Composites Science and Technology

Series Editor

Mohammad Jawaid, Lab of Biocomposite Technology, Universiti Putra Malaysia, INTROP, Serdang, Malaysia

Composites Science and Technology (CST) book series publishes the latest developments in the field of composite science and technology. It aims to publish cutting edge research monographs (both edited and authored volumes) comprehensively covering topics shown below:

- Composites from agricultural biomass/natural fibres include conventional composites-Plywood/MDF/Fiberboard
- Fabrication of Composites/conventional composites from biomass and natural fibers
- Utilization of biomass in polymer composites
- Wood, and Wood based materials
- Chemistry and biology of Composites and Biocomposites
- Modelling of damage of Composites and Biocomposites
- Failure Analysis of Composites and Biocomposites
- Structural Health Monitoring of Composites and Biocomposites
- Durability of Composites and Biocomposites
- Biodegradability of Composites and Biocomposites
- Thermal properties of Composites and Biocomposites
- Flammability of Composites and Biocomposites
- Tribology of Composites and Biocomposites
- Bionanocomposites and Nanocomposites
- Applications of Composites, and Biocomposites

To submit a proposal for a research monograph or have further inquries, please contact springer editor, Ramesh Premnath (ramesh.premnath@springer.com).

More information about this series at http://www.springer.com/series/16333

Mohamad Midani · Naheed Saba ·
Othman Y. Alothman
Editors

Date Palm Fiber Composites

Processing, Properties and Applications

Springer

Editors
Mohamad Midani
Department of Materials Engineering
German University in Cairo
New Cairo City, Egypt

Othman Y. Alothman
Department of Chemical Engineering
College of Engineering
King Saud University
Riyadh, Saudi Arabia

Naheed Saba
Laboratory of Technology Biocomposite
INTROP
Universiti Putra Malaysia
Serdang, Selangor, Malaysia

Composites Science and Technology
ISBN 978-981-15-9338-3 ISBN 978-981-15-9339-0 (eBook)
https://doi.org/10.1007/978-981-15-9339-0

This Springer imprint is published by the registered company Springer Nature Singapore Pte Ltd.
The registered company address is: 152 Beach Road, #21-01/04 Gateway East, Singapore 189721, Singapore

This book is dedicated to the legendary professor Dr. Hamed Elmously, the founding father of date palm byproducts research and development. The work of Prof. Elmously has laid the ground and inspired generations of researchers and practitioners working in the area of utilizing the date palm by products.

Preface

Date Palm (Phoenix dactylifera) is one of the most populous members of the Palmae family and one of the main elements of flora in the Middle East and North Africa. Date palm is a very rich source of cellulosic fibers. Fibers can be obtained from the by-products of annual pruning such as the sheath, spadix stems, midribs, leaflets, and even the trunk at the end of the palm life. The renewed interest in sustainable resources and biobased materials has given a surge to rediscovering those agriculture by-products in new applications with a higher added value. The focus of this book is on the emergence of date palm fiber as a new source of cellulosic fibers that can be used in the reinforcement of polymer composites. Date palm fiber has a unique set of features and benefits, including, abundance, competitive price, in addition to, excellent mechanical, physical and chemical properties. This makes it stand out as an attractive alternative to other fibers currently being used in the natural fiber composites market.

Date Palm Fiber Composites: Processing, Properties and Applications is the first published book title on date palm fiber and its composites. The book acts as a body-of-knowledge or handbook for researchers and industrialists interested in this field. It covers the different aspects of date palm fiber composites with a focus on their processing, properties, and applications and includes up-to-date information on research carried out on date palm fiber composites covering versatile topics such as history of utilization, extraction, treatment, preform formation, composite fabrication, nanofibers, design and modeling, characterization, and properties, in addition to real-life applications in construction and building, wood substitutes, automotive, and other potential future applications.

We are highly thankful to all contributing authors, who provided their valuable ideas and knowledge in this edited book. We attempt to gather all the scattered information of the authors from diverse fields around the world (Egypt, Morocco, Iran, Algeria, Tunisia, Saudi Arabia, Jordan, India, UK, and USA) in the area of date palm fiber composites and finally complete this venture in a fruitful way.

We are highly thankful to Springer Nature team for their generous cooperation at every stage of the book production.

New Cairo City, Egypt — Mohamad Midani
Riyadh, Saudi Arabia — Othman Y. Alothman
Serdang, Malaysia — Naheed Saba

Contents

Editors and Contributors

About the Editors

Dr. Mohamad Midani is an Assistant Professor at the Materials Engineering Department, German University Cairo, Egypt and an Adjunct Assistant Professor at the Wilson College of Textiles, NC State University. He is also the managing partner of inTEXive consulting, Secretary General of the International Association of Palm Byproducts, and member of the Advisory Board of Middle East Company for Textiles. Dr. Midani has long experience in the textile industry, with a focus on technical textiles and nonwovens, with an ongoing research collaboration and consulting to the industry. His research mostly focuses on natural fibers and fiber reinforced composites, and he is the co-inventor of the long textile fibers from the midribs of date palm (PalmFil). He is also the co-founder of the World Conference on Palm Byproducts and their Applications (ByPalma). He published more than 25 articles in international journals and conferences in addition to co-editing one book. Dr. Midani is a reviewer to several international journals from different publishers. He teaches undergraduate and graduate courses on Fiber Reinforced Polymer Composites as well as New Product development and Innovation Management. In 2016, Dr. Midani received a recognition for excellence in teaching from NC State University, and in 2019 from the American University in Cairo. Dr. Midani has his BSc in Mechanical Engineering from Ain Shams University (2006), MT in Textiles & Apparel Technology and Management (2012), and Ph.D. in Fiber and Polymer Science from the College of Textiles NC State University (2016). He is also a certified New Product Development Professional (NPDP) and a member of the Product Development and Management Association (PDMA).

Dr. Naheed Saba completed her PhD (Nanocomposites) with Distinction from Laboratory of Biocomposites Technology, Institute of Tropical Forestry and Forest Products (INTROP), Universiti Putra Malaysia, Serdang, Selangor, Malaysia in 2017. She completed her masters in chemistry and also completed her postgraduate diploma in environment and sustainable development from India. She has published over 50 scientific Peer review Articles in International Journal and 4-5 Articles come under top cited articles in Construction and Building Materials, and Polymers Journal during 2016-2019. She edited six books from Elsevier and also published more than 25 book chapters in Springer, Elsevier, and Wiley publication. She attended few international conferences and presented research papers. Her research interest areas are nanocellulosic materials, fire-retardant materials, natural fiberereinforced polymer composites, biocomposites, hybrid composites, and nanocomposites. She is also recipient of International Graduate Research Fellowship and Graduate on time (GOT) award from Universiti Putra Malaysia, Malaysia. Presently she is Editor-in-Chief of The Journal of Composites and Advance Materials (ISSN 2716-8018) is a peer-reviewed bi-annual journal. She is also Member of Editorial board of Springer-Nature Book Series-Composite Science and Technology. She is also reviewer of several international journals such as Cellulose, Constructions and Building Materials, Composite Part A, Composite Part B, Journal of Polymers and The Environment, Journal of Energy Storage, Journal of Elastomers and Plastics, Journal of Materials Research and Technology, BioResources, and Carbohydrate Polymers. Her H-index=18 (Scopus); H-Index=21 (Google Scholar).

Prof. Othman Y. Alothman is a professor at the department of chemical engineering, King Saud University, with more than 12 years in teaching of undergraduate and graduate levels. His interest is in the field of polymer processing, blends and additives, including processing, aging and characterization of natural fiber/polymer composites, with an interest on the environmental aspects. He has published and presented more than 100 articles in journals and conferences, in addition to co-editing and translating 5 books. He is also a reviewer of a number of high-impact journals from different publishing groups. He is also a member of Polymer Processing Society (PPS) and the Society of Plastics Engineers (SPE).

Alothman holds a B.Sc. degree in Chemical Engineering from King Saud university (1997) and a PhD degree from the University of Manchester in Polymer Science and Technology (2007).

Contributors

S. Abbani EVRNZA Laboratory, University KASDI Merbah Ouargla, Ouargla, Algeria

Wafaa S. Abo Elseoud Cellulose and Paper Department, National Research Centre, Giza, Egypt

Mounir El Achaby Materials Science and Nanoengineering Department (MSN), Mohammed VI Polytechnic University (UM6P), Benguerir, Morocco

Mohammad Saleh Ahmadi Department of Textile Engineering, Yazd University, Yazd, Iran

Faris M. AL-Oqla Department of Mechanical Engineering, Faculty of Engineering, The Hashemite University, Zarqa, Jordan

Nermin M. Aly Spinning and Weaving Engineering Department, Textile Industries Research Division, National Research Centre, Dokki, Cairo, Egypt

Said Awad Department of Civil and Environmental Engineering, College of Engineering, Brunel University London, Uxbridge, UK

Mohamed Naceur Belgacem Université Grenoble Alpes, CNRS, Grenoble INP, Grenoble, France

Rachid Bouhfid Moroccan Foundation for Advanced Science, Innovation and Research (MAScIR), Composites and Nanocomposites Center, Rabat, Morocco

M. Bouziane EVRNZA Laboratory, University KASDI Merbah Ouargla, Ouargla, Algeria

H. Chaib EVRNZA Laboratory, University KASDI Merbah Ouargla, Ouargla, Algeria

Hind Chakchak Unités d'Appui Techniques À la Recherche Scientifique (UATRS), Centre National Pour la Recherche Scientifique et Technique (CNRST), Rabat, Morocco

E. A. Darwish Department of Architecture, Faculty of Engineering, Ain Shams University, Cairo, Egypt

Lamis Darwish Biomedical Engineering Department, Faculty of Engineering, Helwan University, Cairo, Egypt

Zahra Dehghan-Banadaki Department of Textile Engineering, Yazd University, Yazd, Iran

Abdel-Halim M. El-Habbak Mechanical Design & Production Department, Faculty of Engineering, Cairo University, Giza, Egypt

Emad El-Kashif Mechanical Design & Production Department, Faculty of Engineering, Cairo University, Giza, Egypt

Hamed El-Mously Department of Design and Production, Faculty of Engineering, Ain Shams University, Cairo, Egypt

Lobna A. Elseify Department of Materials Engineering, German University in Cairo, New Cairo City, Egypt

Mizi Fan Department of Civil and Environmental Engineering, College of Engineering, Brunel University London, Uxbridge, UK

Mahmoud Farag Mechanical Engineering Department, School of Science and Engineering, American University in Cairo, Cairo, Egypt

Anuradha Gupta College of Textiles, North Carolina State University, Raleigh, USA

F. Hafsi EVRNZA Laboratory, University KASDI Merbah Ouargla, Ouargla, Algeria

Tamer Hamouda Spinning and Weaving Engineering Department, Textile Industries Research Division, National Research Centre, Dokki, Cairo, Egypt

Enas A. Hassan Cellulose and Paper Department, National Research Centre, Giza, Egypt

Mohammad L. Hassan Cellulose and Paper Department, National Research Centre, Giza, Egypt

Fatima-Zahra Semlali Aouragh Hassani Moroccan Foundation for Advanced Science, Innovation and Research (MAScIR), Composites and Nanocomposites Center, Rabat, Morocco;
Faculty of Science, Mechanics of Materials Laboratory, Mohammed V-Rabat University, Rabat, Marocco

Ahmed H. Hassanin Materials Science and Engineering, Egypt Japan University for Science and Technology (EJUST), New Borg El-Arab City, Alexandria, Egypt;
Department of Textile Engineering, Alexandria University, Alexandria, Egypt

Hamdy Ibrahim Mechanical Engineering Department, College of Engineering and Computer Science, University of Tennessee at Chattanooga, Chattanooga, USA

Evina Katsou Department of Civil and Environmental Engineering, College of Engineering, Brunel University London, Uxbridge, UK

Ramzi Khiari Department of Textile, Higher Institute of Technological Studies of Ksar Hellal, Ksar Hellal, Tunisia;
Research Unity of Applied Chemistry & Environment, University of Monastir, Monastir, Tunisia;
Université Grenoble Alpes, CNRS, Grenoble INP, Grenoble, France

A. Kriker EVRNZA Laboratory, University KASDI Merbah Ouargla, Ouargla, Algeria

Sherif Mehanny Mechanical Design & Production Department, Faculty of Engineering, Cairo University, Giza, Egypt

A. Mekhermeche EVRNZA Laboratory, University KASDI Merbah Ouargla, Ouargla, Algeria

Mohamad Midani Department of Materials Engineering, German University in Cairo, New Cairo City, Egypt;
inTEXive Consulting, El Obour City, Egypt;
Wilson College of Textiles, NC State University, Raleigh, NC, USA

A. Mokhtari EVRNZA Laboratory, University KASDI Merbah Ouargla, Ouargla, Algeria

Mehran Nalchian Department of Textile Engineering, Yazd University, Yazd, Iran

Abou El Kacem Qaiss Moroccan Foundation for Advanced Science, Innovation and Research (MAScIR), Composites and Nanocomposites Center, Rabat, Morocco

Shafaat Ahmed Salahudeen Department of Polymer Engineering, B.S.Abdur Rahman Crescent Institute of Science and Technology, Chennai, Tamilnadu, India

Reyhaneh Sarpanahi Department of Textile Engineering, Yazd University, Yazd, Iran

Yonghui Zhou Department of Civil and Environmental Engineering, College of Engineering, Brunel University London, Uxbridge, UK

Date Palm Byproducts (Phoenix Dactylifera L.)

Date Palm Byproducts: History of Utilization and Technical Heritage

Hamed El-Mously and E. A. Darwish

Abstract The date palm abundantly exists in North Africa, The Arab Peninsula and Iran. The first emergence of date palm dates back to 4000 B.C. in Mesopotamia. This long history, along with the high renewability rate, has led to the accumulation of a rich technical heritage associated with the utilization of all the secondary date palm products, including whole leaves (midribs, leaflets), petioles, spadix stems, coir, date kernels and trunks. Midribs have been used in roofing, fencing, furniture making, and manufacturing crates and coops. Leaflets have been used in making mats, baskets and bags. Coir has been used in making ropes, nets, bags, brooms and fly whiskers. Spadix stems have been used in making brooms and household sieves. In addition, fibers obtained from spadix stems have been used for tying agricultural crops. The palm trunk has been used as windows lintels, beams and columns in construction. Moreover, trunks have been used as a wood substitute in furniture making. This chapter reveals the technical heritage associated with several traditional uses of the secondary date palm products to satisfy the human needs in the Arab region. In addition, geometrical description of these secondary products and the procedures of their preparation are included.

Keywords Date palm · Traditional handicrafts · Midribs · Coir · Spadix stem

1 Introduction

The local materials are the material milieu by virtue of which cultures were able to express themselves. Proceeding from the historical perspective, the different cultures of the world were born and developed in company with different materials. Who could deny the relation between the ancient Egyptian culture and papyrus, lotus,

H. El-Mously (✉)
Department of Design and Production Engineering, Faculty of Engineering, Ain Shams University, Cairo, Egypt
e-mail: hamed.elmously@gmail.com

E. A. Darwish
Department of Architecture, Faculty of Engineering, Ain Shams University, Cairo, Egypt
e-mail: eman.atef@eng.asu.edu.eg

M. Midani et al. (eds.), *Date Palm Fiber Composites*, Composites Science and Technology,
https://doi.org/10.1007/978-981-15-9339-0_1

lime stone and granite, nor between the Asian cultures and bamboo, rattan and rice? It is extremely important to capture the relation between culture and local materials as an important asset for development. The linking between development and local materials means that you are building on the existing culture of interaction with these materials, i.e., you are not beginning a development from a zero datum, but with what people—members of each local community—have at hands (the local materials), as well as in minds (psychological familiarity with these materials, and technical heritage, associated with their production, manufacture and use in the different walks of life). In this concern the date palm (Phoenix Dactylifera L.), represents an eloquent example. It is an authentic element of the region's flora, which accompanied our historical march for thousands of years.

As a resource, the date palm could be seen as a system of renewable materials including the primary and secondary products. The date palm is associated with a very rich technical heritage being a product of thousands of years of accumulated expertise of interaction between the diverse local communities and the date palm material system for the satisfaction of various basic material needs.

This technical heritage includes appropriate technologies that may be of contemporary value for peasants in rural areas such as preservation of date in clay jars as food for the whole year and the use of palm midribs and trunks in roofing. In addition, the technical heritage represents a software of its own right, resounding a world of adaptation with the environment and cultural expressiveness, and thus; inspiring to think, imagine and innovate in harmony with the environment and culture.

2 Date Palm: A Basic Element of the Flora of the Arab Region

It may be difficult to record the first emergence in history of the date palm (Phoenix dactylifera L.), but it was well known 4000 years BC, where it was used to build the moon temple near to Ore, south of Iraq (Johnson 2011). The second proof of the deep-rootedness of the date palm comes from the Nile valley, where the date palm was taken as the symbol of the year and the palm midrib as a symbol of the month in the hieroglyphic Egyptian language. But the cultivation of the date palm in Egypt was 2000–3000 years later than Iraq.

The date palm was one of the pivots of economic and, hence, social and cultural life in this region from ancient times. In ancient Egypt the heads of pillars in temples were made resembling the growing top of the date palm. The date palm appeared frequently on walls of temples in different contexts revealing its significance in life in Egypt. The palm leaves were fundamental in ancient Nubian and Upper Egypt houses. The roofs were constructed by split palm trunks and leaves and the interior walls were covered by palm leaves ornaments (Azzam 1960). Until now, date palm constitutes a basic element in several surviving traditions in Nubia and South Egypt, where a palm tree is planted every time a child is born (El-Mously 2001). So when he becomes an adult, the date palm will has grown into many palms that will be the basis of his new life after marriage. As a result, date palm has played a major role in the formation of the culture and heritage in Egypt until the present day (Bekheet 2013).

Economically, date palms are a major part of the life-supporting plantations in every village in Upper Egypt (El-Mously 2001; Bekheet and Elsharabasy 2015). Moreover, the annual products of date palm are being utilized in many traditional crafts by the cultivators and craftsmen in Egypt (Darwish et al. 2019b); thus playing a huge role in sustaining the rural societies against the immigration to urban cities, as date palm related crafts and cultivations support over one million families in Egypt (Bekheet and Elsharabasy 2015).

Thus, the significance of date palm does not only depend on the multiple uses of the fruit in food, spirits, pharmaceutical, cosmetic and medicinal products, but also on the large number of the secondary products that have been widely used in construction and handicrafts(Bekheet and Elsharabasy 2015). Palm midribs and trunks have been used for roofing in a fashion that still survives in the western oases and the poor rural areas in Egypt (Ahmed 2014; Darwish et al. 2019b). Hence, the technical heritage associated with the products of date palm pruning is still thriving as their cheapness and abundance qualified them to be the favorable raw materials for several traditional industries with a know-how that goes back to ancient Egypt (Darwish et al. 2019b).

This technical heritage thrives only because of the high adaptability of date palm in the Arab region environment. The date palm can survive in a wide range of temperature from −15 to 60 C (Barreveld 1993). Direct sunlight helps palm leaves become stronger, taller, and thicker and helps them grow faster (Zaid and de Wet 2002; El-Mously 2018). Moreover, a date palm provides shade and protection for crops and tolerates high levels of heat and salinity as date palm cultivation is found along the seashore in Egypt (El-Mously 2001). In addition, date palms need less water and maintenance and are less prone to diseases than other trees (El-Mously 2001). Thus, the date palm has represented an eloquent example of the sustainable use of renewable material resources as illustrated in Fig. 1.

3 Distribution of Date Palms in the World

The historical roots of date palms cultivation still have a huge impact on the present situation of date palm distribution in the world. Historically, date palm cultivation originated in Iraq (Munier 1973). Now, the Sahara, North Africa, the Arab Peninsula and Iran acquire the most dense date palm plantations in the world as shown in FAO world map of the annual date production (Barreveld 1993). Latest FAO statistics of the number of date palms showed that Saudi Arabia, Algeria, Iran, Iraq and Egypt hold the highest ranks in date palm numbers in the world as shown in Fig. 2 (El-fadda and Abu Ayana 2017).

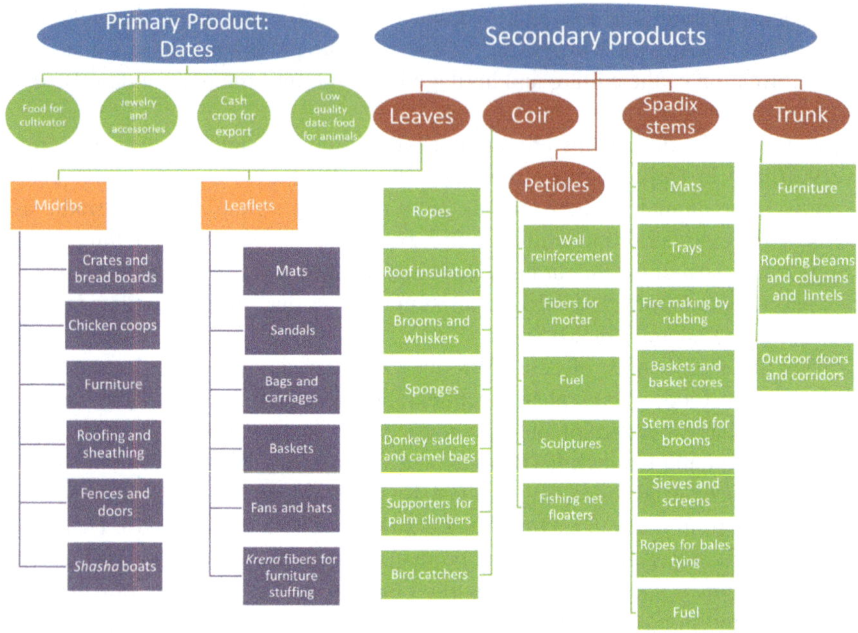

Fig. 1 Traditional Forms of Using Date Palm Byproducts

Fig. 2 Distribution of date palms in the world (El-fadda and Abu Ayana 2017)

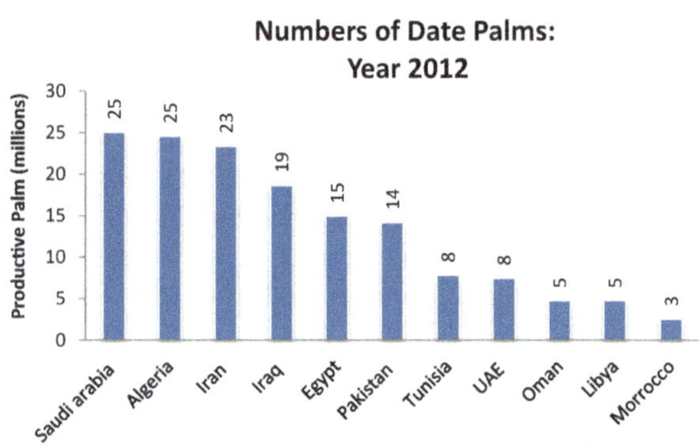

4 Date Palm Byproducts

Date palms (Fig. 3) can live up to 100 years and over, reaching the height of maximum 24 m.

Fig. 3 A date Palm

4.1 Benefits of Pruning

In average, 13 leaves, 13 petioles and 7 bunches are cut per date palm in the annual pruning process (Agoudjil et al. 2011). Annual pruning is necessary for the following (Bekheet 2013):

- Achieving the most suitable symmetry to guarantee the upright standing of the palm.
- Removing abnormal and dead tissues that may take the nutrition from the fruits.
- Stimulating fruit production and flowering necessary for pollination.
- Decreasing the threat of catching fire when the leaves become dry.
- Getting rid of dry and yellow leaves especially if they were infected.
- Removing the thorns and excess leaves that would obstacle the processes of pollination or harvesting.
- Allowing the sunlight reach the fruits for high quality of the photosynthesis process.
- Collecting the products of pruning that represent abundant raw materials for several traditional forms of utilization.

Fig. 4 A climber works on removing excess leaves during the pruning process while supported by a climbing belt made from date palm coir

4.2 Timing and Procedure of Pruning

The annual time of pruning varies from a place to another, but is mainly one of those 3 timings: in autumn after the harvest, in the beginning of spring in the pollination time, and in the ripening time of the leaves in the summer.

Special and trained workers usually perform the annual process as shown in Fig. 4. It begins with the removal of the 3-year old dry leaves using a sharp knife. The cutting should be 10–12 cm above the petiole and the cutting direction should be down-up so the slope of the petiole would expel rainwater.

4.3 Products of Pruning of the Date Palm

The products of the annual pruning process of the date palm are as follows.

4.3.1 Date Palm Leaves

12–15 new leaves are formed annually by a date palm (Barreveld 1993). The life of each leaf ranges from 3 to 7 years (Zaid and de Wet 2002). The length of the palm leaves ranges from 3 to 6 m (Zaid and de Wet 2002). Naturally, each fruit cluster of weight of 8–10 kg is supported on one leaf (Zaid and de Wet 2002). Annual pruning procedures remove the dry leaves in order to provide better access for the crown for harvesting, in addition to save more nutrition for the fruits (Darwish et al. 2019a, b). A whole date palm leaf is shown in Fig. 5.

At the location of the palm leaf near to the trunk, there are sharp spines with lengths that can reach up to 20 cm (Zaid and de Wet 2002). These spines are usually used as sewing needles in traditional weaving (Barreveld 1993; El-Batraoui 2016). The spines taken from three leaves are shown in Fig. 6.

Fig. 5 A Date palm leaf

Fig. 6 Spines taken from 3 leaves

Fig. 7 A date palm midrib

4.3.2 Date Palm Midribs

The dominance of date palm midribs over the total quantities of the products of pruning granted them a well-developed surviving technical heritage in traditional handicrafts and architecture in Egypt and the Arab region (Eldeeb 2017; Darwish et al. 2019b). Midribs are the main ribs of the whole leaves. They extend from their root at the trunk to the last leaflet (Barreveld 1993). The base begins with a triangular shape and the cross section becomes narrower and less triangularly shaped with higher density towards the upper end of the leaf (Barreveld 1993; Elmously 2005). A date palm midrib is shown in Fig. 7. Curved bases of the midribs are often trimmed and used as fuel resource (Barreveld 1993). The curved base and its cross-sections are shown in Figs. 8, 9 and 10.

4.3.3 Date Palm Leaflets

Each leaf contains 120–240 leaflets (Barreveld 1993). Leaflets are used in woven baskets, ropes, mats, fans and sandals. Date palm leaflets are shown in Fig. 11. Date palm leaflets collected from three leaves are shown in Fig. 12.

4.3.4 Date Palm Spadix Stem

Spadix stems are the trimmed stalks of an empty date bunch. The stems grow carrying the relatively heavy weight of the date (El-Mously 2001; Barreveld 1993). As a result, the stems adapt by acquiring notable a high tensile strength and a high fiber ratio (Barreveld 1993). In addition, the fibers in a spadix stem are long and preferred as the main material for several traditional uses (El-Mously 2001). A date palm spadix stems, cross-section of the spadix stem are shown in Fig. 13 and Fig. 14 respectively.

Fig. 8 Base of midrib

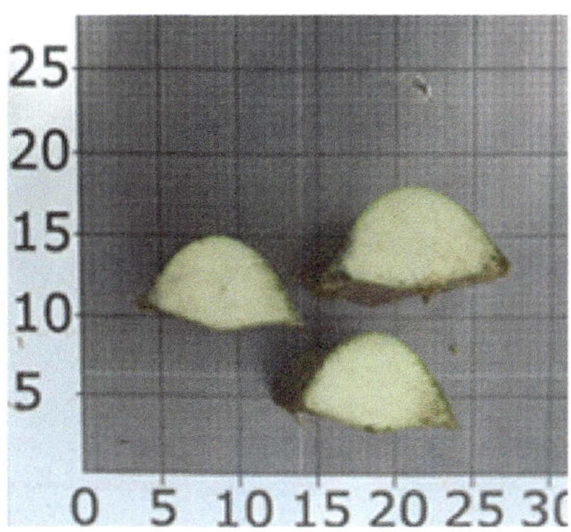

Fig. 9 Cross section at the beginning of the base of the midrib

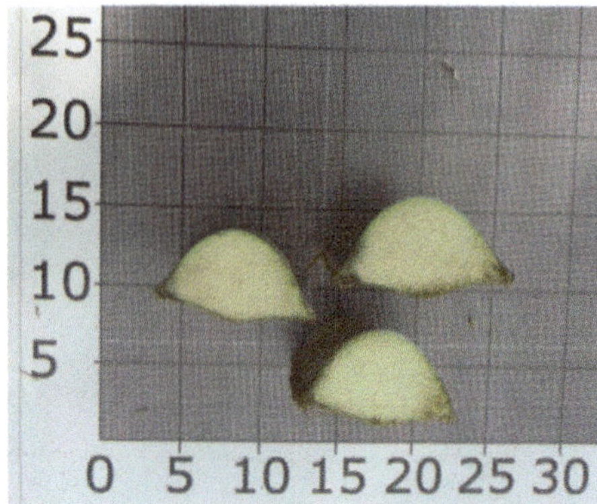

Fig. 10 Cross section at the end of the base of the midrib

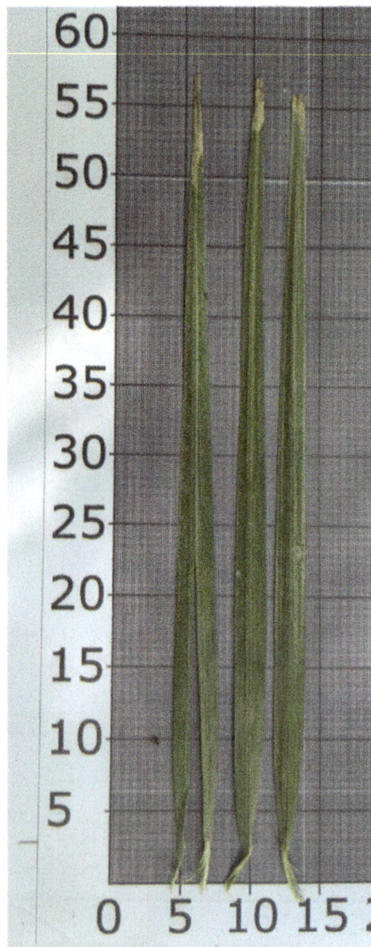

Fig. 11 Date palm leaflets

Fig. 12 Date palm leaflets collected from 3 leaves

4.3.5 Date Palm Coir

Coir originates from the tender tissue that covers the new date palm leaves as they come out and grow (Barreveld 1993). After the growth, the tissue remains attached to the trunk of the palm. This tissue turns into a brownish coarsely—woven fabric, the coir, after drying and can be torn away during the annual pruning (Barreveld 1993). It is used for protecting the newly planted offshoots, shadings, brushes and fishnets. Date palm coir is shown in Fig. 15.

Fig. 13 A date palm spadix stem

4.3.6 Date Palm Petioles

A petiole is the base of the leaf that is left after pruning on the trunk (Barreveld 1993). This leaf base is usually trimmed and removed after drying during the pruning process in the next year. The petioles lack high density that is needed for durable applications (Zaid and de Wet 2002). A date palm petiole and cross-sections are shown in Figs. 16, 17 and 18.

4.3.7 Date Kernel

Date kernels are the pits of the dates. They constitute about 10% of the weight of the fruit (Almana and Mahmoud 1994). Their sizes and colors vary according to the type of cultivar (Barreveld 1993). Fresh seeds are used for breeding and propagation,

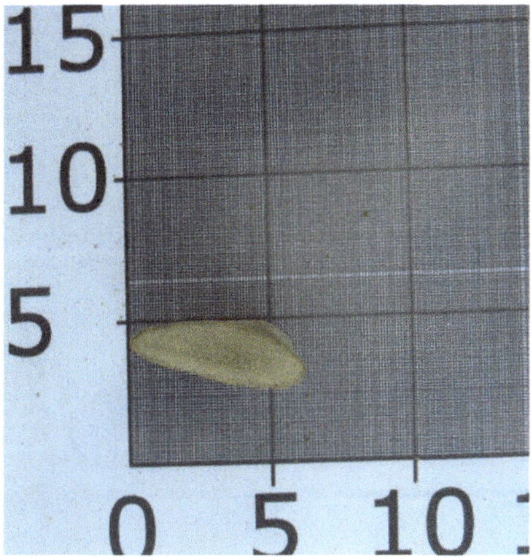

Fig. 14 Cross section of a date palm spadix stem

Fig. 15 Date palm coir

animal feed for their high dietary fiber content such as phenolic acids and flavonoid (Peterson and Dwyer 1998; Mirghani et al. 2012). The palms grown by seeds are of unknown species. They represent approximately 27% of the whole number of palms in Egypt (Bekheet and Elsharabasy 2015). A pair of date kernels is shown in Fig. 19.

Fig. 16 Date palm petiole

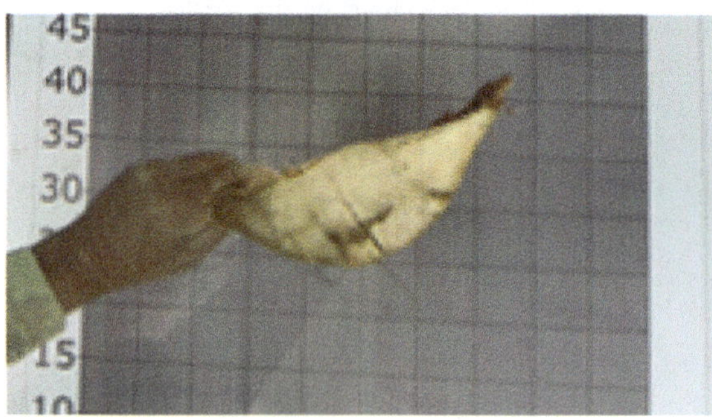

Fig. 17 Cross-section of the beginning of the petiole

4.3.8 Date Palm Trunk

The availability of palm trunks depends on the end of the useful life cycle of the tree. The trunk is the vertical and cylindrical stem of the palm. It consists of tough vascular bundles glued together with cellular tissues (Zaid and de Wet 2002). Hence, the trunk is covered with the bases of the old dry petioles; however the surface of old trunks is mostly softened by the weather (Zaid and de Wet 2002). Date palm trunks are shown in Fig. 20.

Fig. 18 Cross-section of the end of the petiole

Fig. 19 A pair of date kernels

4.4 Estimation of the Quantities of the Annual Pruning of a Date Palm

Date palm products of annual pruning include the midribs, the leaflets, the spadix stems, the coir and the petioles (Elmously 2005). In addition, date palm trunk is considered a valuable byproduct that is mostly used as a substitute of timber (Elmously 2019). In the Arab Gulf region, the annual pruning process produces the average of 6–8 leaves. The weights of the whole leaf, petiole and spadix stem are

Fig. 20 Date palm trunk

0.43 kg, 0.50 kg and 0.50 kg respectively. In Egypt, the quantities of the products of the annual pruning and percentages are shown in Table 1.

Table 1 Products of the annual pruning of date palm (Siwi Species) (10% moisture content-air dried mass) in Egypt (El-Mously 2001)

Quantity available annually	Palm midribs	Palm leaflets	Spadix stems	Coir	Petiole	Total (Kg/palm)
Per palm, dried Kg (mature female)	15	14.6	9	1.56	14	54.2
Percentage	27.6	26.9	16.6	2.8	25.8	100

According to the data shown in Table 1, the quantities of the products of annual pruning of date palms in Egypt pruning are estimated to be approximately 810 thousand tons, which represents a huge material base for a wide spectrum of industries.

5 Traditional Forms of Palm Leaves Utilization

5.1 Traditional Wickerwork Wall Construction

Previous studies predicted that the roofs of the small houses of the workers in ancient Egypt were made of mats of whole palm leaves rows covered with a paste of mud that was so thick that it could be rain-proof (Azzam 1960; Darwish et al. 2019b). This wickerwork technique is still used till now in some rural houses in Egypt.

The plan of the early Egyptian house of a worker was about 3 * 5 m, with walls of a wickerwork of palm leaves that were coated on the inside and outside with mud paste. The maintenance of the gaps that would form as the mud gets old were filled over and over with more layers of mud until the thickness of the walls would reach 20–30 cm as shown in the excavations (Azzam 1960). In a similar manner, the roof was covered with whole palm leaves and straw, with a cover of beaten earth mud (Darwish et al. 2019b). The sole function of the roof was to shelter from heat, sun and dusty winds, regardless of the rain which was considered too scarce to build a more costly type of roof (Azzam 1960; Darwish et al. 2019b). This type of roofing and walling can still be found in some houses in poor rural villages in Egypt.

5.2 Simple Outdoor Sheathing

The roots of using date palm leaves, in layers over a secondary net for roof and wall sheathing, extend back in traditional rural huts that can be seen today in Egypt and UAE (Darwish et al. 2019b).

In that type of wall sheathing, structural nets, made from reeds or date palm midribs are fixed between the structural poles of the hut. Then, the whole date palm leaves are connected together using threads to make mats called *Sedda* or *hassir* (Fig. 21). Finally each mat is fixed by threads to the nets to create a dense sheathing which offers a highly-efficient heat conservation method for the indoor environment of the hut (Darwish et al. 2019a, b).

For roof sheathing, the method relatively resembles thatching technique. The *Sedda* mats are fixed by thread in accumulative layers over the sloped roofing structural grid or a stiff net made from date palm spadix stems or reeds (Fig. 22).

Fig. 21 Whole date palm leaves Sedda

Fig. 22 Sheathing by whole date palm leaves over spadix stem nets supported by wooden poles

5.3 *Sheds and Partitions*

Fences, simple sheds and privacy partitions have been simply built by planting the leaves vertically in the soil and tying them together with two horizontal rows of leaves bundles by ropes as shown in Fig. 23 (Barreveld 1993; Darwish et al. 2019b). In addition, leaves have been used in roofs by laying them across the ceiling beams that are usually made from palm trunks. The thickness of layer of the leaves may reach up to 20–30 cm and then mud is poured above this layer in the present day (Darwish et al. 2019b) (Fig. 24). This method is clearly inspired by the ancient wickerwork walls discussed earlier.

Fig. 23 Traditional palm leaves fence

Fig. 24 Date palm leaves over wooden poles for roofing in a storage house in Menya, Egypt

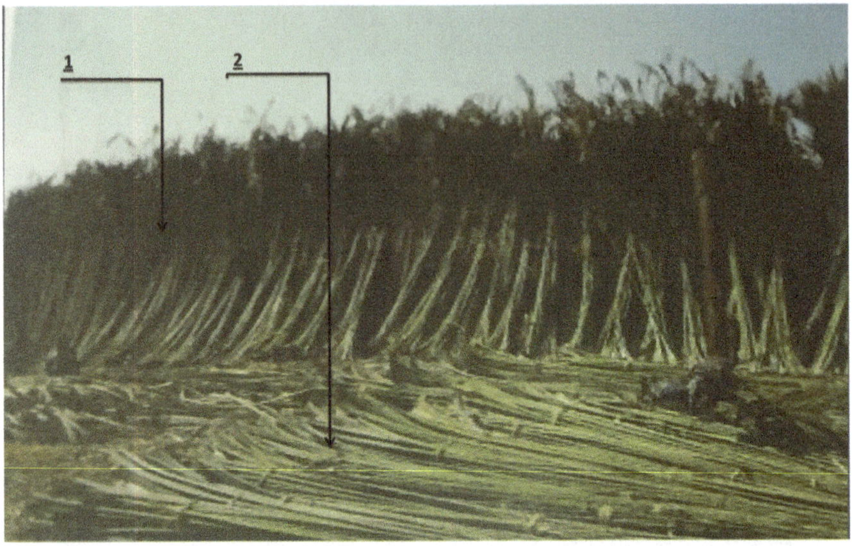

Fig. 25 Drying of date palm midribs. 1: Vertical drying. 2: Horizontal drying

6 Traditional Forms of Palm Midribs Utilization

6.1 Preparation of Midribs

Firstly, the leaflets are stripped from the midribs. The leaflets are to be used later in stuffing furniture. In Egypt, the leaflets are removed from the midribs manually. However in the Arabian Gulf region, the green leaves are laid on the ground, where the goats and sheep would feed on the leaflets, leaving the midribs stripped completely from the leaflets. Then, the midribs are laid vertically for 2–3 weeks to dry with the cut facing down to expel all humidity by gravity. For other purposes where long and straight midribs are needed, the midribs are laid horizontally for 4–6 weeks. Finally, the dried midribs are gathered in bundles: bundles of 5 for handicrafts and bundles of 25 for construction (Darwish et al. 2019b) (Fig. 25).

6.2 Traditional Crates and Bird Coops

Generally the practical quality of the crate is more important for the crate maker than the aesthetic value. The piercing technique used in making crates dates back to the late Roman period in Egypt (Wendrich 2009). A professional crate maker uses his big toe to hold a palm midrib as he punctures and drives a tin tube to create holes into the midrib piece. A light bread board uses only 2 midribs maximum, while heavy-duty chicken coops and fruit crates use 4–5 midribs (El-Batraoui 2016; Darwish et al. 2019b). The majority of the used midribs in crate making in Egypt depend on the Nile Valley in Upper Egypt for the hardness and durability of the palm midribs there.

The used midribs for bread boards, crates and coops are usually only 3 weeks old to have the proper ductility needed for work. Then, they are sprayed lightly with water to gain moderate flexibility and the midribs are shaved lightly by a knife to clean any thorns or bumps (El-Batraoui 2016, p.). Then, the midribs are cut to the desired sizes by a wide sharp knife over a wooden chopping block. Then, the crate maker marks the points where the holes are to be punctured. These holes are driven in the midrib pieces using a sharp thin iron hollow pipe and a mallet (Barreveld 1993; Darwish et al. 2019b). The assembly of the crate is usually from the bottom up, where the punctured horizontal elements are laid on the ground and vertical members are fixed upright those holes (Barreveld 1993; El-Batraoui 2016; Darwish et al. 2019b). Additional punctured horizontal elements are driven down the vertical elements repeatedly as beams of the box to the top (El-Batraoui 2016). Finally, all the edges are hammered to level them (Barreveld 1993). The horizontal elements are almost green, whereas the vertical members are almost dry. With the drying of the horizontal members, their joints with the vertical members become very tight providing rigidity to the crate. Details of making a standard chicken coop and a bread board are shown in Fig. 26 and Fig. 27 respectively.

Fig. 26 Details of a standard date palm midribs chicken coop. 1: The vertical members are fixed through the holes of the bottom horizontal members at the base. 2: Secondary pre-punctured horizontal members are driven down to tie the vertical members. 3: The vertical members are leveled by the top horizontal members

Fig. 27 Details of a date palm bread board. 1: the longitudinal members are cut to standard shapes and punctured at a specific spacing. 2: the transverse members are hammered through the punctured holes of the longitudinal members by friction. 3: additional members are added in the middle to prevent excessive deformation under the loads in the middle

This method is adopted in making crates, coops, cages, and sometimes sliding doors (Barreveld 1993; Darwish et al. 2019b). Sometimes, the crates are lined with palm leaves in the cases of their use for delicate products. This art is developed in more artistic and sophisticated products such as ornaments and furniture.

6.3 Traditional Handmade Furniture

Historically, date palm midribs were used as girders, fixed across the timber frame to build beds in ancient Egypt, above which a woven mat made from spadix stem fibers were fixed to make the mattress as shown in Fig. 28.

Later, furniture makers, often called "artists", give high aesthetic value to their products as the piercing technique becomes more sophisticated (Wendrich 2009; El-Batraoui 2016). Unlike crates, furniture design has no specific standard design and may vary from a place to another. Generally, the midribs are cut according to the desired elements and sizes of the design. The legs of a chair are usually cut from the wide section of the midrib, and the frame and the latticework are made of stiff dry midribs, while the armrests and seats are made from green midribs to facilitate bending (El-Batraoui 2016; Darwish et al. 2019b). Then the elements of the armrest, seat and back are cut respectively.

The same tools of crates making are used here also to make the lattices and the arabesque forms of the armrests and back. Then, all the plates of armrest, seat and back are fixed over the frame that is made of repeated vertical posts and horizontal beams upon which the seat is to be fixed with nails as seen in the figures. 4 cm nails are used to fix the armrests together and 10 cm nail is used to fix the armrests to

Fig. 28 An ancient Egyptian bed, New Kingdom. 1: Date palm midribs girders. 2: date palm spadix stem woven mat, Egyptian Museum in Cairo

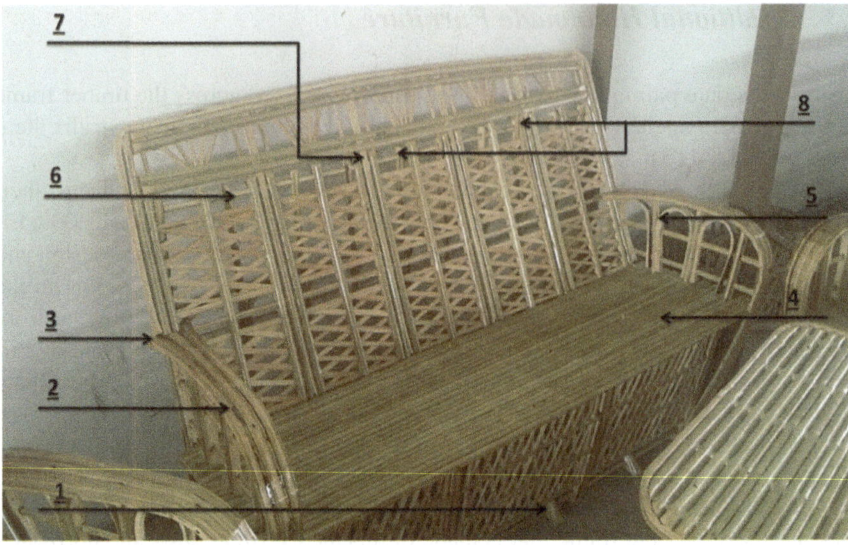

Fig. 29 Details of a standard date palm midribs chair. 1: The legs that bear the chair weight are assembled by horizontal beams through pre-punctured holes along the legs. 2: The armrest members are bent and joined using nails, then fixed to the legs on both ends. 3: Fixation of the armrest to the rear legs using nails. 4: The seat is assembled by nails over cantilever beams protruding from the legs. 5: Lattice with bent midribs to fix the seat to the armrests. 6: Horizontal beam to join main legs and secondary lattice columns. 7: The back is pre-assembled as a lattice using pre-punctured holes. 8: The Secondary lattice columns fix the back to the legs by friction through the lattice

the seat (El-Batraoui 2016; Darwish et al. 2019b). High quality products often use specials keys from the midribs in order to assemble the green members firmly so that they do not disassemble with drying. The whole set of plates are produced in maximum 3 days and are assembled to make a chair or a table in less than a day (El-Batraoui 2016). Details of making a standard chair are shown in Fig. 29.

6.4 *Rural Wall and Roof Sheathing*

Most of the traditional forms of utilizations of date palm midribs focused on their use in sheathing, such as in Arish Houses in UAE (Piesik 2012). Date palm midribs are connected by three rows of ropes to make *Sedda*. This *Sedda* mat has been used to be sheathing between the main structural system elements which used to be made of wooden poles or palm trunks (Darwish et al. 2019b). The natural narrowing geometry of the midribs controls the design of *Seddas*. The assembly of a *Sedda* depends on laying each midrib where its wider end is between the narrower ends of two adjacent midribs.

Fig. 30 Roofing using palm midribs over a series of tree branches as beams, Fayoum, Egypt

The wall sheathing mats are usually fixed by ropes to the columns and side bracings of timber along the mats with maximum spacing of 3 m to ensure the planarity and verticality of the mat (Darwish et al. 2019b). In simple roofing, the midribs are laid in perpendicular layers over a series of local tree branches as beams as shown in Fig. 30. Furthermore in roofing of an outdoor corridor, shown in Fig. 31, these *Seddas* are supported by timber beams by ropes and covered with thick mud layer to increase the thermal and moisture resistance of the roof.

6.5 Doors and Windows

In a manner that is similar to wall sheathing by date palm midribs, date palm midribs have been used in door making using the technique of binding by rope (Wendrich 2009). In ancient Egyptian doors and windows (Fig. 32), the midribs were bound by ropes to form a single layer that was reinforced with diagonal midrib bracings. Loam remains can still be found over the bracings to enhance the coherence between the bracings and the doors. The teeth of the wooden locks were reported to be made from date kernels.

Traditional date palm midribs doors that can still be found now in rural areas in Egypt are clearly inspired by their precedents as shown in Fig. 33. However, the main differences are: wooden posts are used as a frame to which the midribs are fixed by

Fig. 31 Ceiling of an outdoor corridor made of date palm midribs supported by wooden planks, North Sinai, Egypt

Fig. 32 An ancient Egyptian date palm midribs door, Old Kingdom, Egyptian Museum in Cairo. 1: The binding technique using coir ropes. 2: Wooden lock with date kernels teeth. 3: Using loam as a cohesive between the bracings and the door

Fig. 33 A traditional date palm midrib door in Fayoum, Egypt

nails, the bracings are fixed onto the midribs layers by nails instead of ropes, and no loam is used to stick down the bracings to the door.

6.6 *Fencing*

Fencing by date palm midribs *Seddas* is one of the simplest and most spontaneous methods of building fences. Fences made by date palm midribs can be classified into simple and lattice fences. Simple fences depend on the planting of the midribs Seddas in the soil and fixing the fences using ropes to wooden poles at the corners

Fig. 34 A simple date palm midribs fence

(Fig. 34) This kind of fences works only as a temporary outdoor partition element without resisting any loads (Darwish et al. 2019b).

On the other hand, lattice fences are stiffer and sturdier because they consist of date palm midribs lattices that are tied together with ropes. In the UAE, the date palm midribs lattices have a heritage of being used as walls in traditional summer Arish Houses (Piesik 2012; Eldeeb 2017). This lattice is used in relatively heavier weight of fencing between interval structural columns, made of steel or timber, along the walls and in the corners to ensure verticality as shown in Fig. 35. The Egyptian version of lattice fences is heavily inspired by the traditional diagonal lattice of the reed huts in Manzala lake region in Northern Egypt (Helal 1989). The midribs are

Fig. 35 A lattice date palm midribs fence, Ain, UAE

Fig. 36 Diagonal date palm midribs lattice in an outdoor fence, New Valley, Egypt

arranged in a diagonal lattice and fixed with nails to a timber frame as shown in Fig. 36.

This type of fencing can carry the loads of additional wall sheathing and can remain durable and functional for relatively longer periods of time (Eldeeb 2017; Darwish et al. 2019b).

6.7 Boats

Ancient evidences have been found that indicate the use of date palm midribs in boats that go back to the Early Bronze Age in in a fashion that is still common to the present day in the Arab Gulf region, known as the date palm midribs-based *Shasha* raft-boats (Vosmer et al. 2003). Evidences of ancient Magan boats, named after the ancient name of the Arab gulf, were found in the coasts of Kuwait in 1980s. The evidences showed solidified chunks of bitumen, dating back to the Ubaid period (5300–4700 BC) in Kuwait, with the impression of thin parallel longitudinal grooves (Johnson 2016), suggesting that the bitumen was used as a waterproof cover over the date palm midribs or reeds used in building the boats (Carter 2002). Ur III text (dating back to 2100 BC) listed the use of palm-fiber ropes cured with fish oil to tie the reed bundles used in building Mesopotamian boats which are finally also caulked with bitumen (Ray 2003).

Fig. 37 A *Shasha* boat sailing in the sea

Traditional *Shasha* boats (Fig. 37), small raft-boats made from date palm midribs, are still used to the present day for fishing and short distance travelling and racing sports in the UAE (Johnson 2016). The basic material of the boats is date palm midribs after removing the leaflets. The midribs, 150 midribs required for 1 *Shasha* boat, are soaked in saline water for a week until they are manageable and then they are dried overnight (Johnson 2016). Secondly, they are tied together with a date palm coir ropes to make a mat as shown in Fig. 38. Then the mats are fixed tightly to a frame made from local acacia wood with cross beams and side beams to create the hull. Thirdly, the base of the boat is lined with date palm petioles to create buoyancy. Fourthly, more date palm midribs are fixed over the petioles with date palm coir ropes to create the deck of the *Shasha* (Johnson 2016).

Fig. 38 Building a *Shasha* boat from date palm midribs

6.8 Bats and Discs

A traditional household utensil in the Egyptian village is the baking bat, *Matraha*. Baking bats were used to spread the bread dough and bring it into and from the oven. Greco-Roman baking bats depended on the sewing technique which used parallel strings of date palm leaflets by a needle and thread over a network made of shredded and pierced date palm midribs using the piercing technique (Wendrich 2009). The midribs also were used for the bat handles and reinforcement. Such technique was clearly demonstrated in discs, fans and baking bats as shown in Fig. 39. This ancient technique of baking bats is still inherited to the present day. Traditional baking bats now depend on the piercing technique. Rigid midribs are connected by perpendicular midribs that are driven through pierced holes, without the need of sewn leaflets as shown in Fig. 40.

The same piercing technique is used in simple doors and fences, where the vertical midribs are pierced and bounded together by horizontal midribs passing through the pierced holes as shown in Fig. 41.

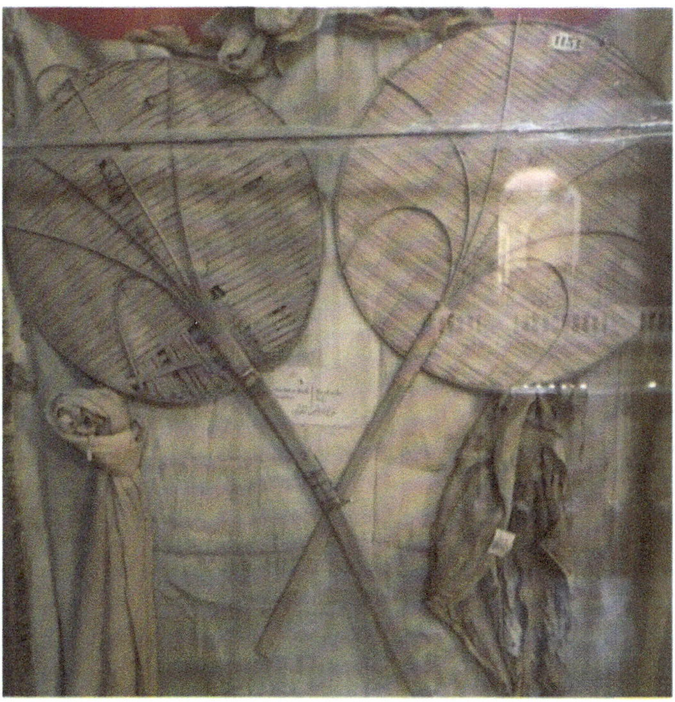

Fig. 39 Baking bats made from sewn date palm leaflets fixed onto a date palm midrib frame, Greco-roman period, Egyptian Museum in Cairo

Fig. 40 A traditional baking bat made from pierced date palm midribs

6.9 *Miscellaneous Uses*

Other miscellaneous uses of midribs are for making fishing rods and supporters for growing grape vines (Barreveld 1993). In addition, bent midribs that cannot be used in crates or furniture have been used traditionally as a source of light charcoal, especially the thick petioles at the base of the midrib (Barreveld 1993).

7 Traditional Forms of Palm Leaflets Utilization

Date palm leaflets are the secondary product of the preparation of the midribs as discussed earlier. As the second most abundant pruning residue, date palm leaflets have acquired a widely spread technical heritage that still thrives to be one of the main sources of income of many families in Fayoum, Sinai and Nubia (El-Batraoui 2016). The well-sustained rich cultural background of Fayoum, Sinai and Nubia has led to the continuity of the date palm leaflets heritage in the field of handicrafts to the present day.

Fig. 41 A simple fence door made form pierced date palm midribs, New Valley, Egypt

Nubia, in South Egypt and North Sudan, is one of the regions where the art of weaving date palm leaflets still thrives. The weaving technique of date palm leaflets have been used since the ancient Egyptian traditions in making many traditional products such as hats, baskets and mats (Barreveld 1993). This technique still thrives because it only needs braiding the leaflets with hand without any other special tools (El-Batraoui 2016).

7.1 Traditional Preparation of Date Palm Leaflets

The leaflets are manually removed from the midribs and are laid in the sun for 2–3 days to get rid of fungus and insects, while in summer it is enough to lay the leaflets in the shade to prevent discoloring (El-Batraoui 2016). Then, each leaflet is split by fingernails into several strips and soaked in saline water for a day to become more flexible. This water may be dyed with vegetable dyes to make various colored strips.

Another method used now for dying the strips is soaking the strips in dyed boiling water with a small amount of salt in it until the desired hue is achieved (El-Batraoui 2016).

7.2 *Bags, Mats and Baskets*

In ancient Egypt, several techniques were employed to use date palm leaflets in making various household accessories as shown in Fig. 42.

After preparation, leaflets are plaited and interwoven together to produce the desired shape according to one or more of the following techniques.

7.2.1 Plaiting Technique

Plaiting is the ancient Egyptian technique where several strands are woven into fabrics by interlacing them with a set of perpendicular strands (Wendrich 2009). The ends of the strands were usually folded into the fabric. This method was widely employed in ancient Egypt basically in making bags and sandals as shown in Fig. 43.

Another traditional form of using date palm leaflets that is deeply rooted in the Bedouin culture in South Sinai in Egypt is the *Sousel, shown in* Fig. 44. The *Sousel* consists of two plaited tubes that are designed to contain two dates, so that every morning, the children would present the *Sousel* with the two dates to their parents as they awaken them in the morning.

This traditional and artistic method in Egypt is more evolved today in making Hassir, depending on weaving the leaflets strips together along their lengths on a

Fig. 42 Ancient Egyptian household utensils made from date palm leaflets, Old Kingdom, Egyptian Museum in Cairo. 1: Protective discs made by the sewing technique. 2: A box made by coiling technique. 3: Protective disc made by looping technique. 4: A basket lid made by looping technique

Fig. 43 Ancient Egyptian sandals from the Middle Kingdom. Egyptian Museum in Cairo

planar desk (El-Batraoui 2016). *Hassir*, a hand-made mat, is made from natural fibers such as reeds and palm leaflets using the traditional plaiting technique. These strips are arranged in 2 diagonal perpendicular grids, then the strips are woven together (Fig. 45). This type of mats depends extensively on high artistic qualities and on high quality leaflets that are preferably just pruned to gain less brittle fibers with high elasticity in work (Barreveld 1993).

In a more modern method, leaflets strips are used after soaking in water to increase their flexibility. Then, every 3 strips are used to make plaited strand that is sewn by machines with thread side to side with the other strands until the wanted area is completed (Barreveld 1993; El-Batraoui 2016). Finally, the edges are bent and sewn to secure the ends of all the strands (Fig. 46). Hence, the used leaflets are not required to be as fresh as in the first type of mats. This type of mat is much lighter than the first type. Therefore, the first type is used for flooring mats and fans as shown in Fig. 47, while the second type is used for ornaments, bags and hats as shown in Fig. 48.

7.2.2 Coiling Technique

The coiling technique depends on creating a coil using a stiff material on which the strands are to be wrapped around (Wendrich 2009). In the ancient Egyptian period, the coiling technique was used mostly in making baskets and plates. Light plates (Fig. 49), probably used as thermal protector below hot pots, were made by wrapping full-size leaflets around coils made from coiled leaflets bundles.

Fig. 44 A *Sousel*, made from plaited date palm leaflets

Traditional trays inherited this technique in making decorative handles that are actually made from spadix stem core on which full-size leaflets are wrapped as shown in Fig. 50.

7.2.3 Looping Technique

When the strands wrapped around the coiled core are linked and intertwined in loops, the technique used is called Looping (Wendrich 2009). This ancient Egyptian technique was widely employed in making sturdy baskets and sandals as shown in Figs. 51, 52, 53 and 54.

This technique is inherited to the present day in making sturdy vase-shaped baskets. Sturdy baskets, shown in Figs. 55 and 56, are made by using dense cores made from date palm spadix stem fibers (Barreveld 1993). These cores are coiled in a spiral form according to the desired shape of the basket. Then, shredded leaflet are wrapped around the spiral cores continuously to link them together while also

Fig. 45 Rolled date palm leaflet Hassir mats

being intertwined. The handles of the baskets are also made from shredded leaflets, wrapped over spadix stem cores, as in the simple coiling technique, in order to provide adequate support while carrying. The bottoms of heavy baskets and heavy-duty plates are made using the same method but with additional sewing in order to fasten the looping of the leaflets over the core made of leaflets bundles as shown in Figs. 57 and 58.

7.2.4 Sewn-Plaits Technique

Tri-plaited strands of date palm leaflets are made according to the needed length (Fig. 59). Then, the strands are sewn together by a large needle using a strong Doum palm *(Hyphaene thebaica)* fiber-based thread or from fibers extracted from the leaflets (Barreveld 1993; Wendrich 2009) (Fig. 60). The skills of the craftsman significantly affect the quality of the product; the stiffness of a bag increases as long as the plaiting is fine and the strands are narrow with tight sewing (Barreveld 1993). Therefore, using freshly pruned leaflets is highly preferred. This technique can also be used in making modern bags (Fig. 61) and mats (Fig. 62).

Fig. 46 Machined date palm leaflet stable mat

Fig. 47 Fans made from woven date palm leaflets with wool embroidery

Finally when the strands are stacked spirally and sewn together, ropes made from date palm coir are added as handles. These bags come in different shapes and volumes, where the largest bags, *Quffa*, (diameter of 50 cm at the bottom and the height of 75 cm) can be used for coal and sand transportation with up to 35 kg capacity (Barreveld 1993). A *Quffa* requires a plaited strand of 10 cm width and 15 m length (Barreveld 1993). Therefore, the bottoms of these heavy duty bags require reinforcement by date palm spadix stem discs made by the coiling technique.

Fig. 48 A hat made from plaited date palm leaflets

Fig. 49 Light date palm leaflet plates, Middle Kingdom, Egyptian Museum in Cairo

7.3 Krena Fibers

Low quality leaflets have been traditionally used as stuffing material for bedding, cushions and mattresses, known as *Krena* (Barreveld 1993). The whole leaves here are dried on the ground and then, the leaflets are collected and soaked in water to soften. The soaked leaflets are then fed into a rippling machine in order to make them into fine threads in order to be dried and baled for later uses (Barreveld 1993). These bales can be used for stuffing of furniture or for thick ropes.

Fig. 50 Coiled decorative date palm leaflets handles

Fig. 51 Date palm leaflets baskets dating back to the New Kingdom, Egyptian Museum in Cairo

Fig. 52 A date palm leaflets basket dating back to the Middle Kingdom, Egyptian Museum in Cairo

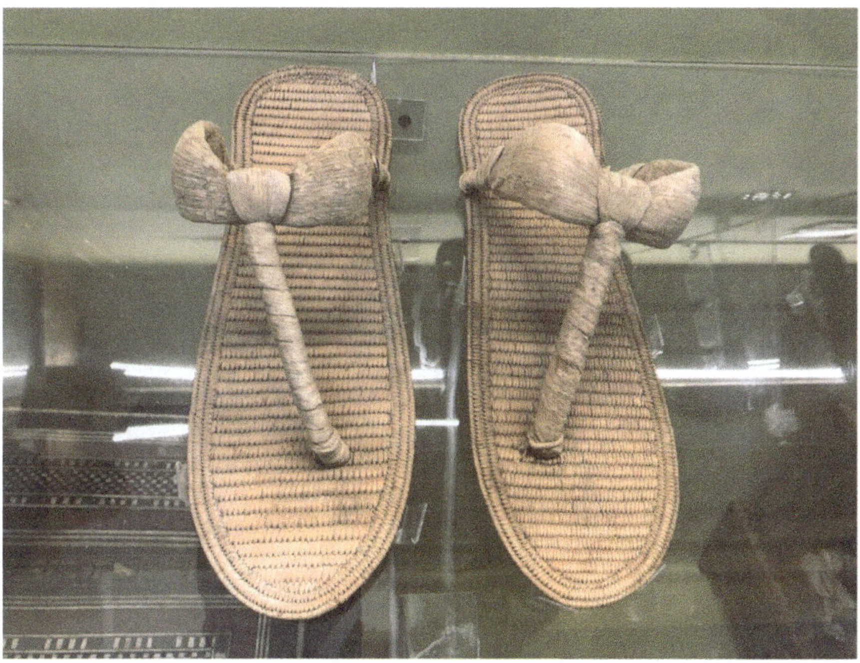

Fig. 53 Ancient Egyptian sandals made by looping technique, New Kingdom, Egyptian Museum in Cairo

Fig. 54 Rolled mat and sandals made from date palm leaflets using the looping technique, Old Kingdom, Egyptian Museum in Cairo

Fig. 55 A reinforced basket made from date palm leaflets and spadix stem cores

7.4 Miscellaneous Uses

Leaflets have been arranged and tied to make simple hand brooms and fly whisks as shown in Fig. 63. Heavy duty ropes are also made using high quality leaflets (Fig. 64). The leaflets here are shredded into 2–3 mm wide strips (Barreveld 1993). Those strips are soaked and made into a strand. Then the strands are plaited to make the final ropes (Barreveld 1993).

Fig. 56 Reinforced plates and trays made from date palm leaflets and spadix stem cores

Fig. 57 Sewing the shredded leaflets around the core in a basket bottom

8 Traditional Forms of Palm Spadix Stem Utilization

Spadix stems acquire recognizable tensile strength because of their natural function of carrying the weight of date through the season. Therefore, several handcrafts products that require durability depend on date palm spadix stems as the main raw material.

Fig. 58 A basket with handles made from shredded leaflets around leaflet bundles core using the looping technique

Fig. 59 Plaiting a palm leaflets strand

Fig. 60 Sewing the spiral strands in a Quffa

Fig. 61 Spiral strands in a date palm leaflet bag

8.1 Preparation of Date Palm Spadix Stem

Spadix stems are firstly soaked to soften the stems and then they are hammered with broad-faced hammers to loosen the fibers. Then, the fibers are stripped away

Fig. 62 Spiral strands in a date palm leaflet mat

Fig. 63 A broom made from shredded leaflets, 4th Century A.D., Egyptian Museum in Cairo

Fig. 64 Date palm leaflets ropes, New Kingdom, Egyptian Museum in Cairo

longitudinally by hand from the basal end of the stalk to the other end (Barreveld 1993).[1]

8.2 *Household Accessories*

The ancient Egyptian twining technique depended on twisting rows of leaflets or ropes around perpendicular sets of spadix stems strips in order to create a stiff disc (Wendrich 2009). Such technique is demonstrated in the sieve shown in Fig. 65.

Utilizing the same technique of making spadix stem *Hassir*, smaller woven patches are made using the shredded fibers from the spadix stems on looms. Those patches are sturdy on their own and can be used to make tissues boxes, table cloths, bags (Fig. 66) and lamp shades (Fig. 67).

[1]A modern method of preparation that is used now in Egypt for faster products is laying the spadix stems on the asphalt roads to be run over by cars and trucks in order to disassemble the fibers of the stems for further use.

Fig. 65 A sieve made from date palm spadix stem, Roman period. Egyptian Museum in Cairo

Fig. 66 Tables cloths, tissues boxes and bags made from woven spadix stem fibers

8.3 Sturdy Baskets

Being stiffer than leaflets, stronger type of baskets (Fig. 68) can be made from thick coiled date palm spadix stems that is sewn and twined with wool threads.

Fig. 67 Lamp shades amp made from woven spadix stem fibers

8.4 *Heavy Duty Mats*

The method used in making mats is inspired by the ancient Egyptian weaving technique that depended on interlacing strands, which were tensioned on a loom, with perpendicular strands (Wendrich 2009). The fibers of the spadix stems are passed and pressed in between the tensioned threads in the machine. Consequently, the fibers are woven in an orthogonal net until the needed area is completed. The resultant mat (Fig. 69) is highly durable and can be used directly over the soil and in the outdoors.

8.5 *Decorative Trays*

The disassembled spadix stem fibers can be gathered to be strips that are woven to create a durable bottom for household trays and saucers (Fig. 70). The sides of the trays are created by wrapping shredded leaflets over fixed decorative spadix stems elements. The method of making the handles for the spadix stem trays, shown in

Fig. 68 A basket made from coiled date palm spadix stems

Fig. 69 Rolled date palm spadix stem *mats*

Fig. 70 A decorative cup saucer

Fig. 71 Decorative trays made from woven spadix stems strips

Fig. 71, is clearly inspired by the ancient Egyptian simple coiling technique discussed earlier in Sect. 7.2.2 Coiling technique.

9 Traditional Forms of Palm Petioles Utilization

Petioles have been used as bordering walls around open wells when the usual brick are not available (Dowson et al. 1978). In this method, the petioles are sharpened at the thinner end and hammered closely until a firm and dense wall is formed. The low density of petioles led to their use as floaters for the fishermen nets and traditional

Fig. 72 A sculpture from date palm petioles

Shasha boats (Popenoe 1973; Barreveld 1993; Johnson 2016). In construction, petioles have been used as vertical sticks that are hammered into the ground to line and stiffen the bond between mortar and mud walls (Popenoe 1973; Barreveld 1993). In handicrafts, date palm petioles offer a suitable soft medium to make distinctive sculptures as shown in Figs. 72 and 73. However, the most prominent use of petioles in the present day is using them as a fuel (Barreveld 1993).

10 Traditional Forms of Palm Coir Utilization

Date palm coir ropes are traditionally known to acquire sufficient strength (Popenoe 1973; Barreveld 1993) as they were the favorite type of ropes for sailing in the UAE (Piesik 2012), although no reliable data has been found regarding the actual mechanical properties of this type of ropes. Coir ropes can be made in different diameters for various uses such as tying, handling and binding (Barreveld 1993). Moreover, ropes can be made into nets that can carry heavy loads for transportation over camels in rural areas (Barreveld 1993).

Other miscellaneous traditional uses of coir include being a fuel source, making fishnet, basket handles, brushes, bedding and shading live plants and offshoots (Barreveld 1993). In addition, raw coir is used as stuffing for the spaces between the date palm trunk beams and midribs in the roofing of traditional rural houses in

Fig. 73 A sculpture from date palm petiole

Egypt (El-Tawil 1989; Ahmed 2014). This coir increases the overall thermal insulation of the roof which enhances the indoor air quality (El-Tawil 1989). Furthermore, date palm coir, among fibers extracted from reeds, is a basic element in the mixture of the traditional wickerwork mortar and plaster in the rural houses in Egypt(Helal 1989). In addition, the poor in Upper Egypt use when making pillows a core made from coir in order to save cotton to decrease the costs of the pillows.

10.1 Plaited Ropes and Bags

Date palm coir was used as the main source of fibers for ropes by plaiting technique since ancient Egypt as shown in Fig. 74 (Wendrich 2009). The same technique is still employed to the present day as shown in Fig. 75.

Plaited coir ropes were employed in miscellaneous uses such as hangers (Fig. 76), fire wicks (Fig. 77), balances (Fig. 78), fishing nets (Fig. 79) and bags (Fig. 80) in ancient Egypt. In the Roman period, the coir fibers were simply rolled by hand to make wigs as shown Fig. 81.

Fig. 74 A coir rope, New Kingdom period, Egyptian Museum in Cairo

Fig. 75 Ropes made from date palm coir

10.2 Cattle Accessories

The well-developed ancient Egyptian expertise in making the coir ropes made them strong enough to be used in bags and agricultural plows. The weaving technique, interlacing strand using a loom, was used in ancient Egypt in saddles and blindfolds for cattle as shown in Fig. 82. The same technique is still used in the present day

Fig. 76 A coir wall hanger, Old Kingdom, Egyptian Museum in Cairo

Fig. 77 A coir wicks with traces of oil, Middle Kingdom, Egyptian Museum in Cairo

Fig. 78 Coir ropes in a balance, Middle Kingdom, Egyptian Museum in Cairo

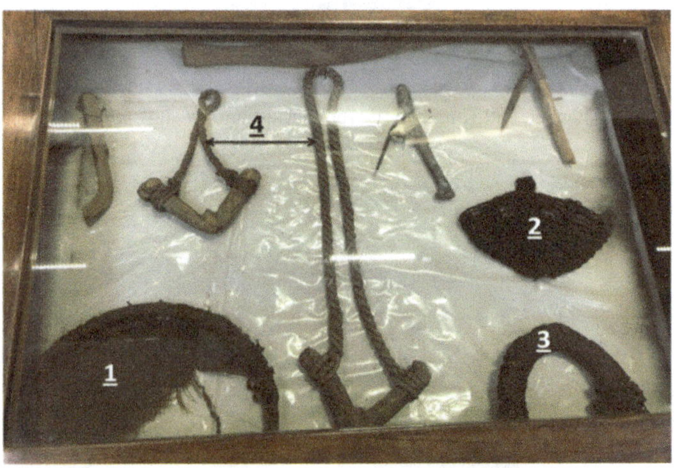

Fig. 79 Date palm coir fishing products, Middle Kingdom, Egyptian Museum in Cairo. 1: Fishing net. 2: Bait bag. 3: safety ropes. 4: Darts

in making camel bags, donkey saddles, supporting belts for palm climbers and bird traps as shown in Fig. 83, Fig. 84, Fig. 85 and Fig. 86 respectively.

Fig. 80 A plaited bag made from coir, Old Kingdom, Egyptian Museum in Cairo

Fig. 81 A coir wig from the Roman Period, Egyptian Museum in Cairo

Fig. 82 Date palm coir cattle accessories, Old Kingdom, Egyptian Museum in Cairo. 1: A buffalo saddle. 2: Blindfold. 3: Bridle. 4: Plaited ropes for collecting water utensil

Fig. 83 Weaving a camel bag from date palm coir using a loom, Fayoum, Egypt

11 Traditional Forms of Palm Date Kernels

Preparation procedures of kernels vary according to the type of use. They can also be pressed to yield edible oil, roasted to make coffee, and heated to make charcoal (Barreveld 1993; Johnson 2016). Prior to using kernels in jewelry, the kernels are extracted, washed, soaked in dyes and dried 48 h (Mirghani et al. 2012). Then, the

Fig. 84 A donkey saddle made from woven date palm coir

Fig. 85 Belts for palm climbers made from woven coir

kernels are holed and stringed together by thread to make jewelry, necklaces, prayer beads (Fig. 87) and bags (Fig. 88) (Popenoe 1973; Johnson 2016).

Fig. 86 A bird catcher made from woven coir and spadix stem

12 Traditional Forms of Palm Trunks Utilization

Date palm trunks have been known to be strong and durable, which qualified the trunks to be used as timber substitutes. In Fig. 89, a worker is confidently ascending to a truck while using a trunk as stairs. Moreover, the worker is carrying a bundle of twenty midribs, with a total weight of approximately 40 Kg. Yet, the natural coarse surface of the trunk offers the needed friction for his bare feet. As a result, date palm trunks were found to be used in the fields which require high durability and stiffness such as construction and furniture elements.

12.1 Traditional Construction

Date palm trunks have also been used as columns and beams in traditional housing in many rural areas in Egypt and the Arab region. The most primitive forms of using

Fig. 87 A prayer bead made from date palm kernels

date palm trunks in construction were as simple door, sheds of outdoor corridors and simple roofing as shown in Figs. 90, 91 and 92.

In Siwa Oasis in the western desert in Egypt, the technical heritage of using date palm trunks in construction evolved to demonstrate a spontaneous cleverness in terms of changing the assembly of the trunks according to the covered span (El-Tawil 1989). In this method, date palm trunks were halved or quartered to work as beams supported by load-bearing walls made from Kershef soil in the Siwa oasis (Ahmed 2014), or from mud bricks in the Nile Valley. The types of roofs were as follows:

1. Primary roof: for rooms with spans of 2–3 m, roofing depended on using planks from the palm trunks supported by the walls in the transverse direction. Then the spaces between the trunks were filled by 10–20 cm thick mortar and date palm coir for intermediate floorings or roofing (El-Tawil 1989).
2. Secondary roof: for rooms with spans of 4–5 m, additional beams are added where each beam consists of two halves of a trunk laid adjacently on the curved side. Above these beams, longitudinal planks of trunks are laid together on which the roofing layers are added as described earlier (El-Tawil 1989).

Fig. 88 A bag made from date palm kernels

3. Tertiary roof: for halls with spans 58 m, main full trunk beams are supported on walls and piers, above which the beams of the secondary roofs and roofing layers are added (El-Tawil 1989).

In a simpler and more recent fashion, date palm trunks are used as visible columns and beams in light huts. In this method, special workers peel the outer tough surface to achieve an organized surface to handle during construction (Fig. 93). Then, the columns are made of whole trunks (Fig. 94), while beams generally consist of quartered trunks that may be arranged back-to-back as in the traditional roofing method in the western oases in Egypt (Fig. 95), or the trunks are shaped into rectangular cross-sections for beams and columns to achieve a modern and sophisticated design as shown in Fig. 96.

12.2 Traditional Furniture

Trunks are shaped into blocks that act as the armrests and supporters of tables and chairs (El-Mously 2001). The fixation methods used are very similar to those used

Fig. 89 A worker ascending on a date palm trunk, carrying his own weight and the weight of a 20-midribs bundle, Asiut, Egypt

Fig. 90 A traditional door made from date palm trunk planks, Fayoum, Egypt

Fig. 91 Using date palm trunks as beams in an outdoor corridor, Arish, Egypt

Fig. 92 Using date palm trunks as beams with date palm midribs ceiling in a simple roof, Arish, Egypt

Fig. 93 Peeling and squaring a date palm trunk

Fig. 94 Using date palm trunks as columns and rafters in a date palm midrib hut

Fig. 95 Quartered date palm trunk beams that support date palm midribs roof

Fig. 96 Squared date palm trunk beams and columns in a modern date palm midribs hut

in timber furniture. The coarse vascular structure of the trunk gives the furniture a rustic look (Fig. 97) that is desirable in various touristic projects.

13 Conclusion

It is clear from the aforementioned that the date palm byproducts enjoyed a long history of utilization in many regions in the world, especially in the Arab region, extending for thousands of years. Relying basically on the periodical pruning (palm

Fig. 97 Furniture made from date palm trunk

service) activity they represented a sustainable material base for the satisfaction of basic material needs of the local populace: in shelter, furniture, agricultural equipment, transportation and household utensils.

The technical heritage, associated with the date palm byproducts reveals a generic perception of the date palm as a whole resource, whereby all the elements of the resource were-according to the available level of technology-efficiently used for the satisfaction of the basic human needs. It also reveals, though implicitly, a huge body of traditional knowledge about the properties and behavior-under different environmental and loading conditions-of these byproducts. The wide spectrum of uses of different palm byproducts reveals a high degree of innovation and, sometimes, high levels of skills seemingly unattainable at our present contexts.

It is also clear from the aforementioned that this technical heritage operated- and still operates-as a force of inspiration to develop new techniques for processing of date palm byproducts and innovative products made from them. But as mentioned in the introduction the most valuable in the technical heritage is that it acts as a software for discovering in future new environmentally and culturally tuned routs of utilization.

References

Agoudjil B, Benchabane A, Boudenne A et al (2011) Renewable materials to reduce building heat loss: characterization of date palm wood. Energy Build 43:491–497. https://doi.org/10.1016/j.enbuild.2010.10.014

Ahmed RM (2014) Lessons Learnt from the vernacular architecture of bedouins in Siwa Oasis, Egypt. In: The 31st international symposium on automation and robotics in construction. Sydney. https://doi.org/10.22260/ISARC2014/0123

Almana HA, Mahmoud RM (1994) Palm date seeds as an alternative source of dietary fiber in Saudi bread. Ecol Food Nutr 32:261–270. https://doi.org/10.1080/03670244.1994.9991406. https://www.research-collection.ethz.ch/bitstream/handle/20.500.11850/134539/1/eth-33671-01.pdf

Azzam OA (1960) The development of urban and rural housing in Egypt. Swiss Federal Institute of Technology, Zurich. https://www.research-collection.ethz.ch/bitstream/handle/20.500.11850/134539/1/eth-33671-01.pdf

Barreveld WH (1993) Date palm products. FAO. https://www.fao.org/docrep/t0681e/t0681e00.htm#con

Bekheet S (2013) Date palm biotechnology in Egypt (Review article). Appl Sci Report 3:144–152. https://www.pscipub.com/Journals/Data/JList/Applied%20Science%20Reports/2013/Volume%203/Issue%203/2.pdf

Bekheet S, Elsharabasy SF (2015) Date palm status and perspective in Egypt. In: Johnson DV, Al-Khayri JM, Jain SM (eds) Date palm genetic resources and utilization. Springer Science+Business Media, Dordrecht. https://doi.org/10.1007/978-94-017-9694-1

Carter R (2002) Ubaid-period boat remains from As-Sabiyah: excavations by the British Archaeological Expedition to Kuwait. In: Proceedings of the seminar for Arabian studies. Archaeopress, Edinburgh, pp 13–30. www.jstor.org/stable/41223721

Darwish EA, Mansour Y, Elmously H, Abdelrahman A (2019) Development of sustainable building components utilizing date palm midribs for light wide-span multi-purpose structures for rural communities in Egypt. J Build Eng 24:100770. https://doi.org/10.1016/j.jobe.2019.100770

Darwish EA, Mansour Y, Elmously H, d Abdelrahman A (2019b) The technical heritage of date palm leaves utilization in traditional handicrafts and architecture in Egypt & the Middle East. In: By-products of palm trees and their applications. Mater Res Proc 11:325. https://doi.org/10.21741/9781644900178-28

Dowson VHW, Aten A, Food and Agriculture Organization of the United Nations (1978) Dates: handling, processing and packing. FAO, Rome

El-Batraoui M (2016) The traditional crafts of Egypt. The American University in Cairo Press, Cairo, New York

Eldeeb A (2017) Recycling agricultural waste as a part of interior design and architectural history in Egypt. Acad Res Community Publ 1:7. https://doi.org/10.21625/archive.v1i1.116

El-fadda S, Abu Ayana R (2017) Palm Secondary Products: Types And Economic Value, second. Al Rajhi Endowment, Riyadh. ISBN: 978-603-01-8810-9. https://www.rajhiawqaf.org/ar/mediacenter/DocLib3/%D8%A7%D9%84%D9%85%D9%86%D8%AA%D8%AC%D8%A7%D8%AA%20%D8%A7%D9%84%D8%AB%D8%A7%D9%86%D9%88%D9%8A%D8%A9%20%D9%84%D9%84%D9%86%D8%AE%D9%8A%D9%84.%20%D8%A7%D9%86%D9%88%D8%A7%D8%B9%D9%87%D8%A7%20%D9%88%D8%A3%D9%87%D9%85%D9%8A%D8%AA%D9%87%D8%A7%20%D8%A7%D9%84%D8%A5%D9%82%D8%AA%D8%B5%D8%A7%D8%AF%D9%8A%D8%A9.pdf

El-Mously H (2001) The industrial use of the date palm residues: an eloquent example of sustainable development. United Arab Emirates University, El Ain, p 21. https://citeseerx.ist.psu.edu/viewdoc/download?doi=10.1.1.588.1154&rep=rep1&type=pdf

El-Mously H (2018) Innovating green products as a mean to alleviate poverty in Upper Egypt. Ain Shams Eng J 9:2039–2056. https://doi.org/10.1016/j.asej.2017.02.001

Elmously H (2019) Rediscovering date palm by-products: an opportunity for sustainable development. In: By-products of palm trees and their applications. Materials research proceedings, Aswan, pp 3–61. https://doi.org/10.21741/9781644900178-1
Elmously HI (2005) The palm fibres for the reinforcement of polymer composites: prospects and challenges. In: The first Ain Shams international conference on environmental engineering. Cairo
El-Tawil H (1989) Environment and architecture in Siwa. Alexandria University, Alexandria, Faculty of Fine Arts
Helal A (1989) Spontaneous architecture in Egypt: analytical study of reed architecture in Manzala Lake region. Helwan University, Cairo, Faculty of Fine Arts
Johnson D (2016) Unusual date palm products: prayer beads, walking sticks and fishing boats. Emir J Food Agric 28:12. https://doi.org/10.9755/ejfa.2015-11-1021
Johnson DV (2011) Introduction: date palm biotechnology. In: Jain SM, Al-Khayri JM (eds) Date palm biotechnology. Springer, Netherlands. ISBN 978-94-007-1318-5
Mirghani MES, Al-Mamun A, Daoud JI, Mustafa SM (2012) Processing of Date Palm Kernel (Dpk) for production of edible jam. Aust J Basic Appl Sci 1:22–29
Munier P (1973) Le palmier-dattier. Maisonneuve Larose, Paris
Peterson J, Dwyer J (1998) Flavonoids: dietary occurrence and biochemical activity. Nutr Res 18:1995–2018. https://doi.org/10.1016/S0271-5317(98)00169-9
Piesik S (2012) Arish: palm-leaf architecture. Thames & Hudson, London. ISBN: 978-0500342800
Popenoe P (1973) The date palm. In: Field h (ed) field research projects. Coconut, Miami
Ray HP (2003) The archaeology of seafaring in ancient South Asia. Cambridge University Press, Cambridge, New York. ISBN: 978-0521011099
Vosmer T, Potts DT, Naboodah HA, Hellyer P (2003) The naval architecture of early Bronze Age reed-built boats of the Arabian Sea. In: Archaeology of the United Arab Emirates. Proceedings of the first international conference on the archaeology of the UAE, pp 152–157
Wendrich WZ (2009) Basketry. In: Nicholson PT, Shaw I (eds) Ancient Egyptian materials and technology. Cambridge University Press, Cambridge. ISBN: 978-0521120982
Zaid A, de Wet PF (2002) Botanical and systematic description of the date palm. In: Zaid A, Arias-Jiménez EJ (ed). Date palm cultivation. Rome, Italy: FAO Plant Production and Protection Paper 156. https://www.fao.org/docrep/006/y4360e/y4360e05.htm

Date Palm Fiber Composites Processing

Date Palm Fiber Extraction and Treatment

Fatima-Zahra Semlali Aouragh Hassani, Hind Chakchak, Mounir El Achaby, Rachid Bouhfid, and Abou El Kacem Qaiss

Abstract Nowadays, the intensive use of petroleum-derived plastics causes many serious environmental issues that are of concern to society as a whole. In order to reduce their use and gradually replace them with sustainable materials, a great deal of research has focused on the development of biopolymer matrix composites and bio-based reinforcements produced from renewable resources. Among the potential bio-reinforcements, the cellulosic fibers contained inside the date palm fibers have the advantage of being inexpensive, not very dense, and have specific properties comparable to synthetic fibers thus making it possible to use them in many sectors. Nevertheless, date palm fibers are currently not used to their full potential. Some parts of these plants are simply not exploited. This chapter summarizes the method of extracting fibers and fibrils from date palm leaves. In order to enhance the fiber/matrix adhesion, many approaches have been presented. Finally, the different stages of date palm fibers examination such as Fourier transform infrared (FTIR), scanning electron microscopy (SEM), thermogravimetric analysis (TGA) and single fiber tensile property were depicts in order to quantify and determine the constituents present in the fibers, the thermal stability, the mechanical properties as well as the morphology of the fibers.

Keywords Date palm fiber · Extraction · Chemical treatment · Morphology · Thermal properties · Mechanical properties

---

F.-Z. S. A. Hassani · R. Bouhfid · A. E. K. Qaiss (✉)
Moroccan Foundation for Advanced Science, Innovation and Research (MAScIR), Composites and Nanocomposites Center, Rabat, Morocco
e-mail: a.qaiss@mascir.com

F.-Z. S. A. Hassani
Faculty of Science, Mechanics of Materials Laboratory, Mohammed V-Rabat University, Rabat, Marocco

H. Chakchak
Unités d'Appui Techniques À la Recherche Scientifique (UATRS), Centre National Pour la Recherche Scientifique et Technique (CNRST), Rabat 10000, Morocco

M. E. Achaby
Materials Science and Nanoengineering Department (MSN), Mohammed VI Polytechnic University (UM6P), Lot 660 – Hay Moulay Rachid, 43150 Benguerir, Morocco

M. Midani et al. (eds.), *Date Palm Fiber Composites*, Composites Science and Technology,
https://doi.org/10.1007/978-981-15-9339-0_2

1 Introduction

The broad awareness of ecological, social and economic imperatives, the search for sustainable green technologies, the increasing problem of waste, environmental legislative standards as well as the depletion of fossil resources are at the origin of the orientation of scientific research towards the development of green materials. In this regard, for over a decade many laboratories and technical centers around the world have been carrying out work with the aim of combining materials of natural origin with plastics of fossil origin. In addition, recent advances in genetic engineering, composite science and the development of natural reinforcements offer significant opportunities for the development of new materials from renewable resources. Materials with various names (bio-composite, biodegradable, biocompatible, etc.) have thus emerged (Alawar et al. 2009; Li et al. 2009; Kalia and Kaith 2011; Ghori et al. 2018). As a result, colossal funds are released in this direction to meet the high market demand.

Cellulose-rich biomass is gaining enormous importance as a raw material for the chemical industry, as it is made up of cellulose, hemicellulose and lignin, which contain many functional groups suitable for chemical treatment. In this context, cellulosic fibers are attracting increasing attention for the reinforcement of composite materials, in particular because of their low cost, their low density, their biodegradability, their availability, their specific interesting module and their capacity to be recycled. As a result, many modern technologies have used composite materials with properties that cannot be obtained by traditional materials (metals, ceramics and polymers). These technologies find applications in areas where light, robust, rigid structures capable of withstanding impact, abrasion and corrosion are required. Among these sectors are the aerospace industry, the automobile industry, the electronics industry, the petroleum industry and the medical industry (Alawar et al. 2009; Li et al. 2009; Kalia and Kaith 2011; Ghori et al. 2018).

Several natural fibers have been proposed as a substitute for conventional fibers including hemp, flax, jute (in Europe) or date palm, alfa and saw palmetto (in Africa). In this chapter, we will focus on date palm fibers, which are one of the most abundant plant resources in Morocco and which today are little exploited. In fact, the annual maintenance work on palm trees generates significant quantities of waste composed mainly of palms which can be used as fiber reinforcement in composites. In the first part of this chapter, a bibliographic study on the date palm as well as current knowledge concerning lignocellulosic fibers and their components, namely cellulose, hemicelluloses and lignins will be presented. Additionally, the different fiber extraction techniques (physical, mechanical and chemical) will also be detailed. While the second part of this chapter, will be devoted to the chemical modification of lignocellulosic fibers. Indeed, due to natural fiber/polymer matrix compatibility a study of the different ways of chemical modification will be investigated. Finally, the effects of different chemical treatment on the morphological, structural, thermal and mechanical behavior of the date palm fiber will be investigated.

2 Date Palm Fiber

The date palm is typically cultivated in Saharan oases. The one native to North Africa, is widely cultivated from Arabia to the Persian Gulf, where it forms the vegetation characteristic of oases. It is also cultivated in the Canaries, in the northern Mediterranean and in the southern part of the United States. It is a cold sensitive plant that grows on any type of soil, provided it is fertile and well drained. In regions with a mild climate, it is grown outdoors, in a sunny position, used mainly as an ornamental plant for its slender appearance and foliage (Alawar et al. 2009; Ghori et al. 2018).

There are more than 2,600 species of palm trees. You would think that it is a tree that has a trunk whereas it is a monocot that does not contain wood or trunk but has a stipe. In addition, it is a dioecious plant therefore containing male palms and female palms. The palm has a very slender trunk, up to 30 m high, visibly covered by the sheaths of the fallen leaves. The leaves, gathered in a number of 20–30 maximum, form a sparse apical crown (Fig. 1). They are pinnate, long up to 6 m; the upper leaves are ascending, the basals curved downward, with leathery, linear, rigid and pungent segments, of green color (Al-Shayeb et al. 1995; Alawar et al. 2009; Ghori et al. 2018).

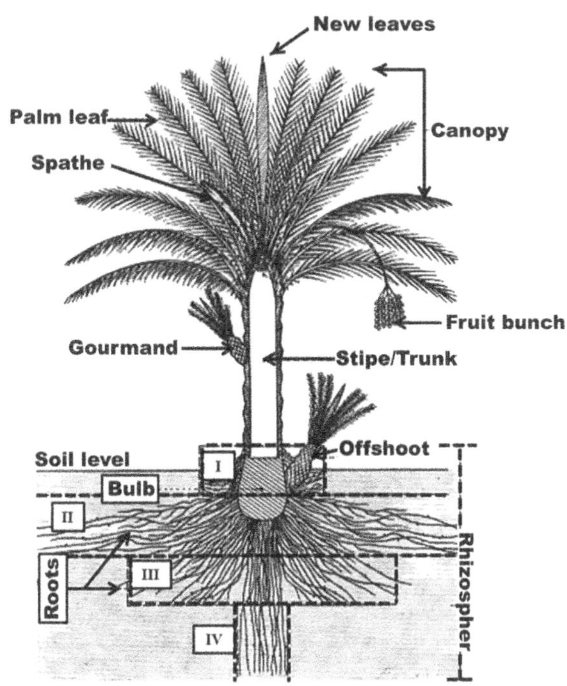

Fig. 1 Schematic presentation of a date palm

2.1 Fiber Structure

In order to better understand the physical structure of the date palm fibers and the influence of the various components and constituents, a diagram of the microstructure of a fiber is presented in Fig. 2 (Ng et al. 2015).

Plant fiber can be represented schematically as a tube, with the lumen in its center, an empty channel allowing the transfer of sap and water, and 4 cell walls superimposed on each other. The primary wall is generally made up of bundles of poorly ordered cellulosic fibers, while the secondary walls S3, S2 and S1 are made up of semi-crystalline bundles oriented helically (Rong et al. 2001; Kalia et al. 2009). These bundles, commonly called microfibrils, provide the mechanical strength of the fiber. In addition, these microfibrils are generally immersed in a matrix of hemicellulose and lignin of variable composition. Lignin is hydrophobic, prevents bacterial attack and stiffens cellulose microfibrils enough to withstand wind and gravity. Hemicellulose acts as a binder between cellulose microfibrils and lignin. Other compounds such as waxes, pectins and even minerals are also crucial in the organization of plant fibers and can also be found in smaller amounts in their structure (Rong et al. 2001; Kalia et al. 2009).

Cellulose Micro-fibrils

In the secondary walls of the fibers and mainly the S2 wall (the most influential due to its large thickness), the cellulose is present in the form of bundles of microfibrils oriented in different helical structures. The level of cellulose in the fibers, the degree of polymerization of the cellulose and the angle of the spirals on each wall vary for each plant and have a direct influence on their mechanical properties. More the cellulose content is higher and the spirals angle is smaller, more the fiber resistance is important (Ghori et al. 2018). Cellulose microfibrils consist of an ordered sequence of amorphous and crystalline phases and are said to be semi-crystalline. Generally for fibers such as cotton, linen or ramie, the crystallinity of cellulose is high (around 65–70%) (Spence et al. 2018). Figure 3 shows schematically the organization and

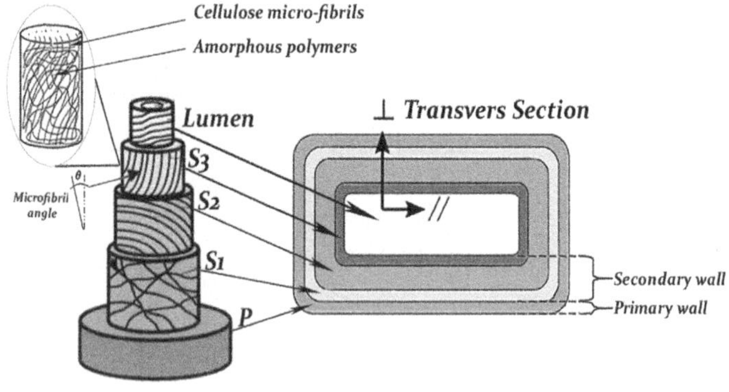

Fig. 2 Structure of a vegetable fiber

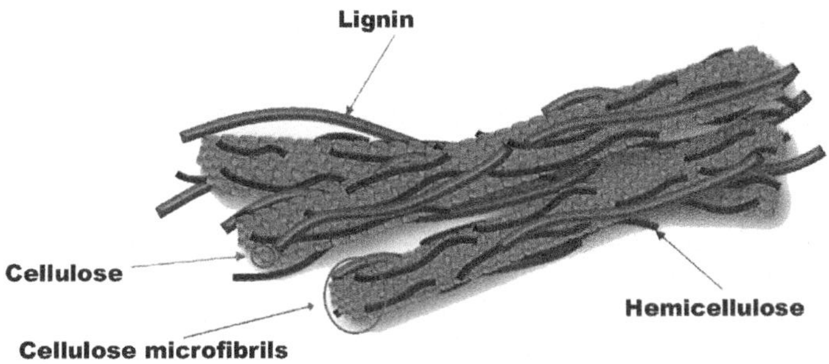

Fig. 3 Diagram showing the structural arrangement of lignin, hemicellulose and cellulose microfibrils

juxtaposition of the different macromolecules within the plant walls.

Hemicellulose

Hemicellulose is a heteropolysaccharide made up of 13 monomeric units classified into two families, furanoses (heterocycles with 5 elements), and pyranoses (cycle with 6 elements). This diversity in the chemical composition generates the formation of many varieties of hemicelluloses (Ebringerová 2006). In addition, hemicellulose is an amorphous polymer, due to numerous ramifications and a generally low degree of polymerization (of the order of 200). Hemicellulose has many accessible hydroxyl groups, in particular thanks to the main chains made up of xyloglucans, which promote adhesion to cellulose microfibrils (Ebringerová 2006).

Lignin

Lignin, like cellulose and hemicellulose, is a hydrocarbon polymer. Unlike these carbohydrate polymers, however, lignin is made up of both aromatic and aliphatic parts. Its biosynthesis takes place by the attack of different enzymes on its precursor, phenylalanine, and an amino acid. Three monomeric units can be obtained during the attack on phenylalanine. Or the units H (hydroxyphenyl), G (guaiacyl) and S (syringyle) corresponding respectively to coumaryl acid, coniferyl alcohol and synapylic alcohol (Abe et al. 2010). Depending on the nature and packaging of each plant, they can be found in different proportions. These monomers then react with each other by radical polymerization, forming a multitude of structural possibilities and preventing precise knowledge of the composition of lignin for each plant. Lignin is a poorly reactive, hydrophobic and fully amorphous polymer. Its properties are close to a plastic material. Until recently, lignin was considered an unwanted by-product that had to be gotten rid of or reduced in content in plants or wood (Abe et al. 2010).

2.2 *Date Palm Fiber Chemical Composition*

The results of the chemical composition of lignocellulosic fibers from the date palm are summarized in the graph in Fig. 4. According to Sbiai et al. (2010) these results represent means of at least 3 tests for each element.

The results obtained confirm that the date palm leaflets are made up of three major components: cellulose, hemicelluloses and lignin. The rest of the composition includes extractables, water-soluble and ash (mineral matter). Cellulose is the first constituent of date palm leaflets, it represents approximately 35% of the dry matter. This percentage is comparable to that of alfa fibers and perennial fibers (flax, kenaf, jute). On the other hand, it is larger than that of sabai fibers. Hemicelluloses represent the second major constituent in date palm leaflets. Their rate (28%) is comparable to that of wheat straw fibers and markedly higher compared to Peribian fibers and sisal leaf fibers. The date palm leaflets are very rich in lignin. The rate measured is around 27%. The latter is close to that determined in the case of bamboo and sugar cane fibers. It is slightly larger than that of jute and sabai and 3 times larger than that measured in the fibers of abaca and sisal leaves where lignin represents only 8% of the dry matter on average (Table 1) (Sbiai et al. 2010).

3 Fiber Extraction Methods

There are four main families of processes for extracting plant fibers; the physical process, the mechanical process, the chemical process and the biological process. The choice of the correct process depends on the type and age of the plant as well as the extracting organ. In some cases the coupling of several processes is required.

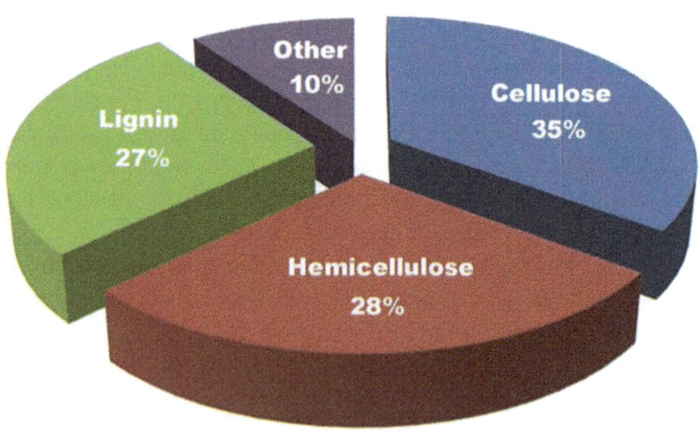

Fig. 4 Chemical composition of date palm fibers (% by weight)

Table 1 Chemical composition of some natural fibers, comparison with palm fibers (PLD) (Gandini and Belgacem 2002; Sbiai et al. 2010)

Type of fibers	Cellulose (%)	Hemicellulose (%)	Lignin (%)	Other (%)
Straw fibers				
Rice	28–36	–	12–16	–
Corn	29–35	27	16–21	11–23
Cane fibers of				
Sugar	32–44	22	19–24	26–8
Bamboo	26–43	15	21–31	9–35
Herbal fibers				
Alfa (Esparto)	3–38	–	17–19	–
Sabai	22	–	23.9	–
Peribial fibers				
Linen	43–47	16	21–23	13–20
kenaf	31–39	19	15–19	23–35
Jute	45–53	15	21–26	5–11
Leaf fibers of				
Palm	**35**	**28**	**27**	**10**
Abaca (Manila)	60.8	20	8.8	10
Sisal (agave)	43–56	12	7–9	23–38

3.1 Physical Process

The pressure explosion by steam (Sotton 1991; Kessler et al. 1998) is a very effective means in the separation of fibers, micro-explosions then occur in the most fragile cells. This cycle is repeated ten times (8–11 times), which results in the explosion of the medium lamellae which is an intercellular adhesive. In the end, the separation of the fibers will be completed by washing with running water (Tung et al. 2004). There are also, ultrasound-based and other microwave-based processes (Raveendran Nair et al. 2013).

3.2 Mechanical Process

This process can be carried out manually or using a machine, in the case of Bamboo the methods used are as follows:

- **The manual method**: Generally the fibers are separated by the use of a knife and a comb after being beaten (Rao and Rao 2007).
- **Method by rolling or pressing**: By crushing in press or by rolling or by combination of the two treatments, the fibers are detached from each other. This is

repeated several times in succession until the fibers are separated as much as possible (Tokoro et al. 2008).

3.3 Chemical Process

The chemical extraction is generally done by alkaline solutions (NaOH, KOH …). The chemical method partially reduces the diameter of the fiber.

3.4 Biological Process

Natural retting, an operation known for centuries, It is based on the activity of microorganisms, which implies a longer extraction time than other processes. Several retting processes are used:

- ***Retting in the field***: After having uprooted the plant from the ground while retaining the roots, the plant is left on the ground (several weeks for flax). It is exposed to different climatic factors (sun, rain, wind) in order to promote the development of microorganisms. The latter attack the pectic matter which connects the fiber bundles together. This action allows easier fiber extraction and refining. Lastly, this step is followed by a washing with water which results in fibers well separated from each other. It is an inexpensive process, which requires little labor. By cons, it requires a large area. Since, the plant is subject to climatic conditions, where we do not have total control of the process (Ali et al. 2018).
- ***Retting in a pond***: This process consists in immersing in water the different parts of the plant from which the extraction will be carried out. This technique is relatively long due to the natural development of bacteria, anaerobic which thanks to a specific enzymatic material are capable of degrading the plant macromolecules which bind the fibrous bundles between them. Although this method has shown excellent results, it is no longer used for health and environmental reasons these days. Indeed this method consumes a significant amount of water (Ali et al. 2018).
- ***Enzymatic retting***: This process aims to replace the classic retting (in the field and in the pond) to avoid water pollution, and reduce the treatment time. It is a low polluting and highly reproducible process. Thanks to enzymology, the fibers are extracted directly by the use of a mixture of enzymes. While touching as little as possible on cellulose, which is interesting for the industry (Ali et al. 2018).

4 Treatment and Surface Modification of Plant Fibers

The pretreatments and surface modifications of cellulose fibers are aimed at reducing the stresses linked to the properties of plant fibers and improving their adhesion with the matrix and reducing the absorption of moisture. Their limited thermal stability and their incompatibility with the matrices of synthetic polymers constitute a major challenge to their introduction to industrial uses (Mohanty et al. 2002; Yao et al. 2009). It has been shown that an appropriate treatment applied to the fibers can lead to compatibility with the polymer matrix, which improves the quality of the composites. Good performance of composite materials requires good dispersion and good cohesion with the polymer matrix in order to achieve an effective transfer of stress inside the structure. For these reasons, several studies have been carried out to circumvent these constraints based on physical, chemical and indeed even biological methods. Chemical modifications include alkaline treatments and the grafting of certain molecules such as silane (functionalized silicon alkoxide) and acetic acid, or molecules based on benzoyl, isocyanate, triazine or imidazolidinone, ect (Bledzki and Gassan 1999; John and Anandjiwala 2008). Their action is either the reduction of hydrogen bonds inside the complex structure of the fiber or by reacting the numerous hydroxyl groups on the surface of the fiber with other active groups of the reagent (Valadez-Gonzalez et al. 1999).

Like chemical methods, physical methods are diverse; explosion by steam, plasma, corona discharge, heat treatment, laser, gamma rays and UV, all are techniques used to modify the structural surface of fibers by causing the separation of the fibrils and their reorganization so that they can be deposited in a uniform direction inside the matrix and thus modifying the structural properties of the fibrillar surface (Bledzki and Gassan 1999; John and Anandjiwala 2008). These treatments are generally used in combination with other basic treatments, whether chemical or biological.

Biological agents such as microorganisms and in particular fungi provided with a specific enzymatic background (Gustavsson et al. 2005; Pietak et al. 2007) are an alternative to chemical and physical methods, (Gulati and Sain 2006; Pickering et al. 2007). Biological modifications offer several advantages over chemical and physical processes. They can selectively separate hydrophilic compounds such as hemicelluloses and pectins with a lower energy supply. Li and Pickering (Pickering et al. 2007) used enzymes and chelators to separate the hemp bundles into individual fibers, and found that the crystallinity (by X-ray diffraction) and thermal properties (thermogravimetric analysis) improved after separation. In another study, Pietak et al. (2007) studied the hydrophilicity of natural fibers using the microscope and the contact angle technique, they found an increase in the adhesion between fiber and matrix. Kardas et al. (2009) also report that the micro-relief of polyester fabric treated with esterase takes on a homogeneous texture during its production. Pectinase also has been proven by Saleem et al. (2008) that it improves the mechanical characteristics of biocomposites; according to their study, hemp fibers treated with 8% pectinase in a solution of maleic anhydride increases the tensile strength, the flexural strength and the elasticity of the composite. The fineness of the fibers is another parameter which can be controlled by the enzymatic treatment, a recent study carried out on bamboo fibers using (xylanase, cellulase, pectin lyase and laccase) revealed an increase in the fineness of the treated fibers, according to Liu et al. (2012) this result is obtained following the removal of the hemicellulosic fraction.

The purpose of the pretreatment, whether chemical, physical or biological of plant fibers differs according to the industrial uses of these fibers, in the case of textiles is the improvement of the fixation and the quality of the coloring of the tissue, in biocomposite uses it is different it is rather improvement of the adhesion of resins with fibers.

Also the pretreatment should guarantee:

- The elimination of impurities present in the fibers to improve their uniformity of structure and color.

The pre-treatment processes and techniques depend on:

- The nature of the fiber to be treated: for raw materials made of natural fibers, such as cotton, wool, linen and silk, mastering the pre-treatment is more difficult than that for raw materials based on synthetic fibers and artificial. In fact, natural fibers are accompanied by a higher quantity of substances liable to interfere with their subsequent transformation, whereas usually artificial fibers contain only preparation agents, soils and synthetic fillers soluble in water.

4.1 Alkaline Treatment

The alkaline treatment with sodium hydroxide NaOH is the most used in the treatment of natural fibers, it aims to increase the contact surface between the fiber and the matrix. It is often used to remove lignin, hemicellulose and to remove residual impurities from the fiber, therefore the interfibrillary region is less dense and less rigid, which allows the fibrils to reorganize in the direction of traction. When the fibers are stretched, there is an arrangement between the fibrils which results in a better distribution of stress. The treatment increases the extraction of fibers from the bundles. On the other hand, it can harm and deteriorate the fiber if the process is not optimized. Indeed, the time, the temperature, and the concentration of the treatment play on the thermal and mechanical properties of the fiber (Bodros and Baley 2008). The alkalization treatment influences the thermal and physical properties of natural fibers as well as the morphology and size of the fibers (Ray et al. 2001; John and Thomas 2008).

4.2 Acetylization Treatment

Acetylization improves the hydrophobic power of the fiber. The principle of this method is to react the hydroxyl groups (–OH) of the fiber with the acetyl groups (CH_3CO–), and therefore to make the surface of the fiber more hydrophobic by causing dimensional stability of the composites, because none Water absorption cannot lead to swelling or removal of the composite material (Baley 2002). Baley et al. carried out treatments on flax fiber and showed that treatment with acetic anhydride improves the inter-facial shear properties of the fiber by 20% (Baley 2002).

4.3 Silane Treatment

The aim of this treatment is to increase the interaction between the fiber/matrix links. This has been confirmed by studies, on flax fibers, on wood fibers as well as on loofah fibers (Ichazo et al. 2001; Demir et al. 2006; George et al. 2014), and which all report that silane increases the compatibility between the fibers and the matrix and thanks to this treatment, they have obtained better mechanical performance.

4.4 Chemical Bleaching

Bleaching process is used on the alkali-treated fibers; it is carried out as a preparatory step to coloring and allows obtaining fibers of a white and uniform color. Chemical bleaching is done by different reagents such as hydrogen peroxide, sodium hypochlorite and sodium chlorite (Semlali Aouragh Hassani et al. 2020). This multi-step bleaching is known to remove the undesired components and thus modifying the surface between matrix and fiber. To conclude, the white fibers obtained imply the increase of the surface ratio and therefore a better fiber-matrix (Ng et al. 2015).

5 Characterization of Date Palm Fiber

5.1 Fourier Transform Infrared (FT-IR)

To analyze the transmittance bands characteristics of lignocellulosic fibers, the FT-IR is used. For this, the IR spectra for the raw, alkaline-treated and bleached date palm fibers are presented in Fig. 5. The peak observed at 3200–3600 is associated to the stretching vibrations of O–H, and the pick at 1027 cm^{-1} to the stretching C–O groups of cellulose (Chollakup et al. 2013). While the alkaline treatment using NaOH is associated to the absorption bands around 2918, 1730, 1650, 1598, 1458, 1247, 1160 and 750 cm^{-1}. The diminution of the lignins, carboxylic esters in hemicellulose, pectins and waxes contents are reflected by the lower peak intensity at 2918 and 1730 cm^{-1} (associated respectively to the C–H stretching of cellulose and C=O stretching of hemicellulose) (Chollakup et al. 2013; Boujmal et al. 2014). Also, the bands at 1650 and 750 cm^{-1} confirms the presence of lignins (Boujmal et al. 2014). The substituted aromatic ringand to the CH_2 symmetric bending in cellulose is represented by the C=C stretching vibration associated to the peaks 1598 and 1458 cm^{-1} (Essabir et al. 2015). The C–O stretching vibration of the acetyl group in hemicelluloses and lignins is represented by the strong peak at 1247 cm^{-1} (Ouarhim et al. 2018). Finally, the band at 1160 cm^{-1} reflects the carbohydrate backbone of cellulose and the C–OH stretching vibration of the cellulose backbone (Boujmal et al. 2014). The complete removal of lignins, pectins, waxes and hemicelluloses

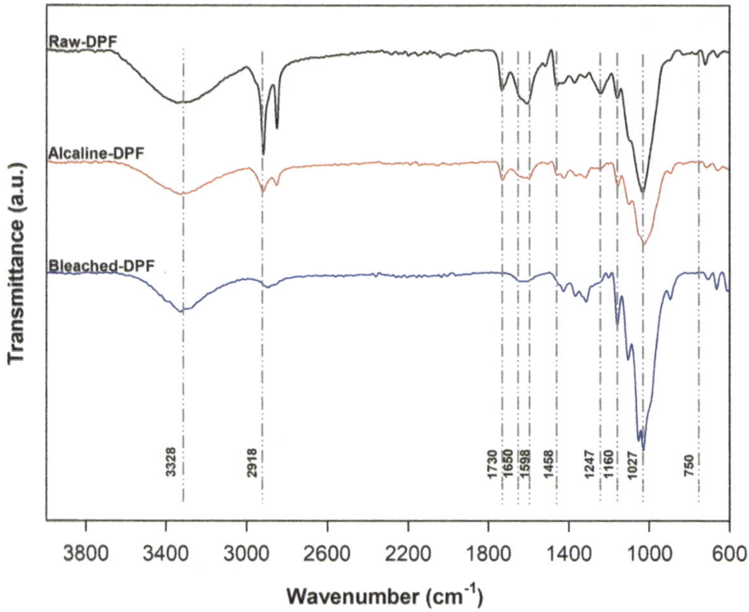

Fig. 5 FTIR spectra for the raw, alkaline-treated and bleached date palm fibers

from the fibers surface during bleaching is confirmed by the absence of this peak for the bleached fibers (Marwah et al. 2014; Semlali Aouragh Hassani et al. 2020).

5.2 Scanning Electron Microscopy (SEM)

SEM is generally used to study the morphology of fibers surface before and after each chemical modification. Figure 6 presents the morphology of the raw date palm fibers (diameter between 100 and 1000 μm and micro-fiber dimension of around 70 μm). In general, raw fibers are constituted by number of micro-fibrils and are covered by different layers. In this case many irregularities in the fiber surface are observed due to the presence of impurities (Alawar et al. 2009; Semlali Aouragh Hassani et al. 2020).

5.3 Thermogravimetric Analysis (TGA)

Generally, natural fibers are known for their low thermal stability which limits their adhesion with the melt thermoplastic polymer (Al-Khanbashi and Al-Kaab 2005). Thus, the hemicelluloses have the lowest thermal stability followed by cellulose,

Fig. 6 SEM morphology for untreated Date palm fiber

while lignins have the highest thermal resistance (Al-Khanbashi and Al-Kaab 2005). Semlali Aouragh Hassani et al. (2020) investigate the effect of alkaline treatment and functionalization on the date palm fiber thermal stability (degradation profiles) as illustrated in Fig. 7. Indeed, the fibers were first bleached, then modified by adding 1% of a novel functionalized dye coupling agent synthetized in their laboratory (dye-IPTMS) (Semlali Aouragh Hassani et al. 2020). The raw DPF shows good stability (260 °C) above which degradation starts, while the onset of thermal decomposition increases from alkaline to functionalized treated fibers, which can be explain by the removal of the impurities covering the raw fiber cell walls (Rosa et al. 2010). The first degradation step observed in DTG curve (around 230 °C) is assigned to hemicelluloses and cellulose decomposition, while the second DTG peak (around 320 °C) is associated to lignins decomposition (Ouarhim et al. 2018). Conversely,

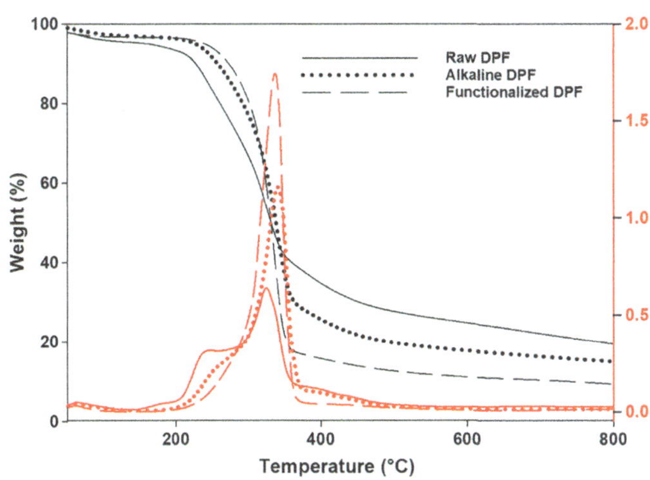

Fig. 7 TGA curve and first derivatives of treated and untreated date palm fiber

Table 2 The mechanical properties of the DPF before treatment and after treatment

Type of treatment	Average tensile strength (MPa)	Average young's modulus (MPa)
Raw	275	845
0.3% HCL	105	3233
0.5% NaOH	458	165,971
1% NaOH	849	153,969
2.5% NaOH	353	142,542
5% NaOH	213	70,474

for the treated fibers, the hemicellulose removal is associated to the disappearance of the shoulder around 230 °C, while the lignins removal is reflected by the lower amount of residues above 370 °C due to the chemical treatment (from alkaline to functionalized treated fibers) (Rosa et al. 2010).

5.4 *Single Fiber Tensile Test*

To measure a single date palm fiber tensile property, the fibers were manually separated to be glued in their both ends (about 20 mm) on stiff galvanic steel. Thus, the tensile strength, Young's modulus, and elongation to break of single date palm fiber can be determined according to ASTM D-3379-75. Table 2 summarizes the average mechanical test results of the un-treated and treated (Acetylation and alkaline) single fibers. It can be observed that acetylation shows a negative effect on fibers tensile strength mainly assigned to the acid attack on the fibers structure and distorted the fiber surface. Conversely, the mechanical performances of the date palm fiber treated with different concentrations of NaOH shows strong improvement. The maximum tensile strength observed at 1% NaOH treatment is due to the impurities removal from fiber surface by soda concentration addition. However, a further weakening in fiber strength was observed, so the tensile strength starts to decrease. Indeed, the soda attacks the main construction components of the fiber and more grooves appear on the surface of the fibers by the increase of soda concentration (Alawar et al. 2009).

6 Conclusion

This chapter focuses on the date palms fibers as a lignocellulosic resource. This plant resource constitutes a renewable resource, naturally biodegradable, and has many high technical qualities. The most used are bast fibers, such as kenaf, jute, flax but these fibers can be replaced by date palm leaflets fibers which occur in large quantities and which are only agricultural waste. Thus, this chapter has extensively detailed four main families of plant fiber extraction processes; the physical process, the mechanical process and the chemical process. In order to better understand the association chemistry of the cellulosic fiber and the polymer matrix, various chemical fiber treatments are proposed in order to facilitate this union within the fiber/matrix composite. The physical, thermal and mechanical results obtained have shown that each treatment has a specific effect on the properties of date palm fibers.

Acknowledgements This work was supported by MAScIR (Moroccan Foundation for Advanced Science, Innovation and Research), MESRSFC and CNRST Morocco grant no. 1970/15.

References

Abe A, Dušek K, Kobayashi SB (eds) (2010) Biopolymers: lignin, proteins, bioactive nanocomposites. Springer, Berlin

Al-Khanbashi A, K. Al-Kaab, Hammami A (2005) Date palm fibers as polymeric matrix reinforcement: fiber characterization. Polym Compos 26:486–497. https://doi.org/10.1002/pc.20118

Al-Shayeb SM, Al-Rajhi MA, Seaward MRD (1995) The date palm (Phoenix dactylifera L.) as a biomonitor of lead and other elements in arid environments. Sci Total Environ 168:1–10. https://doi.org/10.1016/0048-9697(95)04556-G

Alawar A, Hamed AM, Al-kaabi K (2009) Characterization of treated date palm tree fiber as composite reinforcement. Compos Part B 40:601–606. https://doi.org/10.1016/j.compositesb.2009.04.018

Ali A, Shaker K, Nawab Y et al (2018) Hydrophobic treatment of natural fibers and their composites—a review. J Ind Text 47:2153–2183. https://doi.org/10.1177/1528083716654468

Baley C (2002) Analysis of the flax fibres tensile behaviour and analysis of the tensile stiffness increase. Compos Part a Appl Sci Manuf 33:939–948. https://doi.org/10.1016/S1359-835X(02)00040-4

Bledzki AK, Gassan J (1999) Composites reinforced with cellulose based fibres. Prog Polym Sci 24:221–274

Bodros E, Baley C (2008) Study of the tensile properties of stinging nettle fibres (Urtica dioica). Mater Lett 62:2143–2145. https://doi.org/10.1016/j.matlet.2007.11.034

Boujmal R, Essabir H, Nekhlaoui S et al (2014) Composite from polypropylene and henna fiber: structural, mechanical and thermal properties. J Biobased Mater Bioenergy 8:246–252

Chollakup R, Smitthipong W, Kongtud W, Tantatherdtam R (2013) Polyethylene green composites reinforced with cellulose fibers (coir and palm fibers): effect of fiber surface treatment and fiber content. J Adhes Sci Technol 27:1290–1300. https://doi.org/10.1080/01694243.2012.694275

Demir H, Atikler U, Balko D (2006) The effect of fiber surface treatments on the tensile and water sorption properties of polypropylene—luffa fiber composites. Compos Part a 37:447–456. https://doi.org/10.1016/j.compositesa.2005.05.036

Ebringerová A (2006) Structural diversity and application potential of hemicelluloses. Macromol Symp 232:1–12. https://doi.org/10.1002/masy.200551401

Essabir H, Achaby ME, Hilali EM et al (2015) Morphological, structural, thermal and tensile properties of high density polyethylene composites reinforced with treated argan nut shell particles. J Bionic Eng 12:129–141

Gandini A, Belgacem MN (2002) La chimie des substances renouvelables. L'actualité chimique

George M, Mussone PG, Bressler DC (2014) Surface and thermal characterization of natural fibres treated with enzymes. Ind Crops Prod 53:365–373. https://doi.org/10.1016/j.indcrop.2013.12.037

Ghori W, Saba N, Jawaid M, Asim M (2018) A review on date palm (phoenix dactylifera) fibers and its polymer composites. IOP Conf Ser Mater Sci Eng 368. https://doi.org/10.1088/1757-899X/368/1/012009

Gulati D, Sain M (2006) Fungal-modification of natural fibers: a novel method of treating natural fibers for composite reinforcement. J Polym Environ 14:347–352. https://doi.org/10.1007/s10924-006-0030-7

Gustavsson MT, Persson PV, Iversen T et al (2005) Modification of cellulose fiber surfaces by use of a lipase and a xyloglucan endotransglycosylase. Biomacromol 6:196–203. https://doi.org/10.1021/bm049588i

Ichazo MN, Albano C, González J et al (2001) Polypropylene/wood flour composites: treatments and properties. Compos Struct 54:207–214. https://doi.org/10.1016/S0263-8223(01)00089-7

John MJ, Anandjiwala RD (2008) Recent developments in chemical modification and characterization of natural fiber-reinforced composites. Polym Compos 29:187–207. https://doi.org/10.1002/pc.20461

John MJ, Thomas S (2008) Biofibres and biocomposites. Carbohydr Polym 71:343–364. https://doi.org/10.1016/j.carbpol.2007.05.040

Kalia S, Kaith BSK (2011) Cellulose fibers: bio- and nano-polymer composites: green chemistry and technology. Springer, Heidelberg, London

Kalia S, Kaith BS, Inderjeet K (2009) Pretreatments of natural fibers and their application as reinforcing material in polymer composites—a review. Polym Eng Sci 49:1253–1272. https://doi.org/10.1002/pen

Kardas I, Lipp-Symonowicz B, Sztajnowski S (2009) Comparison of the effect of PET fibres' surface modification using enzymes and chemical substances with respect to changes in mechanical properties. Fibres Text East Eur 17:93–97

Kessler RW, Becker U, Kohler R, Goth B (1998) Steam explosion of flax—a superior technique for upgrading fibre value. Biomass Bioenerg 14:237–249. https://doi.org/10.1016/0165-1765(91)90013-B

Li J, Zhang LP, Peng F et al (2009) Microwave-assisted solvent-free acetylation of cellulose with acetic anhydride in the presence of iodine as a catalyst. Molecules 14:3551–3566. https://doi.org/10.3390/molecules14093551

Liu T, Yu F, Yu X et al (2012) Basalt fiber reinforced and elastomer toughened polylactide composites: mechanical properties, rheology, crystallization, and morphology. Appl Polym Sci 125:1292–1301

Marwah R, Ibrahim NA, Zainuddin N et al (2014) The effect of fiber bleaching treatment on the properties of poly(lactic acid)/oil palm empty fruit bunch fiber composites. Int J Mol Sci 15:14728–14742. https://doi.org/10.3390/ijms150814728

Mohanty AK, Misra M, Drzal LT (2002) Sustainable bio-composites from renewable resources in green materials world. J Polym Environ 10:19–26. https://doi.org/10.1023/A:1021013921916

Ng HM, Sin LT, Tee TT et al (2015) Extraction of cellulose nanocrystals from plant sources for application as reinforcing agent in polymers. Compos Part B Eng 75:176–200. https://doi.org/10.1016/j.compositesb.2015.01.008

Ouarhim W, Bensalah M, Rodrigue D et al (2018) Production and characterization of high density polyethylene reinforced by eucalyptus capsule fibers. J Bionic Eng 15:558–566

Pickering KL, Li Y, Farrell RL, Lay M (2007) Interfacial modification of hemp fiber reinforced composites using fungal and alkali treatment. J Biobased Mater Bioenergy 1:109–117. https://doi.org/10.1166/jbmb.2007.012

Pietak A, Korte S, Tan E et al (2007) Atomic force microscopy characterization of the surface wettability of natural fibres. Appl Surf Sci 253:3627–3635. https://doi.org/10.1016/j.apsusc.2006.07.082

Rao KMM, Rao KM (2007) Extraction and tensile properties of natural fibers: Vakka, date and bamboo. Compos Struct 77:288–295. https://doi.org/10.1016/j.compstruct.2005.07.023

Raveendran Nair G, Rho D, Yaylayan V, Raghavan V (2013) Microwave assisted retting—a novel method ofprocessing of flax stems. Biosyst Eng 116:427–435. https://doi.org/10.1016/j.biosystemseng.2013.10.004

Ray D, Sarkar BK, Rana AK, Bose NR (2001) The mechanical properties of vinylester resin matrix composites reinforced with alkali-treated jute fibres. Compos Part a Appl Sci Manuf 32:119–127. https://doi.org/10.1016/S1359-835X(00)00101-9

Rong MZ, Zhang MQ, Liu Y et al (2001) The effect of fiber treatment on the mechanical properties of unidirectional sisal-reinforced epoxy composites. Compos Sci Technol 61:1437–1447. https://doi.org/10.1016/S0266-3538(01)00046-X

Rosa MF, Medeiros ES, Malmonge JA et al (2010) Cellulose nanowhiskers from coconut husk fibers: effect of preparation conditions on their thermal and morphological behavior. Carbohydr Polym 81:83–92. https://doi.org/10.1016/j.carbpol.2010.01.059

Saleem Z, Rennebaum H, Pudel F, Grimm E (2008) Treating bast fibres with pectinase improves mechanical characteristics of reinforced thermoplastic composites. Compos Sci Technol 68:471–476. https://doi.org/10.1016/j.compscitech.2007.06.005

Sbiai A, Abderrahim Maazouz, Fleury E, et al (2010) Short date palm tree fibers/polyepoxy composites prepared using Rtm process: effect of tempo mediated oxydation of the fibers. BioResources 5:672–689. https://doi.org/10.15376/biores.5.2.672-689

Semlali Aouragh Hassani FS, Ouarhim W, Raji M et al (2020) N-silylated benzothiazolium dye as a coupling agent for polylactic acid/sate palm fiber bio-composites. J Polym Environ 21. https://doi.org/10.1007/s10924-019-01585-x

Sotton M. FM (1991) Adjustment of the steam explosion treatment to extract fibres from plants, usable for textile and related end-uses. In: Focher B, Marzetti A, Crescenzi V (eds) Steam explosion techniques: fundamentals and industrial applications. Gordon & Breach

Spence K, Habibi Y, Dufresne A (2011) Nanocellulose-based composites. In: Kalia S, Kaith BS, Kaur I (Eds) Cellulose fibers: bio- and nano-polymer composites, Springer, Berlin, 179–213

Tokoro R, Vu DM, Okubo K et al (2008) How to improve mechanical properties of polylactic acid with bamboo fibers. J Mater Sci 43:775–787. https://doi.org/10.1007/s10853-007-1994-y

Tung NH, Yamamoto H, Matsuoka T, Fujii T (2004) Effect of surface treatment on interfacial strength between bamboo fiber and PP resin. JSME Int Journal, Ser a Solid Mech Mater Eng 47:561–565. https://doi.org/10.1299/jsmea.47.561

Valadez-Gonzalez A, Cervantes-Uc J, Olayo R, Herrera-Franco P (1999) Chemical modification of henequén fibers with an organosilane coupling agent. Compos Part B Eng 30:321–331. https://doi.org/10.1016/S1359-8368(98)00055-9

Yao F, Wu Q, Zhou D (2009) Thermal decomposition of natural fibers: global kinetic modeling with nonisothermal thermogravimetric analysis. J Appl Polym Sci 114:834–842. https://doi.org/10.1002/app

Date Palm Fiber Preform Formation for Composites

Mohammad Saleh Ahmadi, Zahra Dehghan-Banadaki, Mehran Nalchian, and Reyhaneh Sarpanahi

Abstract Preform formation is the process of placement of reinforcing fibers in the desired orientations within an appropriately designed structure utilizing a suitable textile manufacturing technique to shape a reinforcement material required for fabrication of a composite part. Textile preforms can be categorized into one-, two-, and three-dimensional structures. Proper selection of architecture and geometry of the preform has a great influence on the structural performance of the resulting composite part. In this chapter relevant methods for staple yarn formation, manufacturing of two-dimensional fabrics, including woven, knitted, and nonwoven fabrics, and formation methods of three-dimensional fabrics, including 3-D woven, braided, nonwoven, and stitched fabrics are described. Advantages and limitations of each method concerning preform formation from date palm fibers are explained.

1 Introduction

In high performance composite materials, the reinforcing fibers are not only required to have low density and good mechanical properties, but they are also needed to be available in suitable preforms with the desired fiber orientations (Madsen et al. 2013). There are a lot of textile preform architectures available for the designers to select for composite reinforcements. In-plane fiber orientation, existence of through-thickness reinforcements and availability in the near-net-shape form, are the key criteria for the selection of textile preforms. Depending on the end use requirements and the

M. S. Ahmadi (✉) · Z. Dehghan-Banadaki · M. Nalchian · R. Sarpanahi
Department of Textile Engineering, Yazd University, 89168-69511 Yazd, Iran
e-mail: ms.ahmadi@yazd.ac.ir

Z. Dehghan-Banadaki
e-mail: dehghan.z@stu.yazd.ac.ir

M. Nalchian
e-mail: m.nalchian@stu.yazd.ac.ir

R. Sarpanahi
e-mail: r.sarpanahi@stu.yazd.ac.ir

M. Midani et al. (eds.), *Date Palm Fiber Composites*, Composites Science and Technology,
https://doi.org/10.1007/978-981-15-9339-0_3

processing conditions, some or all of these features are required (Ko 1999). Natural fibers have usually low densities and, in some cases, acceptable mechanical properties. In contrast to synthetic fibers, which can be easily found in the continuous form, natural fibers have limited lengths, due to their natural origin. Therefore, in many cases, natural fibers are used as randomly oriented short reinforcements in composite materials. Nonwoven mats are the very common and commercially available kinds of natural fiber preforms, which have already been successfully used for such applications as reinforcing automotive composite panels. However, when composite parts are to be designed with some preferential stress directions, and/or they are needed to withstand higher levels of stress/strain, the fibers are then necessary to be spun into a yarn, so that preforms with specific fiber orientations such as woven, braided and knitted fabrics can be made.

Although suitable textile processes for converting date palm fibers (DPFs) into preforms are not clearly explained in the literature, this chapter aims to describe applicable preform formation methods for this fiber and also to depict the advantages and drawbacks of each method.

2 One-Dimensional Reinforcements

2.1 Yarns

Yarns can been defined as a generic term for continuous strand of textile fibers in a form suitable for knitting, weaving, or otherwise inter-twining to form a textile fabric (Behery 2005). The term spinning may be defined as the process or processes used to produce either fibers or filaments from natural or synthetic polymers, or convert natural or man-made fibers and filaments into yarns by twisting or other means of binding together the fibers or filaments (Lawrence 2010).

Yarns spun from natural staple fibers may be categorized as single, plied and cabled or cord yarns. Single yarns are made by twisting a large number of staple fibers together. Plied yarns are a combination of two or more single yarns (each called a ply) twisted together. When two or more plied yarns are combined and twisted together, cabled or cord yarns are formed. Cord yarns structures are found in some ropes, cordages, sewing threads, and decorative yarns. Single, plied, and cord yarns are illustrated in Fig. 1. A hand-made two-plied yarn from DPF is shown in Fig. 2.

One of the most influential parameters of the yarn, governing its physical and mechanical properties, is the yarn twist. In staple yarns, by increasing the twist up to a particular level called the *optimum twist*, tensile strength increases. Afterwards, further increase in the twist leads to decrease in the strength (Behery 2005). However, in the composites reinforced with staple yarns, the yarn strength, created by fiber-to-fiber friction as a result of yarn twist, is only needed to protect the yarn against the forces applied during preform formation processes. In the final composite part,

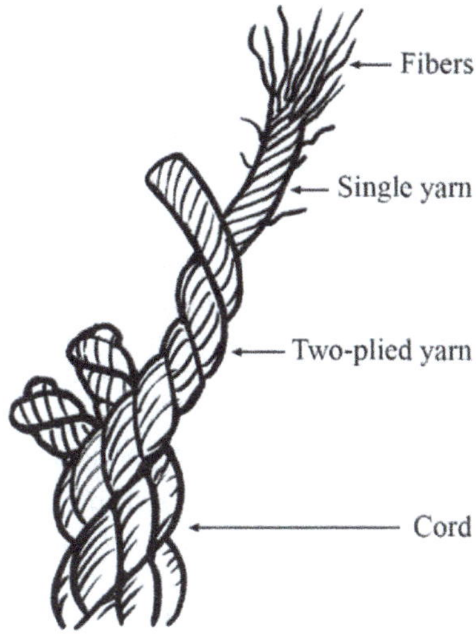

Fig. 1 Single, plied, and cord yarns

Fig. 2 A hand-made two-plied yarn from DPF

instead of fiber-to-fiber friction, the fibers-matrix interfacial adhesion will play the key role in transferring the applied stresses between the fibers. On the other hand, the fiber misalignment, due to the helical configuration of the fibers in the twisted yarn, negatively influences the mechanical properties of the resultant composite. Moreover, at high levels of twist, higher resistance of the yarns to the resin flow during resin impregnation process may result in improper fiber wet-out (Goutianos and Peijs 2003). Goutianos and Peijs (2003) and Goutianos et al. (2006) attempted to optimize the twist level in flax fiber yarns, so that by the selection of the minimum possible level of twist, the processability and the mechanical properties of the final composites can be balanced.

2.1.1 Yarn Formation Methods

Traditionally, in many date palm growing countries, DPFs are spun into yarns by hand twisting or by manually rotating spinning wheels. The products are usually used for making ropes, doormats and handicrafts. Figure 3 shows a pencil holder decorated with date palm yarns.

However, date palm yarn spinning is a field in which mechanization has hardly penetrated. DPF has a diameter range of 100–1,000 μm (Ghori et al. 2018), which is comparatively high among other natural fibers. Thus, there are difficulties in yarn formation from such fibers because of high torsional and flexural rigidity and inadequate inter-fiber friction. As a consequence, producing fine pure yarns from these fibers is very difficult. The pure yarns can be produced with high linear densities, which makes them suitable for making cable yarns and ropes, but not for converting into conventional reinforcement fabrics. Using course yarns in the reinforcement fabrics leads to the composite structures with highly undulating yarns, high resin rich areas, and consequently low mechanical properties. To make yarns with lower linear densities, blending DPF with other finer fibers can be a solution. Therefore, in spite of variety of methods applicable for yarn production from staple yarns, only the methods which seems more appropriate for processing DPF, are presented in this part. Since the shape and structure of DPFs are similar to those for coir fibers (Ghori et al. 2018), some of the methods described herein, are those which have been successfully implemented in processing coir fibers.

Coir spinning

Fig. 3 A pencil holder decorated with date palm yarns

Coir is mainly spun by traditional methods using manually rotated wheels called *ratts*. Nowadays *Motorized Traditional Ratts*, are also available in the market as semi-mechanized spinning wheels, in which a motor is utilized to rotate the wheel. *Automatic coir spinning machines* have been developed following research by the *Coir Board of India* (Mathai 2005). The diagram of a typical double-head automatic coir fiber spinning machine is schematically shown in Fig. 4. The coir fibers, usually in the form of a loose fiber mat or a set of slivers, are fed by the feed lattice or conveyor "A" to the beater "C". The rotating beater opens the fibers and deposits them into four V-shaped collection troughs. One filament yarn "H" is also fed into each trough to be positioned in the core of the yarn. It imparts strength to the yarn and facilitates the spinning process. The yarn is twisted by the false twister "E" and thus picks up the deposited coir fibers as it moves through the length of the V-shaped trough to form a single strand "F". Two adjacent single strands are joint and doubled at the guide "G" and enter the arm twisting and winding unit, which imparts true twist to the two-plied yarn and winds it on the spool "J" (Thanabal and Saravanan 2018).

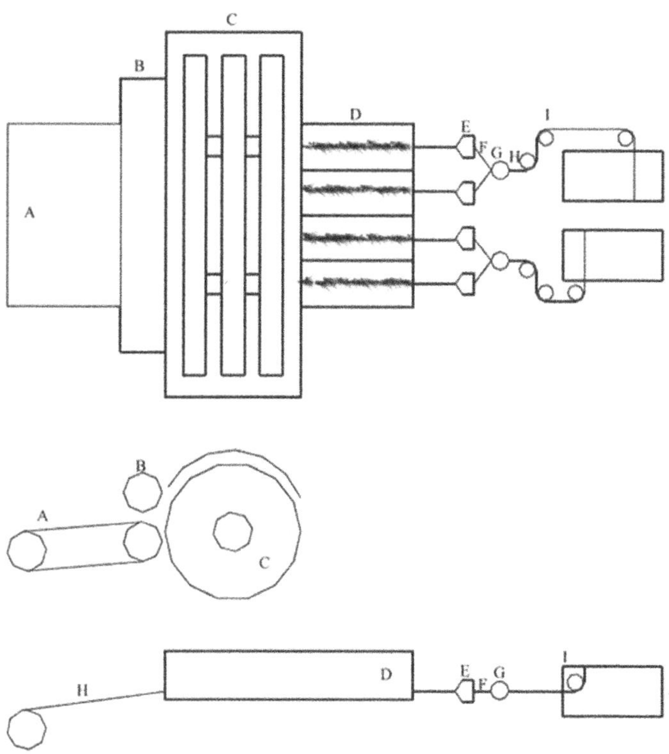

Fig. 4 Schematic diagram of coir fiber spinning process (Reproduced from Thanabal and Saravanan 2018)

In the coir spinning industry, the linear density of the coir yarn is referred to as *runnage*. It represents the length of the yarn in a standard mass of one kilogram (Velayutham and Dhandapani 2019). Runnage of two-plied 100% coir yarns produced by automatic spinning machines ranges from 50 to 300 m/kg (equivalent to about 3.3–20 ktex). The yarns may thus be too coarse to be woven as reinforcement fabrics, but they may be used as 1-D reinforcements in processes such as pultrusion. Nevertheless, the fibers do not possess a high degree of parallelization due to the nature of the spinning process.

Jute spinning

In the jute spinning process, the fibers are commonly prepared by passing two carding and then three drawing stages and are finally spun using a flyer spinning machine. It was shown by some experiments that chemically soften coir fibers blended with jute, sisal, and hemp fibers can be spun using the modified jute spinning system. The blending of coir with jute, sisal, and hemp will improve its spinnability by way of reducing the flexural and torsional rigidity of the fiber assembly. It is reported that the jute spinning process helps to produce coir blended yarns of 25–30% finer as compared to 100% coir yarns produced by the automatic coir spinning machine. Moreover, a higher degree of fiber parallelization in the yarn, owing to the gill drawing stages in the process, will lead to the enhanced mechanical properties in the final composite. It seems that the modified jute spinning system can also be utilized for spinning blended yarns from DPF and other natural fibers for the use in hybrid natural fiber composites.

Hollow spindle spinning

Hollow spindle spinning or *wrap spinning* is a process by which twistless yarns can be produced. In this process, a wrapper yarn, usually a filament one, is wrapped around a bundle of twistless staple fibers. Therefore, instead of being held by the inter-fiber frictional force due to twisting, the staple fibers are held together by the wrapping filament. This yarn has enough strength to withstand stresses applied during the textiles formation or composite fabrication processes. If the final fabric is intended to be formed from pure staple fibers, it is possible to produce the yarn using soluble wrapping filaments, which can be dissolved after the fabric is made (Miao and Finn 2008). In hollow spindle spinning, a sliver of staple fibers is fed to a drafting arrangement, where it is attenuated into the desired linear density. Afterward, by entering into a hollow spindle, the bundle of staple fibers is wrapped by a filament yarn, which is unwound from a bobbin mounted on the rotating hollow spindle (Fig. 5). After leaving the delivery rollers, the wrap-spun yarn is wound on a suitable package (Mankodi 2011). Hybrid yarns can also be produced by using DPF in the core and other yarns in the sheath. Since the twist is not imparted on the core, high torsional and flexural resistance of DPF is not a problem in this method. For making thermoplastic composites, DPF can be wrapped with a thermoplastic filament, which will be melted in subsequent processes to form the matrix material. The topic of comingled preforms for fiber-reinforced thermoplastics is described in Sect. 5.

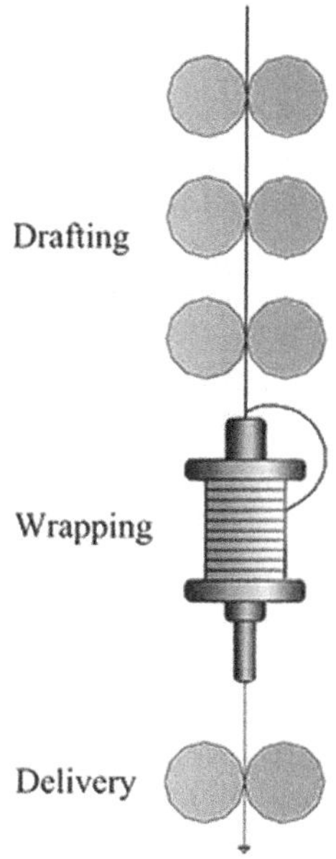

Fig. 5 Wrap spinning process

2.2 Unidirectional (UD) Tapes

UD tapes can be made by positioning DPF yarns parallel to each other in a longitudinal direction. A small amount of another yarn may be used in the transverse direction in order to hold the longitudinal yarns in position. DPF yarns may be used as the warp yarns in narrow woven fabrics, along with some other fine weft yarns with high spacing to hold them in position with the minimum possible crimp level.

UD *prepregs* are resin-impregnated UD reinforcements, manufactured by impregnating fibers with a controlled amount of a thermoset or thermoplastic matrix material. Thermoset prepregs are stored in *B-stage*, i.e., a partially cured state, prior to fabrication. The prepregs will be laid into a closed mold, where the B-stage resin starts to cure after applying pressure and heat.

3 Two-Dimensional Reinforcements

3.1 Woven Fabrics

Weaving is a major textile production method, in which two sets of perpendicular threads, i.e., the warps and the wefts, are interlaced together to form a fabric. This is done on a machine called the weaving *loom*. The main advantage of using woven fabrics as composite preforms is the possibility to pre-orient the yarns in the designed directions (Cristaldi et al. 2010). Moreover, in many composite manufacturing processes such as *resin transfer molding (RTM)*, *spray-* or *hand lay-up* and *vacuum infusion*, for ease of handling, the reinforcements are used in the form of fabrics (Goutianos et al. 2006). Woven fabrics are characterized by the counts or linear densities of warps and wefts, the weave pattern, the number of yarns per unit width (warp and weft densities), the crimp of the yarns, and the areal weight (Lomov and Verpoest 2005).

Figure 6 represents a schematic of a weaving loom. Parallel warp yarns are drawn from the *warp beam* under regulated tension. Each warp passes through the eye of a wire or thin plate, called *heddle*. A set of heddles are mounted on a frame, called *harness*. When a harness moves up and down, the warps connected to its heddles move to the same direction. By moving some harnesses up and the others down, a *shed*, which is a gap between the warps, is formed. The weft (filling) yarn can then be inserted through the shed using a weft insertion device such as a *shuttle*, *projectile*, *air/water jet* or *rapier*. A metal comb, called the *reed*, beats up the weft yarn and pushes it into place. The harnesses then move in opposite directions and close the shed. The fabric is taken up by the cloth roll and the process is repeated (Lomov and Verpoest 2005).

There are three basic weave patterns. In *plain* weave, each warp goes alternately over and under successive wefts. The fabric has the highest stability and the lowest density. Low drapability, and high level of fiber crimp are from the drawbacks of the plain weave compared with the other weave patterns. In *twill* weave, the warps alternately go over and under two or more wefts so that a diagonal pattern is produced

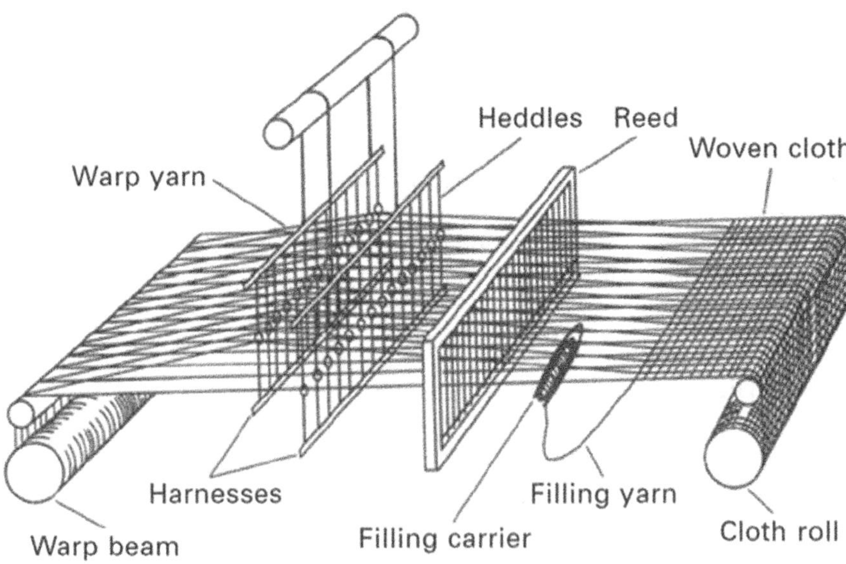

Fig. 6 Basic elements of a weaving loom (Gandhi 2020)

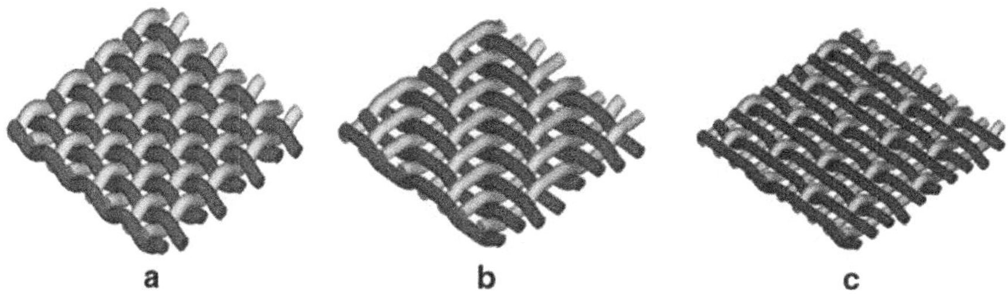

Fig. 7 Three basic weave patterns; a Plain, b Twill 2 × 2, c Satin 5 (Coupé 2011)

on the fabric. Compared with plain weave, twill has a better wet out with matrix resins and higher drapability, but its stability is somewhat lower. Moreover, since the crimp level is lower in twill, the fibers are more straight in the structure and thus the mechanical properties of the resultant composite are slightly improved (John and Thomas 2008). In satin weave, either the warps or the wefts are predominantly floating on the face of the fabric, provided that no two intersecting points of warps and wefts are adjacent to each other. Satin is woven with at least five harnesses. It has the highest drapability, the lowest fiber crimp, and the lowest stability among the basic weaves. Figure 7 shows plain, twill 2 × 2 and satin 5 weave patterns.

It should be noted that woven fabrics from DPF can be efficient reinforcements only when they are woven with fine yarns (which are difficult to be spun with such course fibers), or if possible, spread tows, so that the fiber crimp is minimized. Otherwise, the mechanical properties of the resultant composites are adversely affected.

3.2 *Knitted Fabrics*

Knitted fabrics form one of the main groups of textile structures made from basic construction units called *loops*. Two basic methods for manufacturing knitted fabrics are weft knitting and warp knitting. Although knitted preforms have such advantages as high conformability and productivity, there exist some problems concerning their use as the composite reinforcements, preventing them from being considered for structural applications. The non-linearity of the knitting loops leads to the lower contribution of the reinforcing fibers in the total strength of the composite. The yarns are severely bent during the knitting process, and thus fiber breakage may occur especially for brittle fibers. The knitted fabric has limited packing density, so the fiber volume fraction is low and resin pockets may form within the knitted composite (Hu 2008).

Consequently, the most commonly used knitted preform in the composite industry is the multiaxial warp knitted fabric, in which the reinforcing fibers are positioned straight and in the desired directions (Duhovic and Bhattacharyya 2011). For the production of multiaxial warp knits, which are also called non-crimp fabrics (NCFs),

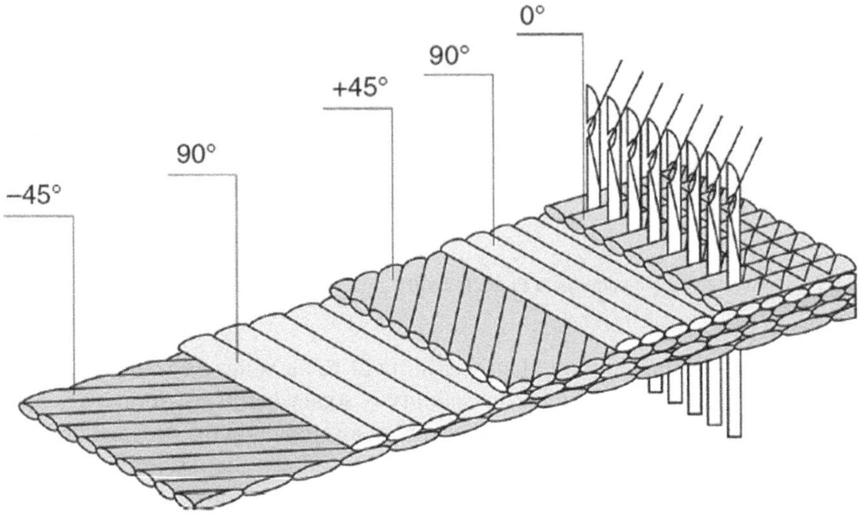

Fig. 8 Schematic structure of a multiaxial NCF (Linke et al. 2012)

several layers of straight reinforcing weft yarns are bonded together by means of knitting yarns in a warp-knitting process. One of the main advantages of this process is that the fibers can be placed with a wide range of orientation angles, approximately from −20° to +20° with respect to the direction of production. Figure 8 shows a schematic structure of a multiaxial NCF. A quadraxial NCF (consisting of four layers with +45°, 0°, 90°, −45° orientation angles) is used to achieve quasi-isotropic behavior in the material (Linke et al. 2012). Since straight and crimp-less yarns are incorporated in the structure of the NCFs, the strength, stiffness and fatigue life of the composite are improved (Meola et al. 2017). DPF yarns can also be used as the weft yarns in this technique, if the resultant reinforcement is cost-effective for the intended application.

3.3 Nonwoven Fabrics

As mentioned earlier, mats or randomly oriented nonwoven fabrics are the very common kinds of preforms for natural fibers. The production of nonwovens fabrics involves the formation of a uniform web and then consolidating the web by mechanical, thermal or chemical means. Compared with other fabric types, nonwoven fabrics are more economical and produced in faster and shorter processes.

3.3.1 Web Formation Methods

Carding

Carding is a process in which mechanical actions are carried out on staple fiber tufts in order to disentangle, clean and intermixes them and produce a continuous web or sliver for subsequent processing. A large rotating metal cylinder is located at the center of the card, whose surface is covered with needles or fine metallic teeth. The fibers are fed to the card using a feed roller. A licker-in roller, located after the feed roller, then opens the fibers into smaller tufts, extracts some of the trash particles and transfers the fibers to the cylinder. On the cylinder, the fibers are brought into contact with pairs of *worker/stripper* rollers. The fibers are separated between the cylinder and the worker with carding action. The stripper roller which revolves faster than the worker, strips the fibers accumulated on the worker's surface and conveys them back to the cylinder. These actions make the fiber tufts to be separated into individual fibers, mainly oriented in the machine direction. A controlled amount of fibers on the cylinder are removed by the *doffer* roller which forms a coherent web. The web is stripped off using stripping rollers or a doffing comb (Wilson 2010). Figure 9 shows a schematic of a standard roller card.

It is usually needed to stack the carded webs in several layers so that the final web with the desired areal weight is produced. The requirements of a web are good mass retention and high lengthwise and widthwise fiber distribution uniformity. In general, the fibers in the carded web are mainly oriented longitudinally, unless the carding machine uses a randomizing mechanism to produce a web with non-emphasized

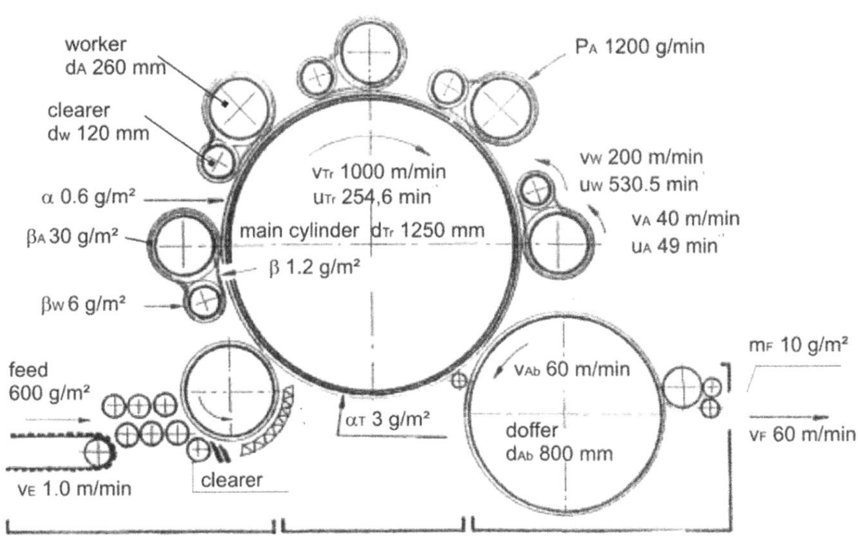

Fig. 9 Standard roller card with technical/technological information (Albrecht et al. 2003)

fiber orientations. The most important types of webs with respect to the web-forming methods are as follow:

Parallel-laid webs: In this type, the webs delivered from several parallel cards, arranged sequentially, are stacked on top of each other on a conveyor belt. In this type of webs, the fibers are mainly positioned in the longitudinal direction.

Cross-laid webs: For this type of webs, a *cross lapper* located after the card takes up the carded web and lays it in several layers with an oscillating carriage movement on a conveyor belt moving perpendicular to the card discretion (Albrecht et al. 2003). Therefore, in cross-laid webs, the fibers are usually positioned in the transverse direction. However, a web with nearly random fibers orientation can also be produced if a *drafter* machine is used after the cross lapper.

One of the problems with web formation from DPF is that the low flexibility and low crimp of the fibers make it difficult to form a coherent web on the card. Blending DPF with other suitable fibers, to form a hybrid web can be a solution to this problem. For the case of thermoplastic matrix composites, DPF can be blended with fibers from the matrix material. The composite fabrication process will then take place on the consolidated web at elevated temperature and pressure.

Air-laying

Air-laying is known as a highly versatile process due to its compatibility with different fiber types and specifications. A schematic of a web formation methods in the air laying process is shown in Fig. 10. In this design, pre-opened fibers are fed to a rotating drum opener using the feed rollers. Then a high-velocity airstream produced by an air blower removes the disentangled fibers from the teeth of the opener and deposits them in the form of a web on an air-permeable conveyor, which allows

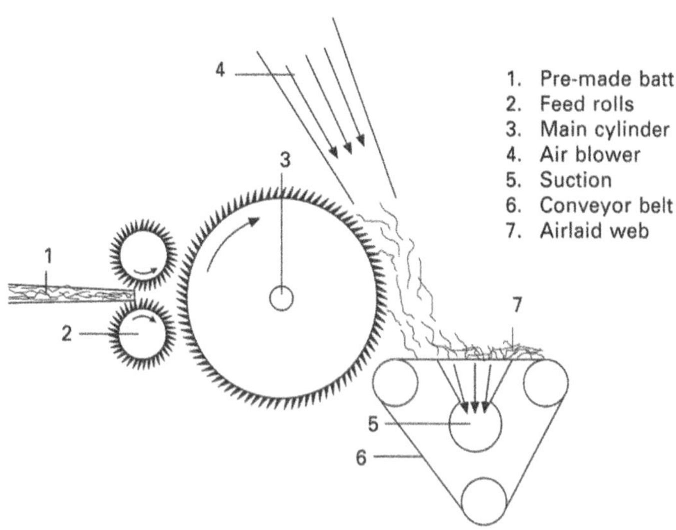

Fig. 10 A simple air-laying process (Brydon and Pourmohammadi 2007)

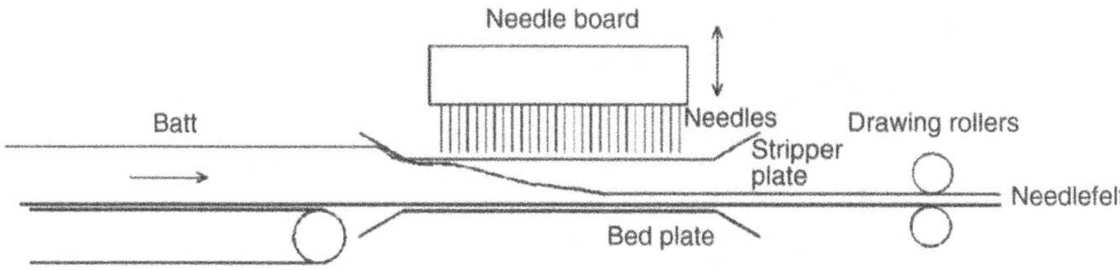

Fig. 11 Basic principle of nonwoven needle punching fabric formation (Smith 2000)

the airstream to pass through. In contrast to carded webs, in which the fibers have mainly lengthwise orientation, in air-laid webs the orientation of fibers is random (Brydon and Pourmohammadi 2007). This feature is important when an isotropic preform for composite fabrication is required. For the bonding of the air-laid webs from DPF, several technologies such as chemical bonding, thermal bonding (when DPF is blended with a thermoplastic binder), mechanical bonding (especially stitch bonding), can be used. The choice of the system depends on the composite fabrication process and the properties required for the final composite.

3.3.2 Web Consolidation Methods

Needle punching

Needle punching or needling is one of the oldest bonding methods in the nonwoven industry. In this method, fibers are mechanically interlocked by an array of barbed needles repeatedly penetrating through the fibrous web to consolidate and densify it (Brydon and Pourmohammadi 2007). The basic principle of needle punching is illustrated in Fig. 11.

The bulkiness, density, and porosity of the final preform which affect its formability in the molding process and permeability to resin flow can be controlled with needling parameters. Two important parameters of the needling process are *depth of needle penetration*, and *punching density*, which the latter refers to the number of times that the layer is penetrated by the needles in the unit area. These parameters also affect the orientation of the fibers in the preform and thus influence the in-plane and through-thickness mechanical properties of the composite. Choosing high levels of the mentioned parameters can also lead to increased fiber breakage during needling, which negatively affects the mechanical properties. In general, besides the effect of fiber type and specifications, mechanical and physical properties of the nonwoven preforms, such as tensile strength and elongation, bursting strength, bending rigidity and porosity will depend on the fabric's areal weight and the machine settings (El Messiry 2017). Figure 12 shows a needle-punched nonwoven from DPF.

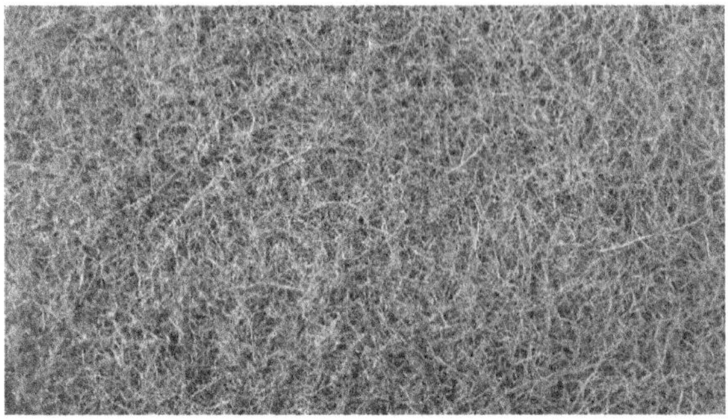

Fig. 12 A needle-punched nonwoven fabric from DPF

Thermal bonding

Thermal bonding or *thermobonding* process is another way to impart cohesion to nonwoven webs. To do this on the webs from non-plastified fiber such as DPF, the fibers must be blended with a specific amount of a thermoplastic fiber as a binder. The binder may be a monocomponent low melt fiber based on polyolefin (polyethylene and polypropylene), polyamides, polyester, polyvinyl chloride, vinyl acetate copolymers, cellulose acetate, etc., or a bicomponent fiber, consisting of both normal and low melt type polymers in the same fiber. As the blend is heated, the low melt fiber component melts and bonds the other fibers. Subsequently, the web is cooled down, so that the binder is solidified and the weld spots are formed. Thermobonding is economically more efficient, and ecologically preferable than chemical bonding, due to the lower cost of machinery and the absence of chemical binders (Tanchis 2008). Thermal bonding may be a more effective bonding method when the nonwoven preform is intended to reinforce thermoplastic polymers. In this instance, the thermoplastic fiber component will be converted into the matrix part of the composite structure. For this purpose, the web may first pass through an oven or heated calender rollers just after the web forming process, to find enough cohesion for handling purposes. It can then be fabricated into composite parts using a hot press, which completely melts the thermoplastic component (refer to Sect. 5). Bradley and Greer (2011) and Huang et al. (2013) reported this technique for fabricating coir fiber-reinforced thermoplastic composites.

Chemical bonding

In chemical bonding, an adhesive material, which is termed *binder*, is used to amalgamate the fibers within the structure of the nonwoven fabric.

The most important binders are emulsions of polymer lattices, which are fine water dispersions of specific polymers. To bond the nonwoven web with this method, the binder must be applied to the web with an appropriate method such as spraying. Since the binder is selected so that its viscosity is close to that of water, it can easily penetrate into even thick and dense webs. After binder application, the nonwoven is dried and the water evaporates, so the binder bonds the fibers at intersections (Tanchis 2008).

An example of using chemical bonding in the nonwoven reinforcement structures is in the production of fiberglass *chopped strand mats* (CSMs). CSM is manufactured by spreading chopped glass fibers randomly in combination with an emulsion binder (e.g., based on vinyl acetate) or a powder one (e.g., based on polyester, copolyester or polypropylene). It should be noted that the binder is designed to dissolve and break down in the matrix material during the resin impregnation process so that it does not adversely affect the fiber-matrix adhesion.

An important advantage of chemical bonding over needle punching in the manufacture of DPF reinforcement structures is that the fibers are not damaged in this technique. Moreover, since it does not change the orientation of the fibers, the resultant composite has higher in-plane mechanical properties. Comparing with thermal bonding, it does not require a second thermoplastic fiber component, and the preform can be easily impregnated with thermoset resins.

Stitch bonding

Stitch bonding is defined as a method of bonding fibers, yarns, fibers and yarns, or fibers and a base fabric together by stitching action with the aid of stitching yarns (Cotterill 1975). In stitch bonding a cross laid web or a nonwoven mat is fed into a special warp knitting machine, in which stitch yarns move through the web from one side to the other and bind the fibrous web (El Messiry 2017).

The *Maliwatt* system is a well-known stitch bonding system successfully employed in the market. Figure 13 shows a schematic of a Maliwatt stitching head. In Fig. 14 stitching action with a compound needle on a web is illustrated. The fiber web is fed to the stitching zone of the machine, where the horizontal compound needles penetrate through the web and the guide bars move the stitching yarns and insert them into the open hooks of the needles. The wire system closes the hooks and the needles exit the web to form the stitches with the aid of knock-over sinkers. On the basic version of the system, which operates with one guide bar, by cam shogging mechanism, stitching can be done with *Pillar* and *tricot* patterns. However, on machines with two guide bars, by pattern disc shogging mechanism, the basic two-guide-bar patterns with a maximum repeat length of sixteen courses can be produced (Anand et al. 2007).

Stitching the webs from DPF can be a good method for consolidation of nonwoven preforms from this special natural fiber. Comparing to needle punching, it is advantageous due to the reduced fiber damages during the process, however, it has less capability for compacting and densifying the web. Therefore, the fabrication process may be carried out under compression to achieve higher fiber volume fractions.

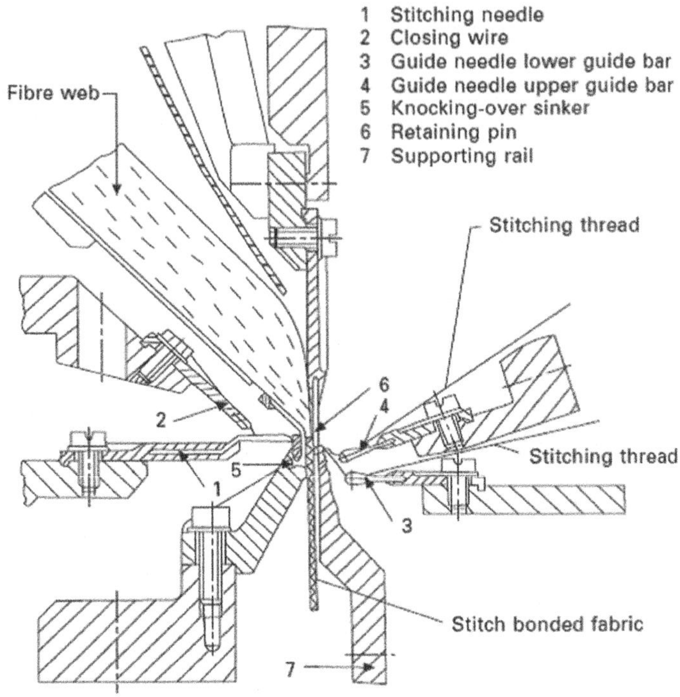

Fig. 13 A schematic of a Maliwatt stitching head (Anand et al. 2007)

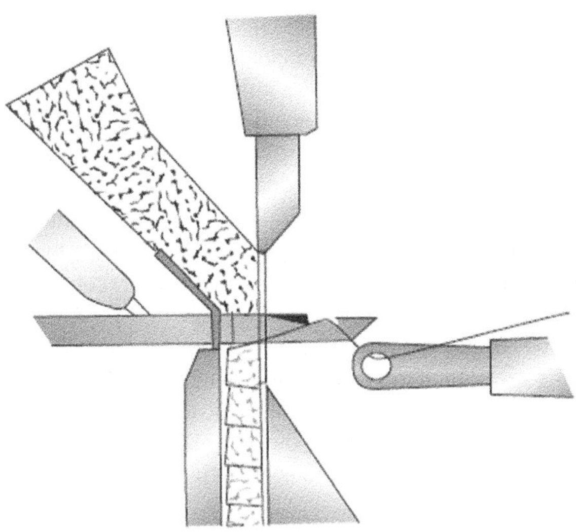

Fig. 14 Stitching action with a compound needle on a web (Anand et al. 2007)

4 Three-Dimensional Reinforcements

In three-dimensional textile preforms, the reinforcing fibers are oriented in both in-plane and out-of-plane directions, so that the final composite is reinforced in all the three mutually perpendicular axes. In-plane fibers provide reinforcement in the x–y plane and the fibers oriented in the z direction, known as z-binders, act as the through-thickness reinforcing elements. Having high interlaminar shear strength and superior resistance to *delamination* are of the unique features of these materials, which cannot be achieved with other 2-D reinforcements. 3-D fiber preforms can be manufactured with various methods and processes, which are usually derivatives of the basic traditional textile procedures.

4.1 3-D Woven Fabrics

3-D woven fabrics comprise in-plane warp and weft yarns and the z-binder yarns interlaced together. It has been found that the composites reinforced with 3-D woven preforms possess higher resistance to damage and delamination compared with the laminates reinforced with 2-D woven fabrics (Hu 2008). The structure of some 3-D weaves are schematically shown in Fig. 15. In some types of 3-D woven fabrics, the z-binder penetrates all the way through the thickness, such as in orthogonal and through-the-thickness angle interlock patterns (Fig. 15a, b), while in the others it travels between adjacent layers as it is seen in Fig. 15c. 3-D woven fabrics may also be classified according to the interlacing pattern of the z-binder in the weave. For instance, it can follow plain, twill or satin patterns. Another basis for classification is the interlacing angle of the z-binder. In some structures the binder is positioned in the vertical direction, making an approximate angle of 90° with the warp yarns. Therefore, the highest through-thickness properties of the composite are achieved, but the binders do not contribute to longitudinal properties. However, in some other

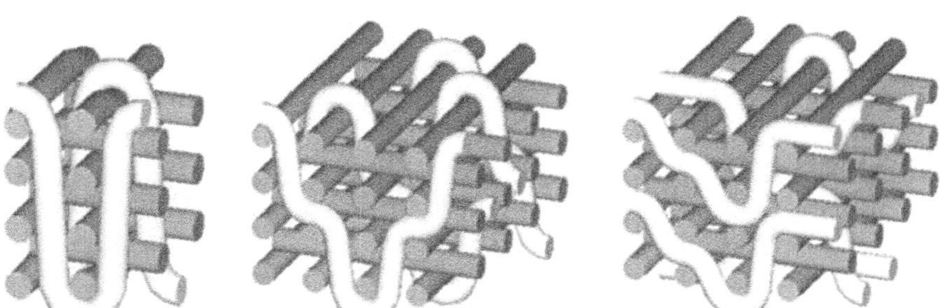

a b c

Fig. 15 Some types of multilayer 3-D woven fabrics, **a** orthogonal, **b** through-the-thickness angle interlock, **c** layer-to-layer angle interlock (Lomov and Verpoest 2005) structures, they lay in a diagonal direction, contributing to both through-thickness and longitudinal properties.

DPF yarns can be used in 3-D woven structures, especially as straight warps and wefts, which are not crimped in the structure. The z-binder yarns may be used from another type of more flexible fibers.

4.2 3-D Braided Fabrics

Three-dimensional braiding can be considered as an extension of the conventional two-dimensional braiding technology. In 2-D braiding, known as *Maypole* braiding, a number of yarn carriers rotate on a pseudo-sinusoidal circular track. The carries, which carry the yarn bobbins are driven by a series of horn gears lying below the track plate. Half the carriers rotate clockwise, and the other half rotate counter-clockwise, and thus, the two groups of yarns are interlaced with each other and a tubular fabric is formed. If longitudinal straight yarns are inserted between the braid yarns, triaxial braids are formed. Longitudinal yarns can also be fed into the center of a tubular braid so that a solid preform with axial core fibers is obtained (Fangueiro and Soutinho 2011). Preforms can be braided directly on a mandrel surface. In this technique, which is termed *overbraiding*, net shape and cost-effective structural reinforcements can be produced. The preform can be braided in several layers, each with a specified braid angle, on the mandrel by reciprocating it through the braiding zone. Figure 16 shows a schematic of overbraiding on a cylindrical mandrel.

In 3-D braids, multiple layers of yarns are interlaced with each other, however, they are also connected in through the thickness direction. In some 3-D braided preforms, the yarns are so interlaced that it is not possible to distinguish separate layers in the final preform.

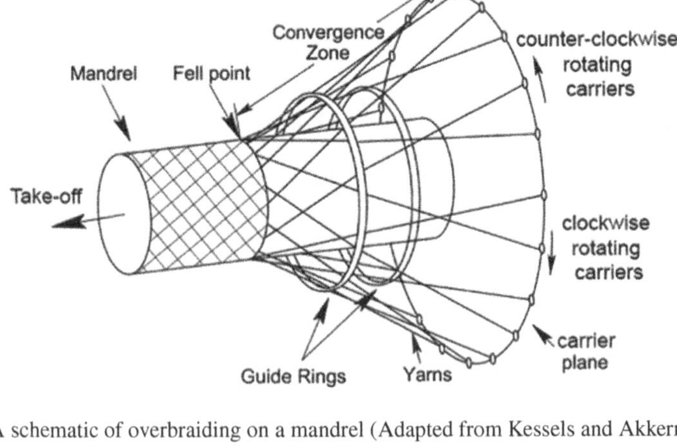

Fig. 16 A schematic of overbraiding on a mandrel (Adapted from Kessels and Akkerman 2002)

3-D braiding machines may be classified as horn gear and Cartesian machines, based on the movement mechanism of the yarn carriers. The *four-step* and *two-step* braiding methods are two important Cartesian 3-D braiding processes, described by Byun and Chou (1996).

In the four-step process (Fig. 17a) the yarn carriers are arranged in an array according to the preform cross section. As implied by the name, one cycle of braiding is comprised of four steps, each involving alternate movements of the rows and columns of yarn carriers. After each cycle, the fabric is *beaten up* to obtain the desired density, and then taken off by one pitch length. In the two-step process (Fig. 17b) a large number of axial yarns are arranged according to the preform cross section, with a smaller number of yarn carriers around them. The carriers are moved

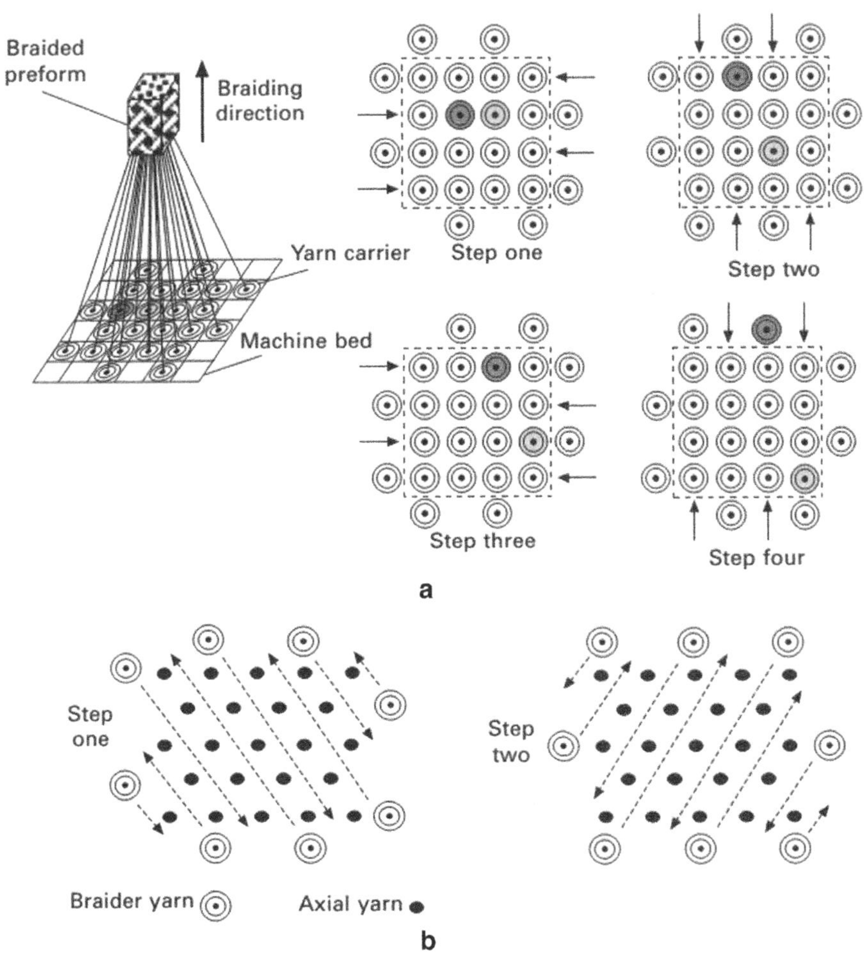

Fig. 17 Process steps in **a** four-step and **b** two-step 3-D braiding (Lomov and Verpoest 2005)

through the array of axial yarns in two alternate directions so that to interlock and hold them in the desired shape (Lomov and Verpoest 2005).

DPF yarns can be easily used as straight axial yarns in both 2-D and 3-D braided structures. However, they do not seem suitable for use as braid carrier yarns due to difficulty in the formation of fine and flexible yarns from DPF. Braid yarns can be chosen from other flexible yarns to make a hybrid reinforcement.

4.3 3-D Nonwoven Fabrics

Nonwoven preforms can be produced in near-net-shape structures with different thicknesses with no need to be made from stacking several laminates as for 2-D woven fabrics (El Messiry 2017). In ideal production conditions, 3-D nonwoven fabrics can be considered as isotropic materials with the same properties in all directions. If an appropriate method is utilized, 3-D nonwoven preforming can be an efficient and advantageous way for making reinforcements from DPF, as the costly and complicated process of yarn formation is omitted.

As previously mentioned, air-laying technology has the potential of positioning fibers in random orientations. Gong et al. (2003) described a technique for manufacturing 3-D nonwovens by air-laying web formation and heat through-air bonding methods (Fig. 18). The fibers were first opened using a modified opening unit of a carding machine, and then transferred by the airflow through a transport duct and deposited in a perforated 3-D mold, moving across the machine. As the web formation step is completed, the 3-D web in the perforated mold, was transferred to the thermal through-air bonding section for consolidation. The distribution of fibers in the 3-D web is an important factor which determines the resultant fabric's performance. Because the angle between the airflow and the deposition surface varies in different parts of a 3-D mold, even at uniform distribution of the airflow, there exists unevenness in fiber distribution in the 3-D fabric. Ravirala and Gong (2003)

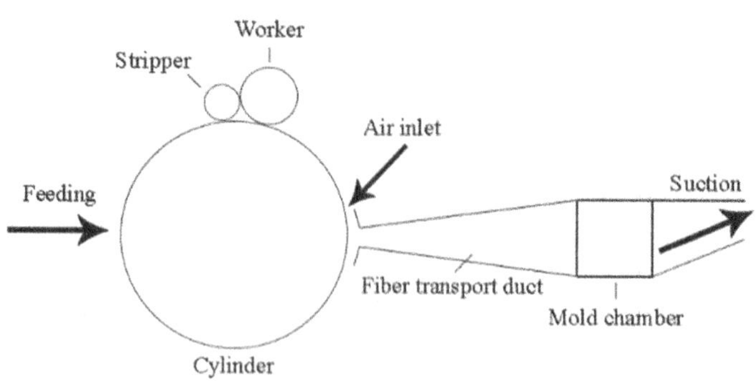

Fig. 18 3-D nonwoven system

suggested that uniform fiber distribution in this technique can be achieved by varying the local porosity of the mold.

3-D nonwoven preforms from DPF can also be produced by the mentioned method. For molding the preform and bonding the fibers, the technique utilized in the *directed fiber preforming* method (Harper et al. 2007) can be used. In this method, which has been developed for the manufacture of 3-D nonwoven preforms from continuous fiber tows or rovings, a robot-mounted mechanical chopper head cuts the fiber into desired lengths and sprays them with a polymeric powdered binder onto a perforated tool face. Then the fibrous structure is compressed by a matched perforated tool, and hot air is cycled through the perforations to consolidate the binder. After binder consolidation the preform is removed and placed in a closed mold, where the resin is injected to fabricate the final composite.

The impregnation of the above-mentioned preform with thermoset resins can be easily done by the RTM technique, in which a liquid catalyzed thermoset resin is pumped into a closed mold containing the 3-D preform under high pressure. Due to the high diameter of DPF and its low flexibility, it seems that even at a highly compacted preforms, the resistance to the resin flow in the mold cavity is low enough for a complete and fast wet-out process. However, in the case of large composite parts, the use of the vacuum-assisted RTM (VARTM) technique can facilitate the flow of resin with the aid of vacuum.

If the production of DPF reinforced thermoplastic composites is intended, the thermoplastic matrix may be blended with DPF in the form of staple fibers, so that the composite is fabricated in the subsequent process under pressure and temperature by melting the thermoplastic fibers.

4.4 3-D Stitched Fabrics

In the stitching process, a stack of 2-D reinforcement fabrics are fed to the stitching machine, where they are sewn together with suitable stitch threads using special sewing needles (Fig. 19) (Tong et al. 2002). If the through-thickness properties of

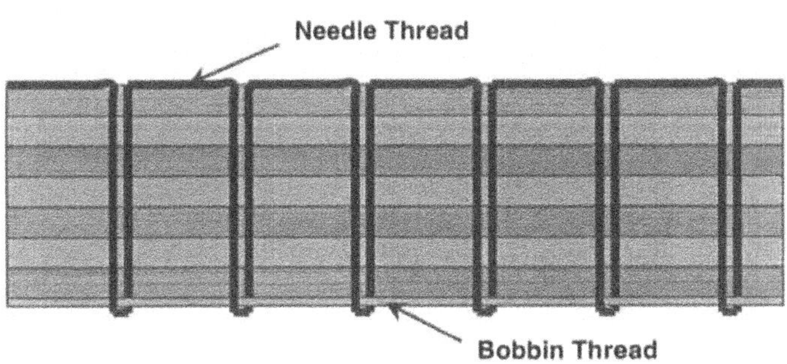

Fig. 19 Schematic of a stitched preform (Tong et al. 2002)

a DPF laminate is required to be improved, the stack of several 2-D fabrics with any structures (woven, nonwoven, etc.), layered with the desired orientations can be sewn with a suitable thread as a z-binder. The stitch bonding process, as described earlier, can also be used in the production of 3-D performs, on the condition that the fibrous web has a suitable thickness, in proportion to the thickness of the final composite, and the bonding threads are sufficiently strong to be considered as the through-thickness reinforcements.

5 Commingled Preforms for Fiber Reinforced Thermoplastics

Recently, thermoplastic composites have attracted a great deal of attention in the automotive industry, due to the recycling considerations, and higher fracture toughness and impact strength. Beside *film stacking*, *powder impregnation*, and *suspension impregnation* techniques, the other important method for the manufacture of fiber-reinforced thermoplastic polymers is to produced commingled preforms and then convert them into composite parts by the application of heat and pressure. Commingling is the process of intimately blending a thermoplastic matrix in the fibrous form with reinforcing fibers. These hybrid yarns may be used as 1-D reinforcements. Moreover, since they are flexible and processable, they can be used in textile processes and converted into hybrid woven, knitted, or braided structures in either 2-D or 3-D forms. As mentioned earlier, hybrid nonwoven fabrics can also be made from a blend of reinforcing and matrix fibers. Comingled preform will subsequently be heated above the melting temperature of the matrix. The pressure is also applied for better impregnation, and then the composite is ultimately cooled down. Since in comingled preforms the flowing distance of the resin for achieving impregnation is reduced, wet-out during consolidation is faster compared with conventional melt impregnation methods (Wakeman et al. 1998). By controlling the reinforcing fiber blend ratio in comingled preforms, the desired fiber volume fraction in the final composite part can be achieved. In co-weaving, the yarns of the two reinforcing and matrix materials are interlaced together in the form of a woven fabric. This technique may also be used in other processes such as knitting and braiding. However, uneven distribution of the reinforcements in the matrix in this method may increase the impregnation time (Wakeman and Månson 2005). Figure 20 shows some fiber/matrix hybrid preforms for DPF reinforced thermoplastic polymers.

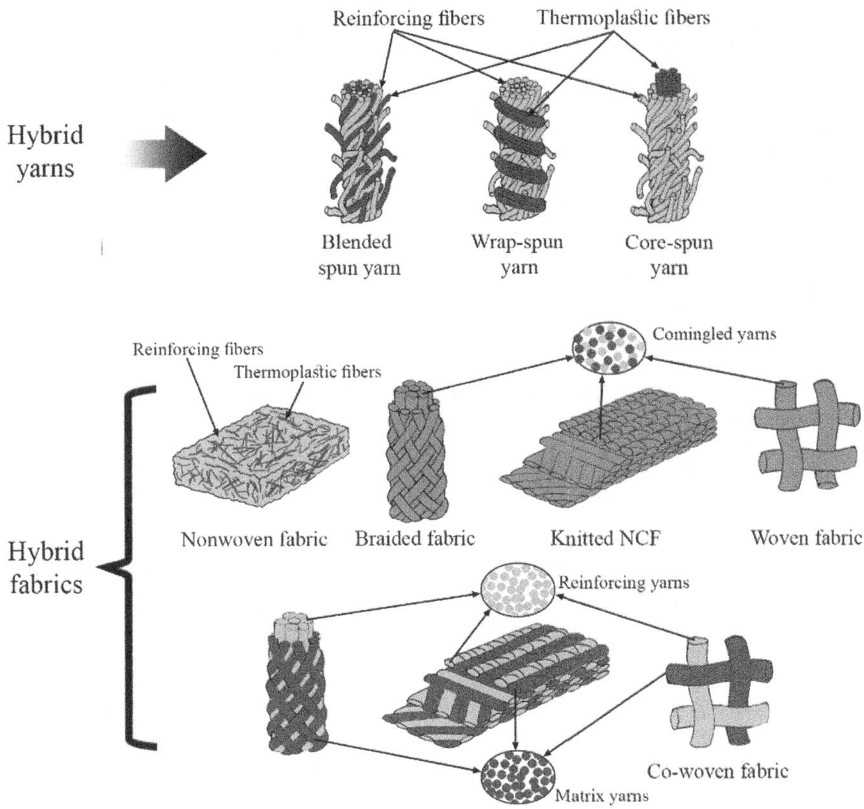

Fig. 20 Some of the fiber/matrix hybrid preforms for DPF reinforced thermoplastic polymers

References

Albrecht W, Fuchs H, Kittelmann W (eds) (2003) Nonwoven fabrics: raw materials, manufacture, applications, characteristics, testing processes. Wiley-VCH, Weinheim

Anand SC, Brunnschweiler D, Swarbrick G, Russell SJ (2007) Mechanical bonding. In: Russell SJ (ed) Handbook of nonwovens, 1st edn. Woodhead Publishing, Cambridge

Behery HM (2005) Effect of yarn factors on fabric hand. In: Behery HM (ed) Effect of mechanical and physical properties on fabric hand, 1st edn. Woodhead Publishing, Cambridge

Bradley W, Greer DS (2011) Non-woven fabric composites from coir fibers. US Patent Application 13/161,316

Brydon AG, Pourmohammadi A (2007) Dry-laid web formation. In: Russell SJ (ed) Handbook of nonwovens, 1st edn. Woodhead Publishing, Cambridge

Byun J-H, Chou T-W (1996) Process-microstructure relationships of 2-step and 4-step braided composites. Compos Sci Technol 56(3):235–251. https://doi.org/10.1016/0266-3538(95)00112-3

Cotterill P (1975) The production and properties of stitch-bonded fabrics. Text Prog 7(2):101–123. https://doi.org/10.1080/00405167508688981

Coupé D (2011) Woven reinforcements for composites. In: Boisse P (ed) Composite reinforcements for optimum performance, 1st edn. Woodhead Publishing, Cambridge

Cristaldi G, Latteri A, Recca G, Cicala G (2010) Composites based on natural fibre fabrics. In: Dubrovski PD (ed) Woven fabric engineering, 1st edn. Intech, Rijeka

Duhovic M, Bhattacharyya D (2011) Knitted fabric composites. In: Au KF (ed) Advances in knitting technology, 1st edn. Woodhead Publishing, Cambridge

El Messiry M (2017) Natural fiber textile composite engineering. Apple Academic Press, New York

Fangueiro R, Soutinho F (2011) Textile structures. In: Fangueiro R (ed) Fibrous and composite materials for civil engineering applications, 1st edn. Woodhead Publishing, Cambridge

Gandhi KL (2020) The fundamentals of weaving technology. In: Gandhi KL (ed) Woven textiles, 2nd edn. Woodhead Publishing, Cambridge

Ghori W., Saba N, Jawaid M, Asim M (2018) A review on date palm (phoenix dactylifera) fibers and its polymer composites. In: IOP conference series: materials science and engineering, vol. 368, no. 1, p. 012009, IOP Publishing

Gong R, Dong Z, Porat I (2003) Novel technology for 3D nonwovens. Text Res J 73(2):120–123. https://doi.org/10.1177/004051750307300205

Goutianos S, Peijs T (2003) The optimisation of flax fibre yarns for the development of high-performance natural fibre composites. Adv Compos Lett 12(6):237–241. https://doi.org/10.1177/096369350301200602

Goutianos S, Peijs T, Nystrom B, Skrifvars M (2006) Development of flax fibre based textile reinforcements for composite applications. Appl Compos Mater 13(4):199–215. https://doi.org/10.1007/s10443-006-9010-2

Harper LT, Turner TA, Warrior NA, Rudd CD (2007) Characterisation of random carbon fibre composites from a directed fibre preforming process: the effect of tow filamentisation. Compos Part A Appl Sci Manuf 38(3):755–770. https://doi.org/10.1016/j.compositesa.2006.09.008

Hu J (2008) Introduction to three-dimensional fibrous assemblies. In: Hu J (ed) 3-D fibrous assemblies, 1st edn. Woodhead Publishing, Cambridge

Huang C-H, Lin J-H, Lou C-W, Tsai Y-T (2013) The efficacy of coconut fibers on the sound-absorbing and thermal-insulating nonwoven composite board. Fibers Polym 14(8):1378–1385. https://doi.org/10.1007/s12221-013-1378-7

John MJ, Thomas S (2008) Biofibres and biocomposites. Carbohydr Polym 71(3):343–364. https://doi.org/10.1016/j.carbpol.2007.05.040

Kessels JFA, Akkerman R (2002) Prediction of the yarn trajectories on complex braided preforms. Compos Part A Appl Sci Manuf 33(8):1073–1081. https://doi.org/10.1016/S1359-835X(02)00075-1

Ko FK (1999) 3-D textile reinforcements in composite materials. In: Miravete A (ed) 3-D textile reinforcements in composite materials, 1st edn. Woodhead Publishing, Cambridge

Lawrence CA (2010) Overview of developments in yarn spinning technology. In: Lawrence CA (ed) Advances in yarn spinning technology, 1st edn. Woodhead Publishing, Cambridge

Linke M, Greb C, Klingele J, Schnabel A, Gries T (2012) Automating textile preforming technology for mass production of fibre-reinforced polymer (FRP) composites. In: Shishoo R (ed) The global textile and clothing industry, 1st edn. Woodhead Publishing, Cambridge

Lomov S, Verpoest I (2005) Manufacturing and internal geometry of textiles. In: Long AC (ed) Design and manufacture of textile composites, 1st edn. Woodhead Publishing, Cambridge

Madsen B, Brøndsted P, Andersen TL (2013) Biobased composites: materials, properties and potential applications as wind turbine blade materials. In: Brøndsted P, Nijssen RPL (eds) Advances in wind turbine blade design and materials, 1st edn. Woodhead Publishing, Cambridge

Mankodi HR (2011) Developments in hybrid yarns. In: Gong RH (ed) Specialist yarn and fabric structures, 1st edn. Woodhead Publishing, Cambridge

Mathai PM (2005) Coir. In: Franck RE (ed) Bast and other plant fibres, 1st edn. Woodhead Publishing, Cambridge

Meola C, Boccardi S, Carlomagno Gm (2017) Composite materials in the aeronautical industry. In: Meola C, Boccardi S, Carlomagno Gm (eds) Infrared thermography in the evaluation of aerospace composite materials, 1st edn. Woodhead Publishing, Cambridge

Miao M, Finn N (2008) Conversion of natural fibres into structural composites. J Text Eng 54:165–177. https://doi.org/10.4188/jte.54.165
Ravirala N, Gong R (2003) Effects of mold porosity on fiber distribution in a 3D nonwoven process. Text Res J 73(7):588–592. https://doi.org/10.1177/004051750307300705
Smith PA (2000) Technical fabric structures—3. Nonwoven fabrics. In: Horrocks AR, Anand SC (eds) Handbook of technical textiles, 1st edn. Woodhead Publishing, Cambridge
Tanchis G (2008) The nonwovens: reference books of textiles technologies. Fondazione ACIMIT, Milano
Thanabal V, Saravanan D (2018) A study on the quality characteristics of yarns made from coir fibres. Int J Recent Technol Eng 7(4):390–393
Tong L, Mouritz AP, Bannister MK (2002) Manufacture of 3D fibre preforms. In: Tong L, Mouritz AP, Bannister MK (eds) 3D fibre reinforced polymer composites, 1st edn. Elsevier Science, Oxford
Velayutham T, Dhandapani S (2019) An investigation on the performance of modified coir spinning machine. Autex Res J 19(3):211–216. https://doi.org/10.1515/aut-2018-0032
Wakeman MD, Cain TA, Rudd CD, Brooks R, Long AC (1998) Compression moulding of glass and polypropylene composites for optimised macro-and micro-mechanical properties—1 commingled glass and polypropylene. Compos Sci Technol 58(12):1879–1898. https://doi.org/10.1016/S0266-3538(98)00011-6
Wakeman MD, Månson JAE (2005) Composites manufacturing—thermoplastics. In: Long AC (ed) Design and manufacture of textile composites, 1st edn. Woodhead Publishing, Cambridge
Wilson A (2010) The formation of dry, wet, spunlaid and other types of nonwovens. In: Chapman RA (ed) Applications of nonwovens in technical textiles, 1st edn. Woodhead Publishing, Cambridge

Polymer Matrix Systems Used for Date Palm Composite Reinforcement

Said Awad, Yonghui Zhou, Evina Katsou, and Mizi Fan

Abstract Date palm fibres have some physical and mechanical properties that are similar to those of synthetic fibres, which make them attractive for developing sustainable composites. However, in common with other natural fibres, date palm fibres have some unique characteristics, such as hydrophilicity and relatively low mechanical properties, which require some specific considerations when they are used for composite formulations. This chapter reviews and discusses various types of polymer matrices that are used to date and can be potentially used with date palm tree fibres to develop sustainable composites in the future, e.g. thermoplastics and thermosets. More importantly, elastomer polymers and bio-derived resins are playing an increasingly critical role in the development of new natural fibre composites, which were often omitted in the currently available book chapters or review articles. Hence, this chapter also pays specific attention on the potentially promising elastomers and bio-resins, such to draw a complete picture of the polymeric matrices. Furthermore, this chapter discusses the effect of various types of date palm fibres and fibre characteristics on the mechanical, physical and thermal properties of the composites developed and provides a database with the reported research done to date on developing date palm fibre reinforced polymeric composites. This paper finishes with some critical suggestions and future perspectives, underlining the issues to be addressed in further in-depth studies and potential industrialisation of the composites.

1 Introduction

Petroleum based polymers (PBP), synthetic polymer matrix systems, play a crucial role in developing various components and structures for wide industrial applications. Synthetic polymers existed since the introduction of Bakelite in early twentieth century, unlike their predecessors which were derived from nature (Feldman 2008; Ali et al. 2018). Polymer production and consumption have proliferated so readily

S. Awad · Y. Zhou · E. Katsou · M. Fan (✉)
Department of Civil and Environmental Engineering, College of Engineering, Brunel University London, Uxbridge, UK
e-mail: mizi.fan@brunel.ac.uk

M. Midani et al. (eds.), *Date Palm Fiber Composites*, Composites Science and Technology,
https://doi.org/10.1007/978-981-15-9339-0_4

in the past 70 years because of their versatility. Polymers can be brittle or resilient, rigid or flexible, coloured or clear, and they can be insulators or conductors, resistant to degradation and also have many other valuable properties, such as superb durability and strength and light weight. Their widely divergent properties allow them to perform many specific applications in packaging, medical, construction, automobile and other industrial sectors.

The global polymer production has increased drastically from 1.5 million tons in 1950 to 359 million tons in 2018 and is expected to continue increasing at a growth rate of 5% per annum, reaching approximately 500 million tons by 2025 ([CSL STYLE ERROR: reference with no printed form.]; Biron 2016). The increase in production of PBP at a large scale resulted in rapidly growing demand of crude oil and serious environmental issues. The production of PBP is energy intensive, requiring approximate 62–108 mega joules (MJ) of energy per kg based on U.S. efficiency average. In addition, the disposal of the plastics is a challenge, considering the extremely slow degradation rate of the deposited plastics in landfills, i.e. from a half-life up to more than 500 years. Moreover, incineration is an expensive method and generates toxic by-products. Although recycling can be done, it alters the properties of the plastic and consumes a lot of time. Also, the presence of additives such as filler, pigments and coatings limits the recycling process. Therefore, to reduce the environmental impact of PBP, an ideal solution is to use bio-based polymers that possess similar physicochemical properties as PBP (Halley et al. 2007; Chan et al. 2018). Bio-based polymers, acting as important alternatives to PBP, are a form of plastics derived from renewable and waste stream biomass sources such as starch, vegetable oil, proteins, wastewater treatment plants. Bio-based polymers provide the dual advantages of conservation of fossil resources and reduction in carbon dioxide (CO_2) emissions, which make them a paramount innovation of sustainable development (Wool and Sun 2011).

The cost of biopolymers has been too high to sustain the industry. Thus, the addition of natural plant fibres (NPF) in polymers matrices develops a bio-composite with sustainable characteristics and improved mechanical performance at a lower cost. Environmentally friendly bio-composites from NPF and bio-based polymers are considered as a new generationmaterial of this century that can solve the growing environmental threat caused by PBP, the uncertainty of petroleum supply and increasing amount of landfill disposal. It was reported that the market size for NPF reinforced composites was projected to reach 5.83 billion U.S dollars in 2019 with a compound annual growth rate of 12.3%, reaching 11.69 billion U.S dollars by 2025 (Ali et al. 2018).

To date, there are numerous NPF that are well recognized through research and suitable for industrial applications. Many researchers have studied the mechanical and physical properties of several NPF, such as sisal, coir, hemp, kenaf, jute and bamboo, reinforced with various polymer matrices to develop bio-composites for a wide range of applications in various industrial sectors (Joseph et al. 1995; Brahmakumar et al. 2005; Dittenber and Gangarao 2012; Węcławski et al. 2014; Al-Oqla and Sapuan 2014; Zhou et al. 2015, 2016; Chimeni et al. 2016; Fan and Fu 2016; Fan et al. 2017). Recently, date palm fibre (DPF) has been explored as a potential reinforcement in polymeric matrix due to its abundancy, especially in the Middle East and North Africa (MENA) region (Abu-Sharkh and Hamid 2004; Al-Kaabi et al. 2005; Mahdavi et al. 2010; Abdal-Hay et al. 2012; AlMaadeed et al. 2012; Shalwan and Yousif 2014; Mirmehdi et al. 2014; Mohanty et al. 2014; Ibrahim et al. 2017; Zadeh et al. 2017; Mohanty 2017; Saleh et al. 2017; Gheith et al. 2019). There are more than 120 million DPTs across the world, each tree grows for more than 100 years, producing fruits and by products on annual harvests (Al-Khanbashi et al. 2005). It was reported that each tree produces approximately 26 kg of waste on annual harvest (Elseify et al. 2019), reaching 3.12 million tonnes annually, majority of which are renewable raw materials that can be used as NPF reinforcement in various polymer matrices for developing bio-composites. However, DPFs are similar to other NPFs, being hydrophilic with many impurities on its surface by nature. Thus, surface modification is commonly considered in order to formulate a composite with stronger interfacial bonding between the DPF and polymer matrix.

This chapter discusses the polymer matrix systems available for DPF reinforcement to develop bio-composites. It also discusses the physical, mechanical and thermal properties of different types of polymer matrices to develop a comprehensive database and platform for researchers and industries to determine how to utilize DPFs in producing more sustainable and renewable bio-composites for various applications.

2 Thermoplastic Matrix Systems for DPF Composites

Thermoplastic polymers are a class of polymers that can be repetitively softened and melted, when subjected to heat under certain conditions, permitting them to be processed in either heat-softened state or liquid state and solidifying again upon cooling without affecting or altering their properties. These properties allow thermoplastics to be recycled at the end of their life, which is considered as the best sustainable technique of non-degradable waste thermoplastic in comparison with the traditional unsustainable methods, incineration and landfill deposition, which have a negative environmental impact due to the formation of dust, fumes and toxic gases (Van De Velde and Kiekens 2001; Grigore 2017). The recycling of thermoplastic can be achieved through several approaches. Firstly, traditional recycling also known as primary recycling refers to the reutilisation of products in their original structure, and it is simple and cheap. The drawback of this method is that each material has a limited number of recycling cycles. Moreover, mechanical recycling is another method that can be applied to recycle thermoplastic polymers. Although this method significantly reduces the quantity of waste plastics, it deteriorates the properties of the polymer with each recycling cycle.

Additionally, another method of recycling is feedstock or more commonly known as chemical recycling. It can be used as a complementation to the mechanical recycling, which involves chemically transforming or decomposing the polymer into monomers or partially depolymerize to oligomers through chemical reactions that alters the structure of the polymer. The monomers can then be polymerized to produce the same or a related product. Polymers can be decomposed into monomers through many different chemical reactions like hydrogenation, glycolysis, gasification, hydrolysis, pyrolysis and many more. This method is still not fully developed because of its high costs and the requirement of highly skilled experts to perform the process. Many of the chemical reactions are still being studied.

Furthermore, for better understanding and selection of the most suitable thermoplastic matrix for developing composites reinforced with DPF, the physical, thermal and mechanical properties should be well understood and taken into consideration.

2.1 Synthetic Thermoplastic Polymers for DPF Composites

Synthetic thermoplastic polymers are polymers that are produced from crude oil, and represented by a wide range of plastic materials, which can be divided into three groups, semi-crystalline, amorphous and elastomeric thermoplastic polymers, each with its own unique characteristics. Crystalline thermoplastics, mostly translucent with molecules arranged regularly, are composed of partly amorphous and partly crystalline polymers, thus the term crystalline polymer always implies as semi-crystalline polymers (Sastri 2010). Polymers classified in this group are high density polyethylene (HDPE), medium density polyethylene (MDPE), low density polyethylene (LDPE), linear low density polyethylene (LLDPE), polypropylene (PP), polyamide—6 (PA—6), polyamide—4,6, (PA—4,6), polyamide—6,6, (PA—6,6), polyethylene terephthalate (PET), polyoxy-methylene (POM), polyvinyl chloride (PVC), chlorinated polyvinyl chloride (CPVC), polyvinylidene chloride (PVDC), polytetrafluoroethylene (PTFE), polybutylene terephthalate (PBT) and polyphenylene sulphide (PPS). However, amorphous thermoplastic polymers are usually transparent with molecules arranged in a random orientation. Polymers classified in this group are polystyrene (PS), styrene acrylonitrile (SAN), acrylonitrile butadiene styrene (ABS), polycarbonate (PC), plasticised PVC (uPVC), polymethyl methacrylate (PMMA), styrene maleic anhydride (SMA) and polyethylene terephthalate (PET). Furthermore, elastomeric thermoplastics are materials that behave like elastomeric rubbers at room temperature but can be processed similarly to thermoplastics when heated up. They are considered as another class of hybrid composites, with better functional performance than elastomer polymers since their properties are altered by varying the ratio of elastomer to thermoplastic in the composite (Zhou et al. 2015).

Elastomeric thermoplastic polymers can be divided into five groups; styrene-diene block copolymers (TPE-S), elastomeric alloys, thermoplastic urethane elastomers, thermoplastic ester-ether copolymers (TPE-E), and thermoplastic amide copolymer (TPE-A). Styrene-diene block copolymers are thermoplastic elastomers based on styrene and are block copolymers in which a polydiene unit divides polystyrene blocks. SBS (Styrene/Butadiene Copolymer), SIS

(Styrene/Isoprene Copolymer), SEBS (Styrene/Ethylene-Butylene Copolymer), and SEPS (Styrene/EthylenePropylene Copolymer) are examples of styrene-diene block copolymers. Moreover, elastomeric alloys are combinations of thermoplastics and elastomers that are processed similarly to thermoplastic processing methods. Examples of elastomeric alloys are Thermoplastic Olefin Elastomers (TPO/TOE), which are blends of PP and ethylene propylene rubber (EPM)/ethylene propylene diene monomer rubber (EPDM). Thermoplastic vulcanizates (TPE-V/TPV/DVR) combines both elastomers and thermoplastics that have been dynamically vulcanized during their mixing. Additionally, Melt Processible Rubbers (MPR), which are rubbery materials, have the same texture and appearance of traditional rubbers, but can be processed like thermoplastics. However, they differ from other elastomeric thermoplastics by having only one phase structure rather than two (Fig. 1).

Physical and thermal properties

Physical properties include density, moisture uptake and melt flow index (MFI), while thermal properties include heat deflection temperature and coefficient of thermal expansion (Table 1). Density is a crucial parameter in determining the weight of the end composite developed and affects the implementation of the composite. It is mainly determined using ASTM D1505, BS EN ISO 1183-2:2019, and ASTM D792 (Van De Velde and Kiekens 2001; BS EN ISO 2019). Thermoplastic polymers density can range from 0.9 to 2.2 g/cm^3 depending on the type of polymer. Thus, it can be stated that the implementation of a lighter composite, polymers with relatively low density, becomes less hazardous and easier to assemble.

Water absorption or moisture uptake of polymers after 24 h of immersion in water demonstrates the percentage of migration of water into the polymer, which

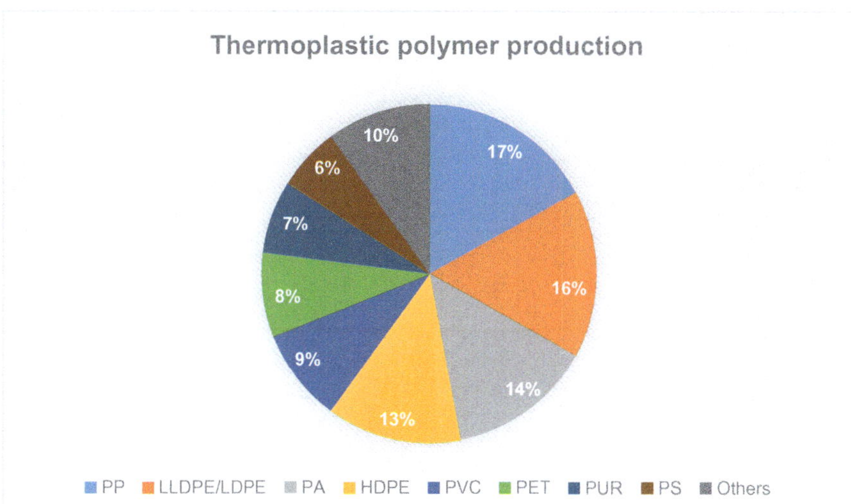

Fig. 1 Types of thermoplastics produced worldwide as percentage in year 2015 (Yeo et al. 2018)

Table 1 Physical and thermal properties of synthetic thermoplastic polymers (Da Silva and Coutinho 1996; Oksman et al. 2003; Cheung et al. 2009; Russo et al. 2014; Olabisi and Adewale 2016; Koo and Koo 2016)

Polymer	Density (g/cm^3)	Moisture uptake (%)	MFI (g/10 min)	Melting temperature (°C)	T_g (°C)	Processing temperature (°C)	HDT @ 1.82 MPa (°C)	α_T (10^{-5}/°C)	Thermal conductivity (W/m/°C)	Specific heat (kJ/kg/°C)	Mold shrinkage (%)
Semi-crystalline thermoplastics											
HDPE	0.94–0.97	0.01–0.20	1.30–20.00	120–140	−140 to −90	150–290	43–82	12–13	0.49	1.57–2.28	2.0–6.0
MDPE	0.93–0.97	–	5.00	126–130	−120	210	–	–	–	–	–
LDPE	0.91–0.93	<0.015	4.44–35.00	105–120	−125 to −40	150–230	32–50	10	0.32	1.90–1.99	1.5–5.0
LLDPE	0.89–0.94	0.01	26.30	112–124	−80 to −68	160–230	55–65	–	–	–	–
PP	0.90–1.40	0.01–0.035	12.00–45.00	152–176	−23 to 16	200–290	50–63	6.8–13.5	1.70–1.90	1.70–1.90	1.0–2.5
PA—6	1.12–1.15	1.3–1.8	–	215–228	40–58	215–270	56–80	8.0–8.6	0.17	1.60	0.5–2.6
PA—4,6	1.18	–	–	295	75		160		0.30	–	–
PA—6,6	1.13–1.60	1.0–1.6	–	250–269	50–80	250–320	55–105	7.2–9.0	0.24	1.68	0.8–2.0
PET	1.33–1.34	0.07	20.00	212–265	67–125	100–256	100	5.9–6.5	1.10–1.15	1.10–1.15	2.5
POM	1.41	0.21		160–180	−85 to −50	–	110–125	–	–	1.37	2.0–3.0
PVC	1.3–1.5	0.04–0.75	7.27	170–220	68–110	–		–	0.21	1.01	1.0–2.5
CPVC	1.54	0.03	–	–	100	–	91	–	0.31	0.78	0.3–0.7

(continued)

Table 1 (continued)

Polymer	Density (g/cm^3)	Moisture uptake (%)	MFI (g/10 min)	Melting temperature (°C)	T_g (°C)	Processing temperature (°C)	HDT @ 1.82 MPa (°C)	α_T (10^{-5}/°C)	Thermal conductivity (W/m/°C)	Specific heat (kJ/kg/°C)	Mold shrinkage (%)
PVDC	1.75	–	–	205	−18	–		–			–
PTFE	2.20	0.0093	0.60–15	332	−73 to 97	370–395	70	–	0.24–0.25	1.05	–
PBT	1.31	0.08–0.10	16.00	220–267	20–52	246–290	49–85	7.0–7.2	0.20	1.20	1.6–2.3
PPS	1.340	0.03–0.05	24.60	270–290	85–95	300–340	105–138	4.1–5.4	0.29	–	0.8–1.2
PEEK	1.320	0.30	2.5	334–343	139–153	370–400	150–204	4.7–5.5	0.25	1.1	0.7–1.9
PB	0.92	–	–	124–126	–	–	–	–	0.19–0.22	–	–
PVDF	1.76	–	–	169–180	−40	180	85	–	0.17–0.19	1.12	–
Amorphous thermoplastics											
PS	1.05–1.08	0.04	11.57	–	85–110	–	90	2.14	0.14–0.15	1.25	0.4–0.7
SAN	1.07–1.08	0.26	–	–	100–127	–	99	–	0.15	1.38	0.6
ABS	0.99–1.10	0.30	23.00	–	88–120	–	85–95	–	0.23	1.40	0.4–0.9
PC	1.20–1.24	0.12–0.70	14.20	–	140.5–150	260–330	127–140	6.5–7.0	0.19	1.17–1.26	0.5–0.8
PVCu	1.41	0.24	–	180	80–85	–	69	–	0.16	0.88	–
PMMA	1.170–1.23	–	2.00	180	43–160	–	65–100	2.41	0.19	1.26	0.1–0.4
SMA	1.05–1.15	–	–	–	114	–	95–130	–	–	–	0.4–0.8

(continued)

Table 1 (continued)

Polymer	Density (g/cm^3)	Moisture uptake (%)	MFI (g/10 min)	Melting temperature (°C)	T_g (°C)	Processing temperature (°C)	HDT @ 1.82 MPa (°C)	α_T (10^{-5}/°C)	Thermal conductivity (W/m/°C)	Specific heat (kJ/kg/°C)	Mold shrinkage (%)
PVP	1.05	–	–	150–180	–	–	–	–	–	–	–
PAI	1.38–1.45	0.22–0.28	–	–	244–290	340–400	260–275	2.9–3.6	0.20–0.73	1.00	–
PEI	1.27–1.28	0.25	–	–	215–225	330–420	190–216	5.6–6.1	0.18	1.20	0.5–0.7
Elastomeric thermoplastics											
EPDM	1.15–1.45	–	1.6	149	–54	160–210	–	16	–	–	–
TPU	1.10–1.30	–	–	–	–	–	–	–	–	–	0.80–0.81
TPE	1.10–1.20	–	–	–	–	–	–	–	–	–	1.5–2.5
SBS	0.92–1.00	–	–	–	–	160–190	–	–	–	–	–
PEBA	–	–	–	180–271	–	–	–		–	–	–

can reduce the overall strength of the composite developed. It is mainly determined by ASTM D570 and BS EN ISO 62:2008 (BS EN ISO 2008). In the case of DPF or any NPF reinforcement, water adsorption is a major drawback due to the tendency of NPF in absorbing huge amount of water which results in fibre swelling, affecting the fibre/polymer interface and significantly reducing the overall strength of the developed composite. Therefore, this parameter is very crucial when choosing a suitable polymer matrix for a DPF based composite. According to the data demonstrated in Table 1, thermoplastic polymers have the tendency to absorb water up to 0.28% and PTFE as the lowest moisture uptake polymer, 0.0093%. However, PA polymers and homopolymers accounted for the highest moisture uptake being greater than 1%.

MFI which is a determination of polymers viscosity by measuring its ability to flow when melted, is used in the manufacturing of polymer composites for understanding the workability of the composite while developing. Calculations of MFI are conducted based on BS EN ISO 1133-2:2011 and ASTM D1238 (BS EN ISO 2011). Generally, a lower MFI indicates a higher polymer viscosity, and when comparing polymers of the same type, a higher MFI corresponds to a lower molecular weight and/or more branching (Shenoy and Saini 1986). For semi-crystalline polymers to flow, their temperature must be above the crystalline melting temperature, indicating that all the crystalline regions have disappeared (Aumnate et al. 2019). Also, as an example, according to Fazeli and his colleagues investigating the effect of branching characteristics of LLDPE on the MFI. Results showed that with the increase of the degree of branching in LLDPE samples, the MFI reaches minimum. Before the minimum point, the increase in short chain branches (SCB) will cause more entanglements between the chains (inter-molecular entanglement) and, therefore, will impede the flow, increasing the MFI (Fazeli et al. 2006). Thus, knowing the branching characteristics of the polymer matrix can predict the MFI which is crucial for controlling its processing.

Heat deflection temperature (HDT) is the temperature at which a polymer resists to deform by losing its dimensional stability, when subjected to a specified load, generally 0.45 or 1.8 mega pascals (MPa) at an elevated temperature. It is done according to ASTM D648 and BS EN ISO 75 standards (BS EN ISO 2013). The HDT is considered to depend on the thermostability and rigidity of the polymer which can be influenced by the molecular weight, chemical linkages, presence of thermostable and rigid moiety of polymer, crystallinity and degree of branching. Thus, thermoplastic polymers have HDT varying from 32 to 275 °C.

The glass transition temperature (T_g) is another important characteristic when studying physical properties, which depends on the rigidity and flexibility of the polymer. It is the transition temperature, when the polymer becomes soft but not fluid after subjecting the polymer to heat from a very low temperature, at its glassy and hard conditions, until the polymer is completely melted and flows similarly to a liquid with high viscosity (Van De Velde and Kiekens 2001). T_g of polymer is influenced by several factors which include the crystallinity of polymers, chemical linkages, molecular weight, branching and cross-linking structure and the presence of a filler or plasticizer . The flexibility of amorphous polymers is reduced drastically, when they are cooled below T_g. At these temperatures, there are no dimensional changes or segmental motion in the polymer (Grigore 2017). Semi-crystalline thermoplastic polymers have Tg that varies from −140 °C to 153 °C. On the other hand, amorphous thermoplastic polymers have Tg that varies from 43 to 225 °C.

The coefficient of thermal expansion (α_T) expresses the dimensional change of the polymer (volume, length, etc.) in response to change in temperature. Thermoplastic polymers undergo a linear thermal expansion where the values of the linear coefficient of thermal expansion is shown in Table 1. The expansion of polymers under heating depends on mostly intermolecular forces, because the bond length between atoms is virtually independent of temperature. Semi-crystalline polymers have higher coefficient of thermal expansion than amorphous polymers, varying from 4.1 to 13.0 (10^{-5}/°C).

Mould shrinkage of polymers can be calculated using BS EN ISO 294-4:1998, which indicates the volume contraction of polymers during the cooling step while processing (BS EN ISO 1998). This shrinkage is partly due to the difference of density of polymers from the melt state and the cooled, rigid state. Most of the moulded polymers experience shrinkage through the cooling phase at various rates depending on the polymer compositions, thermal properties, processing conditions and the geometry of the mould. Semi-crystalline polymers show higher mould shrinkage than amorphous polymers. This is because semi-crystalline polymers, when cooled down, have part of their macromolecular chains re-arranged to form crystallite that is a well-organized structure, leading to less space needed for the same number of atoms. As demonstrated in Table 1, the mould shrinkage for semi-crystalline thermoplastic polymers varies from 0.3 to 6.0%. Meanwhile, the mould shrinkage for amorphous thermoplastic polymers varies from 0.1 to 0.9%.

The thermal conductivity, is determined by ASTM C177 and BS EN ISO 22007-1:2017 and expressed in W/m/°C as given in Table 1 (BS EN ISO 2012), is a significant characteristic, when considering the use of a polymer for thermal insulation applications. However, most of the polymers demonstrated in Table 1 have the same thermal conductivity, varying from 0.15 to 0.75 W/m/°C, other than PET and PP where there is a significant difference.

Mechanical properties

Polymers tend to exhibit a wide variation of behavior in their mechanical properties ranging from hard to brittle to ductile which are defined under tensile, flexural and impact strength properties. Tensile properties include both tensile strength and tensile modulus, which are determined according to BS EN 527-1:2019 (Table 2) (British Standards Institutions 2019). It is crucial to understand the tensile properties of polymers to predict their performance under stress, especially when used in structural applications. Flexural properties include both flexural strength and flexural modulus which are determined according to BS EN 178:2019 (Table 2) (ISO 178 2010). Flexural properties of polymers demonstrate the polymers rigidity, stiffness, which denotes the ability of a polymer to bend. Moroever, Impact properties are determined according to BS EN ISO 180:2019 (2019). The impact strength can be related to the ultimate tensile strain of the different polymers. More flexible polymers tend to better withstand an impact than the more brittle, less flexible polymers.

Table 2 Mechanical properties of synthetic thermoplastic polymers (Oksman et al. 2003; Cheung et al. 2009; Olabisi and Adewale 2016)

Polymer	Tensile strength (MPa)	Tensile modulus (GPa)	Flexural strength (MPa)	Flexural modulus (GPa)	Young's modulus (GPa)	Notched Izod impact (J/m)	Elongation (%)
Semi-crystalline thermoplastics							
HDPE	14.5–38	0.41–1.49	13.8–48.3	0.41–1.38	0.6–1.1	20–1,068	12–800
MDPE	14.8	0.6	–	0.55	0.6	–	150
LDPE	4.0–78.6	0.05–0.38	25.0	–	0.1–0.3	>854	90–800
LLDPE	7.45–21.0	–	–	0.2–0.48	0.48	443–500	40–800
PP	26–41.4	0.95–1.78	55.2	0.83–1.73	0.3–1.7	20–267	15–700
PA—6	43–80	2.5–2.9	69–117.3	1.9–2.8	1.5–3.0	42.7–160	20–150
PA—4,6	99	3.0–3.3	150.0	3.0	2.0–3.0	98	40–45
PA—6,6	12.4–94	2.5–3.9	89.7–131.1	1.1–3.5	3.3	16–854	30–300
PET	50–70	2.0–4.0	110.4–112.3	2.8	2.7–4.1	26.7–45	100
POM	61–70	2.0	86.87	2.55	3.0	65–85	25–75
PVC	48.25	–	–	–	2.5–4.0	61	21
CPVC	48.81–56.74	2.24–2.94	82.05–94.46	2.41–3.38	2.41–3.38	80.01–320.27	6–120
PVDC	25–110	0.3–0.55	29–42.7	0.5	0.35–0.55	16–53	30–80
PTFE	6.89–31	0.55	13.8–15.2	0.5	0.50	80–320	6.5–300
PBT	51.8–55.9	1.7–2.37	82.8–96	1.9–2.6	2.65	48.1–70	100–300
PPS	65.6–90	2.6–3.9	96–151	3.4–4.1	3.90	10.7–133	1.1–6.0
PEEK	70–103.5	3.1–3.8	110–110.4	2.8–3.9	3.70	120	50
PB	29–36.5	0.29–0.30	–	–	0.53	–	215
PVDF	40.0	1.1	49.0	0.90	1.2–8.3	160	250
Amorphous thermoplastics							

(continued)

Table 2 (continued)

Polymer	Tensile strength (MPa)	Tensile modulus (GPa)	Flexural strength (MPa)	Flexural modulus (GPa)	Young's modulus (GPa)	Notched Izod impact (J/m)	Elongation (%)
PS	46.0	2.9	75.0	3.2	2.4–3.5	19.0	3.4
SAN	59.57–79.29	3.5–4.71	86.18–131.69	2.6–3.77	3.9	9.61–29.89	3.5
ABS	40.0	2.2	97.0	2.1	2.59	138	90–130
PC	53–72	2.3–3.0	81.4–93.2	2.14–2.38	2.2–2.4	534–960	90–130
PVCu	47.0	3.7	80.0	3.5	3.7	360	58.0
PMMA	50–170	2.8	134.0	110.0	2.2–3.8	15–20	2.0–6.0
SMA	–	–	–	–	–	25–650	2.0–30.0
PAI	90–192	2.8–4.4	185.6–240.8	3.6–6.6	–	58.7–133	12.0
PEI	103.5–104.9	3.0	144.9–151.8	3.0–3.5	–	53.4–133	6–60
Elastomeric thermoplastics							
EPDM	5–25	–	–	–	–	–	100–700
TPU	19.58–40.61	0.09–0.81		0.01–0.19	–	–	–
TPO	17	0.9	22.0	0.65–0.83		50 kJ/m^2	–
SBS	24.8–27.9	–	–	–	–	–	820–860
PEBA	9–56	0.01–4.03	–	0.0120–2.69	–	53.4	1–1000

2.2 Bio-based Thermoplastic Polymers for DPF Composites

Bio-based thermoplastic polymers are polymers that can be reformed into new biomass, water and CO_2 by living organisms or by bacteria through biological activity. Bio-based thermoplastic polymers can be synthesized from sustainable resources in plants, polylactide or polylactic acid (PLA) fermented from starch, and bacteria, polyhydroxyalkanoates (PHA), as well as from non-renewable petroleum, synthetic polyesters. Most frequently studied biodegradable polymers have been either aliphatic polyesters, PHA, Polycaprolactone (PCL), polybutylene (PBS) and PLA, or polysaccharides, producing thermoplastic starch (TPS) from starch, which have been introduced into the market through various industrial sectors. The typical thermal and mechanical properties of these bio-based polymers are shown in Tables 3 and 4. It can be noticed that biodegradable aliphatic polyesters have similar mechanical properties to those of PP and PE, with PLA and P(3HB) possessing higher brittleness (Table 3). Starch based polymers posses lower tensile strength and elongation at break when compared to PP and PE which is improved by the addition of plasticizers and various polymer blends (Jiménez et al. 2012; Chan et al. 2018). The material properties of P(3HB), PHA homopolymer, can also be tailored through the inclusion of other monomers, such as 3-hydroxyvalerate. The wide range of properties of the resulting random copolymer, poly(3-hydroxybutyrate-co-3-hydroxyvalerate) (P(3HB-co-3HV)) and several other copolymers that are also presented in Tables 3 and 4 (Chan et al. 2018). A brief description of several bio-based thermoplastic polymers will be discussed in this section.

PLA polymers or polylactides are polyesters of lactic acid which have been recently commercialized for developing biodegradable. PLA is a versatile polymer made from renewable agricultural raw materials, which are fermented to lactic acid which is then polymerised to the wanted polylactic acid via a cyclic dilactone, lactide and ring opening. The polymer is modified through certain techniques, which intensify the temperature stability of the polymer and lowers the residual monomer content. The resulting PLA can be processed similar to polyolefines and other thermoplastics although the thermal stability could be better. PLA are distinguished as brittle and stiff polymers, therefore it is necessary to use plasticizers to improve the impact properties and elongation of PLA. Moreover, PLA polymer is fully biodegradable where the degradation occurs by hydrolysis to lactic acid, which is metabolised by micro-organisms to carbon monoxide and water. By composting together with other biomass the biodegradation occurs within two weeks, and the material will fully disappear within 3–4 weeks (Meinander et al. 1997; Oksman et al. 2003; Chan et al. 2018).

PHA are a family of polyesters that are synthesized intracellularly by more than 298 species of bacteria in the presence of excess carbon source with one of the elements necessary for growth in limitation, such as nitrogen, oxygen or phosphorus. The most common homopolymer and copolymer of PHAs are P(3HB) and P(3HB-co-3HV) respectively. PHA exhibits several properties that make them promising candidates for natural fibre composites. They have higher melt flow indices than

Table 3 Physical and thermal properties of bio-based thermoplastic polymers (Oksman et al. 2003; Ray et al. 2007; Kim et al. 2008; Pivsa-Art et al. 2011; Vallejos et al. 2011; Teixeira et al. 2012; Chuah et al. 2013; Rudnik 2013, 2019; Chang et al. 2014; Hossain et al. 2014; Mizuno et al. 2014; Sirin et al. 2014; Zhang et al. 2014, 2017; Bugnicourt et al. 2014; Janošová and Lenfeld 2015; Sharma et al. 2015; Singh et al. 2015; Farah et al. 2016; Nofar et al. 2017; Huang et al. 2018; Malik et al. 2018; Suleiman 2018; Xu et al. 2018; Chan et al. 2018, 2019; Gierej et al. 2019; Lule and Kim 2019; Seggiani et al. 2019)

Polymer	Density (g/cm^3)	Moisture uptake (%)	Tm (°C)	T_g (°C)	Processing temperature (°C)	HDT @ 1.82 MPa (°C)	Thermal conductivity (W/m/°C)	Specific heat (kJ/kg/°C)	Mold shrinkage (%)
PLA									
PLA	1.21–1.43	0–2.2	150–175	45–64	110	55	0.11–0.20	1.59–6.5	0–10
PLLA	1.24–1.30	0–2	170–200	55–65	123.3	61	0.16–0.42	1.55	0–4
PDLLA	1.25–1.27	0–4	100–180	50–60	130	50	0.1–0.5	1.2	1.3–2.8
PBS									
PBS	1.26	0.06–5	112–114	−32 to −18	76.5–80	97	0.2	1.2–3.3	2–7
PBSA	1.08–1.13	2.4–5	49–96	−36 to −45	220	69	0.12–0.38	01.0–2.7	5
TPS									
Corn based	1.4–1.65	15–30	114–140	20–25	40–70	75	0.19	0.5–0.8	0.6–1.5
Pea based	0.19–1.3	5–20	126	−75 to −19	40–70	81	0.403	0.3–1.0	0.6–0.90
Rice based	1.34–1.38	6.8–14	117.4		40–70	79	0.245	1.2–2.0	0.89
Cassava based	1.38	10–19	169.6	−0.5 to 13	40–70	90	0.189	0.28–0.4	0.65–1.2
Potato based	1.22	8–15	118.7		40–70	78	0.2	0.9–1.2	0.7
Maize based	1.47	5–11	125.5	−21	40–70	65	0.27	1.0–1.6	0.5–0.8
PHA									
P(3HB)	1.18–1.39	1.0–1.2	162–181	−4 to 18.0	50–180	59	0.15	1.4–5.2	11–16
P(3HB-*co*-3HV)	1.18–1.26	4.8	64–171	−13 to 10	52–170	46	0.21	1.1–3.6	12.7
P(3HB-co-4HB)	1.17–1.40	3.2–4.1	49–178	−48 to 4	131	61	0.23	1.0–4.2	6.7

(continued)

Table 3 (continued)

Polymer	Density (g/cm^3)	Moisture uptake (%)	Tm (°C)	T_g (°C)	Processing temperature (°C)	HDT @ 1.82 MPa (°C)	Thermal conductivity (W/m/°C)	Specific heat (kJ/kg/°C)	Mold shrinkage (%)
P(3HHx-*co*-3HO)	0.2–0.9	1.9–2.8	60–135	−5 to 3	105	73	0.19	1.3–5.0	5–11.2
P(3HB-co-3HA)	0.8–1.2	0.9–2.1	133	−8	160	68	0.21	1.1–4.6	12–6.9
P(3HB-*co*-HP)	0.4–1.0	1.3–1.9	44	−19	112–140	85	0.18	1.2–5.4	6.7–9.1
P(3HB-*co*-3HHx)	0.1–0.8	1.0–1.7	52	−4	123–150	79	0.22	1.0–4.9	5.9–10.3
P(3HO)	0.5–1.3	0.7–1.4	61	−10	104–131	53	0.21	1.4–5.1	7.1–11
PCL	1.11–1.15	0.1–1.5	58–65	−65 to −60	60	59	0.19	0.9–6.0	1.0–9.0
PGA	1.50–1.71	0.21–1.8	220–233	35–45	30–62	71	0.24	1.2–7.1	2.1–10

Table 4 Mechanical properties of bio-based thermoplastic polymers (Oksman et al. 2003; Zhao et al. 2003; Coskun et al. 2005; Kim et al. 2008; Pivsa-Art et al. 2011; Vallejos et al. 2011; Teixeira et al. 2012; Rudnik 2013, 2019; Chuah et al. 2013; Hossain et al. 2014; Sirin et al. 2014; Zhang et al. 2014, 2017; Bugnicourt et al. 2014; Chang et al. 2014; Mohd Ishak and Mat Taib 2015; Sharma et al. 2015; Singh et al. 2015; Farah et al. 2016; Nofar et al. 2017; Das et al. 2018; Xu et al. 2018; Chan et al. 2018, 2019; García-Quiles et al. 2019; Seggiani et al. 2019)

Polymer	Tensile strength (MPa)	Tensile modulus (GPa)	Flexural strength (MPa)	Flexural modulus (GPa)	Young's modulus (GPa)	Notched Izod impact (J/m)	Elongation (%)
PLA							
PLA	37–71	2.7–16	83	3.4–4.8	1.2–8.9	11–26	1–3.1
PLLA	15–59	2.7–4.1	190–202	4.8	2.7–4.14	66	4–7
PDLLA	27.6–50	1.0–3.5	90–119	4.9	1–3.45	18	2.0–10.0
PBS							
PBS	34–56	2.5–3.6	36–49	2.5–4.0	1.61	45–101	2.5–19.6
PBSA	19–30	2.12–3.34	13.5–20.1	0.3–3.6	0.27–1.7	2.99–164	382–997
TPS							
Corn based	2.5–3.6	0.01–1	3.1	1.0–1.80	0.97–1.31	No break	48–63
Pea based	1.4–5.8	0.34–0.89	3.5	0.9–2.1	0.8–3.00	No break	38–51
Rice based	1.6–11	0.49–0.7	4.8	0.4–1.3	0.63–0.72	No break	3–60
Cassava based	1.4–1.6	0.3–0.79	4.2	1.1–2.4	0.56–1.28	No break	83
Potato based	3	0.1–0.6	2.6	0.8–1.9	0.4–0.98	No break	3–47
Maize based	0.2–5.8	0.23–0.81	5.4	0.7–1.5	0.6–1.6	No break	30–45
PHA							
P(3HB)	19–44	1.2–4	78.28	1.1–2.1	3.5	–	5–6
P(3HB-*co*-3HV)	30–38	0.14–8.7	60	1.0–3.2	0.7–2.9	–	50
P(3HB-co-4HB)	17–65	1.23–3.5	63.45	1.2–3.0	0.4–5.9	–	444
P(3HHx-*co*-3HO)	9–10	0.39–0.9	62	0.8–1.4	0.008–0.84	–	380
P(3HB-co-3HA)	17	0.95	65–67	0.6–1.1	0.2	–	680
P(3HB-*co*-HP)	10		64	0.9–1.6	0.2–1.14	–	620
P(3HB-*co*-3HHx)	7	0.39	70–71	0.66–1.7	0.135–1.8	–	107.7
P(3HO)	6–10	0.2–3	945	0.4–1.5	2.8–3.5	–	300–450
PCL	21–42	0.21–0.44	9.7–17.2	5.75–7.0	0.21–0.44	No break	300–1000
PGA	60.0–99.7	6.0–7.0	21–108	1–4.6	6.0–7.0		

most of thermoplastic polymers which allows them to have a better distribution and contact with natural fibres and they are 100% biodegradable. However, tensile strength of PHA copolymers show to be constant upon the loss in molecular weight until below 100 kg/g-mol at which point it decreases dramatically, as the degree of chain entanglement decreases. Also, PHAs havelow thermal stability while melted and while processing some of the PHAs show slow crystallization rate which affects the mechanical properties. Moreover, the utilization of PHA polymer in developing composites with natural fibre are not yet feasible on a large scale due to the high costs for sterilizing equipment, in combination with a refined carbon source, the productivity limitations for biological processes and the need for an extraction stage, which lead to a high market price of PHA when compared to the costs of conventional polymers such as PVC and PE (Chan et al. 2018).

Furthermore, starch is a natural polysaccharide, which is known to be the most abundantly available renewable material worldwide. It can be obtained from a variety of crops, such as maize, potato, corn, pea, cassava and rice. It is 100% biodegradable and edible, which makes it an attractive material for food packaging and biodegradable composites development. However, the processing of starch is challenging as it is not a truly thermoplastic, but investigations and studies showed that starch can be altered and by blending with other polymers, by grafting with vinyl monomers or by processing with plasticizers. It can perform thermoplastic properties in the presence of plasticizers, such as glycerol and water at elevated temperatures (90–180 °C) and under shear. Thus, TPS is a potential candidate to replace synthetic PBP. However, its low mechanical stability and sensitivity to water are considered as the main disadvantages of TPS when compared to other bio-based polymers. The tensile modulus and strength of TPS is in the range of 0.01–1 GPa and 0.20–22.0 MPa respectively, which are much lower than those of PLA and PHA polymers. Upon storage, the stiffness and mechanical characteristics of TPS are weakened due to an aging effect occurred through hydrolysis, known as retrogradation due the diffusion of water into the polymer matrix, and to the recrystallization of amylopectin(Nafchi et al. 2013; Zhang et al. 2014; Chan et al. 2018).

Moreover, PCL is a semi-crystalline linear polyester prepared through the ring-opening polymerization of E-caprolactone that is completely biodegradable in compost environments and aerobic soils (Goldberg 1995). PCL is a semi-rigid and tough polymer that has a T_g of 65–60 °C and a low T_m of 58–65 °C. At room temperature, PCL has a modulus of that between LDPE and HDPE. Also, PCL is compatible with many polymers and organic materials, thus it is used in many polymer formulations as compatibilizers (Sreekumar and Thomas 2008).

2.3 *Thermoplastic Matrix Systems in Date Palm Fibre Composites*

Research on DPF reinforced thermoplastic composite was initially reported by Abu-Sharkh and Hamid (2004), who investigated the degradation and stabilization of PP/DPF composite under artificial and natural weathering conditions. Results showed that PP/DPF reinforced composite possessed higher stability than PP itself under both accelerated and weathering exposures, and the compatibilized PP/DPF reinforced composite showed substantial degradation compared to uncompatibilized PP/DPF reinforced composites (Abu-Sharkh and Hamid 2004). The effect of DPF rachis, trunk and petiole, at different loading concentrations, 20%, 30% and 40%, on the mechanical properties of HDPE reinforced composites was also investigated, and the results showed that the addition of more than 20 wt% of DPF from the different parts of DPF produced better mechanical properties than pure HDPE. The optimal fibre content that attained the highest strengths was 40 wt% for the trunk DPF, increasing both the tensile and flexural strength by 17% and 37% respectively. On the other hand, 30 wt% was the optimal fibre loading for both rachis and petiole DPFs, increasing the tensile strength by 15.5% and 6%, and the flexural strength by 8.5% and 60% respectively (Mahdavi et al. 2010). Additionally, AlMaadeed and her colleagues investigated the mechanical and thermal properties of recycled PP (RPP) based hybrid composites of date palm wood flour (DPWF)/glass fibre although only one composite was made from RPP reinforced with 30% DPWF, and found that RPP reinforced DPWF composite has 10% higher tensile strength than pure RPP, where the tensile strength increased from 14.8 to 16.5 MPa. However, the melt flow index (MFI) decreased by 45%, from 1.78 to 1.23 g/10 min (AlMaadeed et al. 2012). The addition of the treated DPF with 5% NaOH aqueous solution to reinforce recycled PET matrix was reported to increase both the tensile and flexural strength with the increase of the concentration of DPF, 5, 10 and 15% in the composite. This is due to the increase in interfacial bonding between DPF and the polymer matrix after treating the DPF with NaOH which removes the impurities from the surface and exposes a larger area of the fibre to be bonded with the polymer matrix. The incorporation of DPF lowered the thermal stability, but increased the overall degree of crystallinity of the composites (Dehghani et al. 2013). AlMaadeed and her fellow researchers (2014a; b) also investigated the mechanical properties of date palm wood powder (DPWP) as a filler for recycled LLDPE composite with the fillers concentration ranging between 10 and 70% at 10% increment. The Young's modulus of the composites increased significantly with increasing DPWP concentration, reaching 466% higher at 70% filler than that for the neat RLLDPE. The stress at break of the composites decreased sharply with increasing DPWP concentration and the composite became brittle when filled with more than 10 wt % DPWP (AlMaadeed et al. 2014b).

Research done on TPS as a source of biodegradable matrices to develop hybrid composites with DPF reinforcement showed that the thermal stability increased as the DPF content increases. The tensile strength and modulus also increased by 760% and 600% respectively for 50 wt% DPF reinforcement. Flexural strength and modulus showed similar behavior. At 60 wt% fiber content and above, the mechanical properties started to deteriorate as a result of the increase in the composites porosity (Ibrahim et al. 2014). Afterwards, Saleh et al. (2017) reported similar results as Ibrahim et al. (2014) that DPF reinforced TPS matrices had the highest fatigue and flexural strengths at 50 wt% DPF content. Furthermore, using a mixture of recycled HDPE, LDPE, and PP with 1% maleic anhydride as a compatibilizing agent was effective for chemical modification of the DPFRC, which promoted dispersion and better interfacial adhesion between the polymer matrices and DPF. Thus, the mechanical properties of the composites were improved where results showed a noticeable increase in both tensile strain and modulus but a large decrease in the percentage of elongation at break. The physical properties of the composites showed good resistance to alkalies and acids. The polymer melting and crystallization temperature were not affected by the addition of DPF or maleic anhydride, indicating that the additives did not affect the thermal behavior of the polymer matrices.

A latest study reported (2019) that the biodegradation of linear low density PE (LLDPE) reinforced with powder DPF composites is affected by the fibre treatment, UV stabilizers, antiblock additives and polymer additives. Results showed that the composite containing UV stabilizer degraded faster than the composite containing antiblock additive, which is due to the difference in the chemical additives within them. Finally, a summary on reported studies of DPF reinforced thermoplastic composites are shown in Table 5.

3 Thermoset Matrix Systems for DPF Composites

3.1 Synthetic Resins for DPF Composites

Phenolic resins

Phenolic resins are one of the first polymeric products discovered in early 1900s to be composed from simple low molecular weight compounds, being the first legitimately synthetic resins to be used. They are produced from phenol, or phenol derivatives, and formaldehyde by step-growth polymerisation using base or acid catalyst. In the presence of an acid catalyst, the reaction of phenol with less than equimolar proportions of formaldehyde produce novolac resins that contain aromatic phenol units combined mostly by methylene bridges. Additionally, in the presence of a base catalyst, the reaction of phenol or phenol derivative with an excess amount of formaldehyde produces resole resins. Novolac resins are thermally stable and cured by cross-linking with formaldehyde donors, such as hexamethylenetetramine. Moreover, when developing composites using phenolic resins, resoles are the most extensively used phenolic resins for composite development as they are considered less viscous and easier to process than novolac resins. Furthermore, phenolic resins are mostly used in structural applications, due to their built-in fire-resistant properties. However, they are not tough enough and their curing reaction generates water, which remains trapped in the composite and transforms into steam in the case of a fire which could damage the structure of the material (Ratna 2009).

Unsaturated polyesters (UP) resins

Unsaturated polyester (UP) resin is the second thermoset resin discovered after phenolic resins, in early 1940s. UP resins consist of an unsaturated polyester, a monomer and an inhibitor. With the increasing environmental concerns, low styrene co-reagent additives are being used in UP resins, usually 50 wt% added to obtain a satisfactory flow behaviour. In some cases, these additives compromise the adhesion by dividing and sealing the surface in the final stages of cure. When cured, UP resins HDT can range from 60 to 230 °C having the lowest glass transition temperature and thermal coefficient expansion of 60 °C and 3.0 °C respectively. Furthermore, UP

Table 5 Summary on exclusively reported studies on DPF reinforced thermoplastic composites

Polymer matrix	Reinforcements	References
Synthetic thermoplastic polymer matrices		
PP	Leaf	Abu-Sharkh and Hamid (2004)
HDPE	Petiole/Rachis/Trunk	Mahdavi et al. (2010)
R-PP	Date palm wood flour	AlMaadeed et al. (2012)
PP-EPDM	N.S	Asadzadeh et al. (2012)
R-HDPE	Mesh	Aldousiri et al. (2013)
R-PET	Leaf	Dehghani et al. (2013)
PP	Rachis	Mahmoudi (2013)
R-LLDPE	Date palm wood flour	AlMaadeed et al. (2014a)
R-HDPE+ R-LDPE+ R-PP	Leaf	Noorunnisa Khanam and AlMaadeed (2014)
LLDPE	Leaf/Leaflet/Rachis flour	Mirmehdi et al. (2014)
PVP	Leaf	Mohanty et al. (2014)
PVP	Leaf	Mohanty et al. (2014)
PP-EPDM	Mesh	Eslami-Farsani (2015)
PP/Flax	Leaf	Aly and ElNashar (2016)
ABS	Leaf	Neher et al. (2016)
LDPE	Frond	Alzebdeh et al. (2017)
HDPE+ LDPE+ R-PP	Leaf	Zadeh et al. (2017)
EPS	Leaflets	Masri et al. (2018)
R-LLDPE	Mesh flour	Alshabanat (2019)
Bio-based thermoplastic matrices		
PLA	N.S	Amirou et al. (2013)
TPS	Spadix stems	Ibrahim et al. (2014)
TPS	Spadix stems	Ibrahim et al. (2017)
TPS	Mesh	Saleh et al. (2017)
PCL	Mesh	Dhakal et al. (2018)

resins are considered very commercially feasible due to their relatively cheap prices as well as their practical nature, easy processability and rapid cure schedules, which allowed them to serve the civil sectors, such as electronic gear, pipes, and tanks. The major use of UP resin is as a matrix for fibre—reinforced composites for various applications and industries, including aerospace, automotive, construction industries as well as household, electrical appliances and military applications (Ratna 2009).

Epoxy resins

Epoxy resins are one of the most important polymer matrices, which play an important role in developing composites. They are created using different reinforcing phases and processing techniques (Paluvai et al. 2014). However, curing comprises polyaddition with a multipurpose co-reagent identified as "hardener" to create a three-dimensional network that can resist hard chemical environments. The epoxy resin has excellent adhesion and presents low cure shrinkage (Plummer et al. 2016). Using conventional resins has drawbacks like longer curing times, moisture uptake and higher viscosities than for UPs. There is a big range of resins and co-reagents that can be used. However, the most common epoxy resin is diglycidyl ether of bisphenol A (DGEBA), and aliphatic amine hardeners are used (Ratna 2009). They cure at room temperature but the reaction is greatly exothermic, which causes issues in thick mouldings. Aromatic amine hardeners on the other hand, need a higher curing temperature, but are more suitable for big parts and give high HDTs reaching 230 °C. Anhydride hardeners are not as used as amines, but are less toxic and need an accelerator. With the convenient catalysis, epoxies can be cured thermally or photolytically without any co-reagent (Plummer et al. 2016). Thus, epoxy resins can range from hard low-temperature epoxies to more breakable high-temperature epoxies, having good mechanical strength, chemical resistance, very low creep, corrosion and weather resistance, fire retardance, good wettability, and adequate electrical properties, allowing it to be used in various industrial applications such as aerospace, automotive, construction and commodity industries (Ratna 2009; Mohan 2013).

Vinyl Ester (VE) resins

Vinyl Ester (VE) resins are produced by the esterification of an epoxy resin with an unsaturated carboxylic acid using a catalyst such as triphenyl phosphine. VE resin exhibits desirable mechanical properties like epoxy and simultaneously offers processability like a polyester resin. VE resins offer much better flexibility and corrosion resistance compared with general purpose UP resins. VE resins are more enduring and resistant to water and chemically threatening environments which results in them being much more expensive than UPs. Composite applications of UPs and VEs are found in several industries like the construction and transport industries as well as in furniture and chemical storage tanks (Ratna 2009).

Polyimide resins

Polyimides are a class of high-temperature resin with a –CO–NR–CO– backbone that predominantly consists of a ring structure (Setianto et al. 2009; Shalwan and Yousif 2014; Sharath et al. 2016; Sharma et al. 2015). In general, polyimides are made by reacting a dianhydride with an aromatic amine. Because of this ring structure, polyimides offer outstanding thermal stability. The demand of polyimides originates from their outstanding thermal properties and thermo-oxidative resistance in combination with excellent mechanical properties. Polyimides burn but they have self-extinguishing properties. They have a very low dielectric constant and are resistant to ionising radiation. Among the low-cost thermoset resins, phenolic resin has a thermal stability comparable with polyimides. However, phenolic resin is extremely brittle and generally needs blending with other resins like epoxy/polyurethane or rubber for augmenting toughness. Such modifications significantly reduce their thermal stability and they cannot be used for high-temperature structural applications. Polyimides, due to their high strength and high heat resistance, often replace glass and metals (e.g., steel) in many demanding industrial applications. The thermosetting polyimides have aroused tremendous interest as advanced materials in civilian and defence applications (Ratna 2009). Thus, the properties that polyimide resins possess make them suitable as a polymer matrix for developing DPF composites that can be used in structural applications.

Amino resins

Amino resins are produced by reacting compounds that contain amino group and formaldehyde. Amino resins are used as curing agents for the resins containing carboxyl, hydroxyl and amide groups. The most popular amino resins are urea–formaldehyde (UF) resins and melamine–formaldehyde (MF) resins which are produced by reacting formaldehyde with urea and melamine, respectively (Ibeh 1999). They are considered to supplement and complement phenolic resins. Also, they can replace phenolic resins applications. MF resin shows the best performance in terms of toughness and water absorption. However, the use of MF resin is restricted due to its high cost (Ratna 2009) (Tables 6 and 7).

3.2 Bio-based Thermoset Resins for DPF Composites

Bio-based thermoset resins can standout when compared to synthetic thermoset resins by having many advantages, such as low price, universal availability, less hazardous, and is mostly preferred by chemical industries as an alternative to conventional synthetic thermoset resins. There is an extensive demand for bio-based resins in the industries to develop sustainable composites that can perform similarly to synthetic polymers (Tschan et al. 2012; Voirin et al. 2014; Zini and Scandola 2016). This section discusses several types of bio-based thermoset resins, their properties and applications.

Bio-based PU resins

Vegetable oils like castor oil for example can be utilized to produce bio-based thermoset PU resins, which have also been exploited in natural fibre composites. The castor oil triglyceride is distinguished by the existence of ricinoleic fatty acid that contains double bonds as well as hydroxyl groups on its backbone (Crosky et al. 2013). Castor oil- based polyurethane (COPU) can be formed by reacting hydroxyl groups with isocyanates. The produced COPU matrix resin has a tensile strength of 2.5 MPa, much lower than the tensile strength of synthetic thermoset resins, but has an extremely high elongation to break of 31% (Milanese et al. 2011). PU can also be

Table 6 Physical and thermal properties of synthetic thermoset resins (Sreekumar et al. 2007; Biron 2012, 2016; Crosky et al. 2013)

Polymer	Density (g/cm^3)	Tm (°C)	T_g (°C)	Processing temperature (°C)	HDT @ 1.82 MPa (°C)	α_T (10^{-5}/°C)	Thermal conductivity (W/m/°C)	Specific heat (kJ/kg/°C)	Mold shrinkage (%)
Unsaturated Polyester	1.09–2.00	260	60	230–300	60–230	3.0	0.70	1.20	4–8
Vinyl Ester	1.03–1.32		110	104–121	95–100		0.23–0.24	–	1–2
Epoxies	1.11–1.30	350	120	−250 to 175	54	5.5	0.20	1.50	–
Phenolics	1.20–1.46	220	170	−250 to 260	153	4.0	0.25	1.40	–
Polyimides	1.40	330	267–400	280–300	240	5.4	0.29	1.10	–

Table 7 Mechanical properties of thermoset resins (Sreekumar et al. 2007; Biron 2012, 2016; Crosky et al. 2013)

Polymer	Tensile strength (MPa)	Tensile modulus (GPa)	Flexural strength (MPa)	Flexural modulus (GPa)	Young's modulus (GPa)	Notched Izod impact (J/m)	Elongation (%)
Polyester	20–105	2.1–3.5	53.8–265	0.36–16	1–16	0.30–0.45	0.5–3.3
Vinyl Ester	73–85	3.0–3.5	125–150	3.5	2.7	0.16	5.0–6.0
Epoxies	55–130	2.7–4.1	80	1.5	2.4	0.18	4.5
Phenolics	35–60	2.7–4.1	42	0.06–0.08	3	0.30–0.35	2.0
Polyimides	72–186	3.3	83–211	3.1	3.3	0.43	5.9

made using rubber seed oil (RSO), however, they have very low mechanical properties when compared to synthetic thermoset resins. They have a tensile modulus of 0.4 GPa and a strength of 4.8 MPa (Bakare et al. 2010). A research conducted by Fiorelli and few other researchers to develop particleboard using coconut fibre and bio-based PU resins showed superior results in producing a composite that can be used in housing, farm buildings and structural applications when compared to producing the same composite from urea–formaldehyde resin (Fiorelli et al. 2012). Thus, bio-based PU resins can be applied in developing DPF that can be used in housing and structural applications, however extensive research is required.

Bio-based phenolic resins

Bio-based phenols extracted from cashew nut shell liquid (CNSL), are considered as good and sustainable natural alternative to synthetic or petroleum-derived phenols (Kim 2015). It is a low-cost material that is abundantly available across the globe, e.g. India, Brazil, and tropical region of Africa. Thus, CNSL can be used as starting material in polymer industry due to its low cost, abundant availability and chemically reactive nature. Several researchers have studied and monitored the extraction, chemistry and composition of CNSL. It is extracted from cashew nut using different processes, namely, hot oil process, roasting process, solvent and supercritical fluid extraction process (Setianto et al. 2009). CNSL-based phenolic resins were used in coating, laminate and friction materials industries. The CNSL-based resins show high rigidity and thermal stability due to the presence of aromatic rings on their backbone (Menon et al. 1996). Jaillet and his colleagues reported with isophorone diamine (IPDA) as curing agent, Tg value for cardanol-based resin was found to be 45 °C, while for diglycidyl ether of bisphenol A (DGEBA), it was found to be 73 °C (Jaillet et al. 2014). Furthermore, bio-based phenolic resins has also been used as the matrix resin for developing 100% biodegradable bio-composites when reinforced with natural fibres (Barreto et al. 2010; Crosky et al. 2013). Njoku and his colleagues reported that alkali treated coir fibres reinforced with CNSL possess higher tensile strength than neat CNSL composites. Their results showed that 30% of coir fibres, weight fraction, added as a reinforcement increases the tensile strength by 70.9% (Njoku et al. 2012). Thus, it is presumed that these resins can also be used in developing DPF composites that possess enhanced mechanical properties with higher tensile strength and being 100% biodegradable.

Protein based resins

Soy protein products can also be used to make thermoset resins. They include soy protein isolate (SPI), soy flour (SF) and soy protein concentrate (SPC). SPI can be synthesized at a pH of 4.5 by precipitation and consists of more than 90% protein. The SPC is prepared by elutriating soluble units from defatted SF and contains almost 18% carbohydrate and more than 70% protein. The SF has a different constitution, it consists of almost 56% protein and about 34% carbohydrate (Paetau et al. 1994). The mentioned soy protein materials have been utilized as the matrix material in natural fibre thermoset matrix composites. To give an enhancement to its tensile properties, SF can be modified to improve its protein content. The modified SF is recognized as MSF. SF can also be modified by cross-linking SF with the use of glutaraldehyde (GA) to give cross-linked soy flower (CSF). In addition to that, the SPI can also be modified using stearic acid to enhance the properties of the composites (Bobade et al. 2016).

Epoxidized plant oil-based resins

Plant oils are mostly composed of triglycerides that are made up of 3 unsaturated fatty acids joined at glycerol juncture. The fatty acids are generally 14–22 carbon atoms with 0–3 double bonds per fatty acid (Wool and Sun 2011). There are many methods that can be used to epoxidize fatty acids to convert the unsaturation to epoxy (Biermann et al. 2008). Soybean oil and linseed oil also referred to as flaxseed oil that are usually used for epoxidation to produce epoxidized linseed oil (ELO) and epoxidized soybean oil (ESO). ELO-based resin utilized as the matrix for natural fibre composites has showed a flexural modulus of 1.8GPa and a strength of 60 MPa (Boquillon 2006), which are very close to the strength and flexural modulus of synthetic resins. ESO can be modified using acrylic acid and become acrylated epoxidized soybean oil (AESO) (Akesson et al. 2009), which is also used as the matrix resin in natural fibre thermoset matrix composites. In addition to that, ESO can be transformed into methacrylated soybean oil (MSO) (Ramamoorthy et al. 2012) and the MSO can also be modified to methacrylic anhydride-modified soybean oil (MMSO) using methacrylic anhydride (Adekunle et al. 2011). Bio-based epoxy resins have been proven to be environmentally friendly. However, their high water absorption and their swelling present limitations when compared to the traditional epoxy resins (Masoodi and Pillai 2012). Akesson and his colleagues reported that hemp reinforced AESO resin shows an increase in mechanical properties which can be also applied to DPF to develop sustainable biodegradable composites (Akesson et al. 2009).

3.3 *Thermoset Matrix Systems in Date Palm Fibre Composites*

Research on DPF reinforced thermoset composites are done using synthetic thermoset polymer matrix and to date no bio-based thermoset matrix has been used for developing DPF composites. Al Kaabi and his colleagues are the first researchers who investigated the effect of different DPF loading, 6–10%, and DPF length treated with three different chemicals on the mechanical properties of the composite. Their results showed that the optimal fibre loading was 9% treated with 5% NaOH solution for 2 h. Where the date palm fibre reinforced composite (DPFRC) had higher flexural strength, flexural modulus and impact strength than the thermoset polymer by approximately 66%, 50%, 475% respectively (Al-Kaabi et al. 2005). The effect of a higher fibre loading, 20–60%, and the fibres orientation, woven or unidirectional, on the mechanical properties of DPFRC without any prior treatment of the DPF was also reported, showing that both unidirectional and woven orientation DPF reinforced polyester resin had greater mechanical properties than neat polyester resin, however, unidirectional orientation had higher mechanical properties than the woven orientation reinforced DPF, where the flexural strength and elongation increased with the increase of fibre loading. On the other hand, the impact strength decreased with the increase of DPF loading. The flexural strength and elongation of the optimum unidirectional orientation DPF (60% loading) reinforced polyester composites increased approximately by 55% and 10% respectively compared to neat polyester composites. However, the inter-laminar failure and delamination always occurred along the fibre/matrix interface (Wazzan 2005). Using poly epoxy as a matrix with varying DPF loadings, 5%, 10%, and 15%, subjected to oxidation using soxhlent extraction, was also investigated and the results showed that there was a slight increase in flexural modulus of the reinforced composites with the increase of DPF loading by 19%, and a decrease in the stress at break or elongation and impact strength with the increase of DPF loading by 71% and 87% respectively (Sbiai et al. 2010).

The effect of various diameters of DPF treated in 6% NaOH solution on the strength of reinforced epoxy composites was reported. It was found that the mechanical properties increased with the decrease in diameter of treated DPF. DPF with diameter size ranging from 200 to 400 μm has 120% and 13% higher tensile strength and Young's Modulus respectively than DPF diameter size ranging between 600 and 800 μm. The elongation strain of untreated DPFRC is higher than the treated DPFRC by 65%, 77%, and 15% for DPF diameters ranging between 200 and 400, 400 and 600, and 600 and 800 μm respectively (Abdal-Hay et al. 2012). A recent investigation on the effect of different DPFs from different parts of DPT on epoxy matrix composites showed an increase in mechanical properties of epoxy composites, with the tensile strength, tensile modulus, flexural strength, flexural modulus and impact strength showing an increase using different types of DPTF. The highest increase was with the reinforcement of date palm fruit bunch stalk, having higher tensile strength, tensile modulus, flexural strength, flexural modulus and impact strength than neat epoxy resin by 95%, 405%, 243%, 179% and 119% respectively. However, the incorporation of DPF in epoxy resins resulted in an increase in the water absorption with time which can be explained due to the hydrophilic nature of DPF (Alshammari et al. 2019). Likewise, different DPF loadings, 45, 50 and 60% showed the epoxy DPF composites resulted in an increase in flexural strength and flexural modules by approximately 25% when loaded with 50% DPF (Gheith et al. 2019). Finally, a summary on reported studies of DPF reinforced thermoset composites are shown in Table 8.

4 Elastomers Matrix Systems for DPF Composites

Elastomers are synthetic or natural amorphous polymers with high molecular weight (10^5–10^6 g/mol), low glass transition temperature and high chain flexibility, which allow them to exhibit large reversible elongation and extension, when subjected to relatively low stress (Council 1994). The well-known elastomeric material is rubber and therefore elastomeric composites are usually named as rubber composites. Elastomers differ from thermosets and thermoplastics with their highly elastic mechanical behaviour (Sim and Arof 2017). However, originally, elastomers were thermosets where their reprocessing and recycling were limited. Afterwards, blending elastomers with thermoplastics formed a better composite that can be recycled and processed by heating it above the glass transition temperature of the blended thermoplastic (Council 1994). Elastomers can be classified as natural material when considering natural rubber that can be obtained from more than 200 different species and synthetic elastomers that are produced using crude oil. This section will discuss the four largest elastomers produced by volume and their physical, thermal and mechanical properties as shown in Tables 9 and 10. Also, a comparison between various types of elastomers' properties are shown in Table 11. Styrene-butadiene rubber (SBR), one of the cheapest and largest volume elastomers, is a synthetic rubber that is derived from styrene and butadiene, and is produced by either emulsion polymerization or by free-radical solution polymerization. SBRs exhibit good abrasion resistance and their properties are influenced by the styrene/butadiene ratio, normally 10–25% styrene. As the styrene content increases, the polymer gets harder and less rubbery, thus improving the strength and abrasion resistance, this also reduces the price and improves the compatibility. Moreover, polybutadiene (PBD), the second largest elastomer polymer matrix, have a very low T_g of around 90 °C and very high resilience, and is accounted for the highest elasticity between all rubber polymers. Furthermore, nitrile-butadiene rubber (NBR), a polymer with good resistance to oil, chemicals and fuel that has better temperature resistance compared to PBD, withstanding 33–107 °C. The higher the nitrile content, the greater the resistance to oils is, but its flexibility decreases. To achieve the optimum chemical resistance and flexibility, 30–45% nitrile content is utilised. Polychloroprene rubber (CR) accounts for being the fourth largest volume elastomer polymer, and possesses a good UV stability and has good chemical resistance to oil, oxygen and ozone, fuel and other chemicals. CR

Table 8 Summary on exclusively reported studies on DPF reinforced thermoset composites

Polymer Matrix	Reinforcements	References
Synthetic thermoset polymer matrices		
Phenolic (phenol formaldehyde) resin	Leaf	Al-Sulaiman (2002, 2003)
Bisphenol resin with amine-based slow curing agent		
Polyester resin (8120TEC)	Mesh	Al-Kaabi et al. (2005)
Polyester resin (8420-P) Hardned (MEKP)	N.S	Wazzan (2005)
Epoxy Resin	Leaf	Kaddami et al. (2006)
Unsaturated polyester resin (G154TB)		
Polyepoxy (epoxy with amine curing agent)	Leaf	Sbiai et al. (2010)
Epoxy Resin (YD-128) Hardend (D-230)	Mesh	Abdal-Hay et al. (2012)
Polyester resin (Siropol 7440)	Frond	Anyakora (2013)
Epoxy Resin (R246TX) Kinetix (H160)	Mesh	Shalwan and Yousif (2014)
Polyester Resin Hardened (Methyl Ethyl Keton Peroxide Mekp)	Mesh	Hammood (2015)
Epoxy Resin (R246TX) Kinetix (H160)	Mesh	Alajmi and Shalwan (2015)
Epoxy Resin (A–B)	Leaf	Mahdi et al. (2015)
Polyester Resin (SIR RESIN)	Date palm seeds powder	Ra (2015)
Epoxy Resin	N.S	Sharath et al. (2016)
Epoxy Resin	Spadix stems	Tripathy et al. (2016)
Polyester Resin	N.S	Salih et al. (2015)
Epoxy Resin (LY 556) Hanrdend (HY951) + E-Glass fibre	Mesh	Swain et al. (2018)
Epoxy Resin Hardened Jointmine 905-3S	Leaf/Bunch/Mesh/Trunk	Alshammari et al. (2019)
Epoxy Resin (Jointmine 905-3S)	Mesh	Gheith et al. (2019)
Polyester Resin	Midrib	Raghavendra and Lokesh (2019)
Epoxy Resin (D.E.R 331) Hardened Jointmine 905-3S	Spadix stems	Saba et al. (2019)

Table 9 Physical and thermal properties of elastomer polymers (Hanhi et al. 2007; Laoui 2013; Database 2015; Wu et al. 2016; Harandi et al. 2017)

Polymer	Density (g/cm^3)	Tm (°C)	T_g (°C)	Processing temperature (°C)	Thermal conductivity (W/m/°C)	Specific heat (kJ/kg/°C)
NR	0.93–0.95	−25	−70	−55 to 70	0.2	0.44–0.45
SBR	0.98	−29	−65 to −70	−25 to 100	0.148–0.16	
NBR	0.952	−23	−68 to −71	−40 to 125	0.157	
Cis-1, 4-polyisoprene	0.9–1.0	70	−73	−56 to 82	0.15	1.8
Butyl Rubber	0.92–2.0		−75 to 67	−45 to 150	0.13–0.23	1.95
Cis-1, 4-polychloroprene	1.23–1.25		−20	−43 to 107	0.19	2.17
CR	1.5		−45	10	0.19	1.1
Epichlorohydrin Copolymer Rubber (ECO)	1.4	−40 to −80	−30 to 20	−35 to 125	0.17–0.22	–
ACM	1.2–1.6		−57 to −18	−10 to 150	0.2–0.4	–
Polyurethane rubbers	0.8–1.3		−40 to 25	−50 to 100	0.022–0.028	–
Fluorocarbon Rubbers	0.9–1.1		−46 to −18	−22 to 120	0.025–0.15	–

polymers are known with the superior toughness and abrasion resistance. Generally, CR's mechanical properties are lower than those of natural rubber but it has excellent chemical resistance. Butyl rubber, a copolymer of isobutylene and minor amounts of isoprene, has a low gas permeability and superior resistance to ozone, oxygen UV attack. Due to the low content of unsaturated bonds, it has superior thermal stability and chemical resistance (Database 2015).

Moreover, to date, there is no reported work on using elastomer polymer matrix on its own for DPF composites. However, it was reported by both Asadzadeh and his colleagues and Eslami-Farsani of blending EPDM with PP with DPF as a reinforcement to develop bio-composites (Asadzadeh et al. 2012; Eslami-Farsani 2015).

Table 10 Mechanical properties of elastomer polymers (Gent et al. 1998; Hanhi et al. 2007; Tonpheng et al. 2010; Renner and Pék 2011; Laoui 2013; Wang et al. 2013; Xu and He 2015; Arsada et al. 2018; Wei et al. 2019)

Polymer	Tensile strength (MPa)	Tensile modulus (GPa)	Flexural strength (MPa)	Flexural modulus (GPa)	Young's modulus (GPa)	Elongation (%)
NR	4–27.6	1.5–2	52–61.353	1.814–2	0.01–0.12	100–700
SBR	4–36	0.01–0.1	65.18	2.26	0.002–0.01	450–770.9
NBR	0.5–15	0.002–0.15	20–34	2.9	0.004	100–400
Cis-1, 4-polyisoprene	29	0.5–1.5	28–41	1.7- 2.5	0.22–5.9	660–850
Butyl Rubber	4–17	0.0002–0.001	31–45	2.1–2.9	0.3–3.4	100–800
Cis-1, 4-polychloroprene	3.4–24.1	0.0085	20–29	1.78	0.7–20.1	100–800
CR	4–20	0.001–0.1	23.5	1.6	1.7	100–500
Epichlorohydrin Copolymer Rubber (ECO)	4–18	0.0075	–	–	0.1–0.12	100–500
ACM	8.6–17.2	0.0007–0.01	17.9	2.7	0.01–0.1	600–900
Polyurethane rubbers (AU/EU/PUR)	6.9–69	0.0002–0.034	16.8	2.9	0.04–0.09	250–900
Fluorocarbon Rubbers (FKM/FPM)	3–20.7	0.0031–0.0034	19.6	2.8–2.9	0.11–0.2	100–500

5 Conclusion

DPF is a potentially suitable reinforcement in sustainable and bio-degradable composites, owing to its great properties, low cost, abundancy, bio-degradability and low density. Various types of available polymer matrix that could be used in combination with DPF to develop such composites have been comprehensively reviewed and discussed. The polymer properties reported in this chapter provide a database for researchers and industries to decide the most suitable polymer for developing an ideal date palm fibre reinforced composite for various applications. It can be concluded that thermoplastics possess several advantages (e.g. high recyclability, high impact resistance, good chemical resistance and reshaping/remoulding capabilities) and few disadvantages (e.g. highest cost compared to thermoset and elastomers, melt when heated, high temperature processing window, undergo fracture rather than deforming under high stress level, and suffer from creep). The advantages that thermoset resins possess include low cost, higher resistance to temperature compared to thermoplastics, high flexibility design, and excellent aesthetic appearance, and the disadvantages that thermoset resins possess are difficult to recyclable and to be reshaped or remoulded. Elastomers have several advantages (e.g. low cost, recyclable, abundant

Table 11 Advantages and disadvantages of elastomer polymer based on their properties (Hanhi et al. 2007)

Elastomer polymer type	Advantages	Disadvantages
Natural rubber	• Good processability • Excellent elastic properties • Good tensile strength • High elongation • Good wear and tear resistance • Little dissipation factor—low heat build-up in dynamic stress • Excellent cold resistance • High resistance to water and acids (not to oxidizing acids) • Good electrical insulator	• Poor weather and ozone resistance • Restricted high temperature resistance (short-time maximum temperature 100 °C) • Low fuel and oil resistance • Unsuitable for use with organic liquids in general
Styrene-Butadiene Rubber (SBR)	• Good abrasion and aging resistance • Good elasticity • Low price	• Low mechanical properties • Poor ozone and oil resistance • Do not resist aromatic, aliphatic or halogenated solvents • Low elongation
PBD	• Excellent cold resistance and heat resistance • Good elasticity • Excellent low temperature resilience and flexibility • Good abrasion resistance	• Poor processability • Weak mechanical properties
NBR	• High oil and heat resistance • Good resistance to oil, aliphatic and aromatic hydrocarbons and vegetable oils • Good water and abrasion resistance • High tensile strength	• Low ozone resistance • High swelling with some oils and solvents (e.g. esters and ketones)
CR	• Non-flammable • Good abrasion, ozone, oils and solvents resistance • Good tear strength • Increased hardness at elevated temperature environments	• High swelling in hot water, acids and some oils and organic solvents

(continued)

Table 11 (continued)

Elastomer polymer type	Advantages	Disadvantages
Isoprene Rubber, Polyisoprene (IR)	• Good toughness and abrasion resistance • Competitive price • Good processability and adherence • High tensile strength and resilience • Good resistance to many inorganic chemicals	• Restricted life time at high temperatures and in oxidative conditions • Poor oil resistance needs protection against oxygen, ozone and light • Unsuitable for use with organic liquids
Fluorocarbon rubbers	• Excellent heat, oxygen, ozone and weather resistance • Good solvent, chemical and abrasion resistance • Good compression-set resistance at elevated temperatures	• Low alkali resistance • Poor mechanical properties • High cost
Butyl rubbers	• Heat stability • Excellent weathering and ozone resistance • Low water absorption and gas permeability • Resistant to oxidizing agents, polar solvents, vegetable and animal fats	• Poor wear resistance • Not resistant to oils and hydrocarbon solvent • Relatively low elasticity
ACM	• Excellent weathering and ozone resistance • Good elasticity and oil resistance • Excellent flexural properties • Low gas permeability	• Poor acid, alkali and water resistance
Polysulphide rubbers	• Excellent solvent and oil resistance • Good ozone and weather resistance	• Narrow processing temperature window, difficult to process • Bad smell

availability of raw materials, easily processed, high resistance to environment, and light weight) and disadvantages (e.g. low durability, low Young's modulus, low heat resistance, and susceptibility to degradation under the influence of UV radiation). The critical issue for developing a composite with superior properties is the interfacial bonding between DPF and the polymer matrix, denoting the importance of the interface and surface modifications of fibres to enhance the physical and mechanical properties of DPF reinforced composites. Challenges still exist in the development of DPF composite, especially its industrialization as other natural fibre composites for various industrial applications.

References

Abdal-Hay A, Suardana NPG, Jung DY et al (2012) Effect of diameters and alkali treatment on the tensile properties of date palm fiber reinforced epoxy composites. Int J Precis Eng Manuf 13:1199–1206. https://doi.org/10.1007/s12541-012-0159-3

Abu-Sharkh BF, Hamid H (2004) Degradation study of date palm fibre/polypropylene composites in natural and artificial weathering: mechanical and thermal analysis. In: Polymer degradation and stability. Elsevier Ltd., pp 967–973

Adekunle K, Patzelt C, Kalantar A, Skrifvars M (2011) mechanical and viscoelastic properties of soybean oil thermoset reinforced with jute fabrics and carded Lyocell fiber. J Appl Polym Sci 2855–2863. https://doi.org/10.1002/app.34360

Akesson D, Skrifvars M, Walkenstrom P (2009) Preparation of thermoset composites from natural fibres and acrylate modified soybean oil resins. J Appl Polym Sci 114:2502–2508. https://doi.org/10.1002/app.30773

Alajmi M, Shalwan A (2015) Correlation between mechanical properties with specific wear rate and the coefficient of friction of graphite/epoxy composites. Materials (Basel) 8:4162–4175. https://doi.org/10.3390/ma8074162

Aldousiri B, Alajmi M, Shalwan A (2013) Mechanical properties of palm fibre reinforced recycled HDPE. Adv Mater Sci Eng 2013. https://doi.org/10.1155/2013/508179

Ali A, Shaker K, Nawab Y et al (2018) Hydrophobic treatment of natural fibers and their composites—a review. J Ind Text 47:2153–2183. https://doi.org/10.1177/1528083716654468

Al-Kaabi K, Al-Khanbashi A, Hammami A (2005) Date palm fibers as polymeric matrix reinforcement: DPF/polyester composite properties. Polym Compos 26:604–613. https://doi.org/10.1002/pc.20130

Al-Khanbashi A, Al-Kaabi K, Hammami A (2005) Date palm fibers as polymeric matrix reinforcement: fiber characterization. Polym Compos 26:486–497. https://doi.org/10.1002/pc.20118

AlMaadeed MA, Kahraman R, Noorunnisa Khanam P, Madi N (2012) Date palm wood flour/glass fibre reinforced hybrid composites of recycled polypropylene: mechanical and thermal properties. Mater Des 42:289–294. https://doi.org/10.1016/j.matdes.2012.05.055

AlMaadeed MA, Nógellová Z, Mičušík M et al (2014) Mechanical, sorption and adhesive properties of composites based on low density polyethylene filled with date palm wood powder. Mater Des 53:29–37. https://doi.org/10.1016/j.matdes.2013.05.093

AlMaadeed MA, Nógellová Z, Janigová I, Krupa I (2014) Improved mechanical properties of recycled linear low-density polyethylene composites filled with date palm wood powder. Mater Des 58:209–216. https://doi.org/10.1016/j.matdes.2014.01.051

Al-Oqla FM, Sapuan SM (2014) Natural fiber reinforced polymer composites in industrial applications: feasibility of date palm fibers for sustainable automotive industry. J Clean Prod 66:347–354. https://doi.org/10.1016/j.jclepro.2013.10.050

Alshabanat M (2019) Morphological, thermal, and biodegradation properties of LLDPE/treated date palm waste composite buried in a soil environment. J Saudi Chem Soc 23:355–364. https://doi.org/10.1016/j.jscs.2018.08.008

Alshammari BA, Saba N, Alotaibi MD et al (2019) Evaluation of mechanical, physical, and morphological properties of epoxy composites reinforced with different date palm fillers. Materials (Basel) 12. https://doi.org/10.3390/ma12132145

Al-Sulaiman FA (2002) Mechanical properties of date palm fiber reinforced composites. Appl Compos Mater 9:369–377. https://doi.org/10.1023/A:1020216906846

Al-Sulaiman FA (2003) Date palm fibre reinforced composite as a new insulating material. Int J Energy Res 27:1293–1297. https://doi.org/10.1002/er.957

Aly NM, ElNashar DE (2016) Polyester/flax biocomposites reinforced using date palm leaves and wood flour as fillers. Int J Eng Technol 8

Alzebdeh KI, Nassar MM, Al-Hadhrami MA et al (2017) Characterization of mechanical properties of aligned date palm frond fiber-reinforced low density polyethylene. J Eng Res 14:115–123. https://doi.org/10.24200/tjer.vol.14iss2pp115-123

Amirou S, Zerizer A, Haddadou I, Merlin A (2013) Effects of corona discharge treatment on the mechanical properties of biocomposites from polylactic acid and Algerian date palm fibres. Sci Res Essays 8:946–952. https://doi.org/10.5897/SRE2013.5507

Anyakora AN (2013) Investigation of impact strength properties of oil and date palm frond fiber reinforced polyester composites

Arsada R, Surojo E, Ariawan D et al (2018) Effect of NBR (Nitrile Butadiene Rubber) on flexural strength of composite friction brake. IOP Conf Ser Mater Sci Eng 420. https://doi.org/10.1088/1757-899X/420/1/012057

Asadzadeh M, Khalili SMR, EslamiFarsani R, Rafizadeh S (2012) Bending properties of date palm fiber and jute fiber reinforced polymeric composite

Aumnate C, Rudolph N, Sarmadi M (2019) Recycling of polypropylene/polyethylene blends: effect of chain structure on the crystallization behaviors. Polymers (Basel) 11. https://doi.org/10.3390/polym11091456

Bakare IO, Okieimen FE, Pavithran C et al (2010) Mechanical and thermal properties of sisal fiber-reinforced rubber seed oil-based polyurethane composites. Mater Des 31:4274–4280. https://doi.org/10.1016/j.matdes.2010.04.013

Barreto ACH, Esmeraldo MA, Rosa DS et al (2010) Cardanol biocomposites reinforced with jute fiber: Microstructure, biodegradability, and mechanical properties. Polym Compos 37:915–924. https://doi.org/10.1002/pc.20990

Biermann U, Friedt W, Lang S et al (2008) New Syntheses with oils and fats as renewable raw materials for the chemical industry. Biorefineries Ind Process Prod Status Quo Futur Dir 2:253–289. https://doi.org/10.1002/9783527619849.ch25

Biron M (2012) Thermoplastics and thermoplastic composites. William Andrew

Biron M (2016) Thermoplastics. In: Material selection for thermoplastic parts. William Andrew, pp 77–111

Bobade SK, Paluvai NR, Mohanty S, Nayak SK (2016) Bio-based thermosetting resins for future generation: a review. Polym Plast Technol Eng 55:1863–1896. https://doi.org/10.1080/03602559.2016.1185624

Boquillon N (2006) Use of an epoxidized oil-based resin as matrix in vegetable fibers-reinforced composites. J Appl Polym Sci 101:4037–4043. https://doi.org/10.1002/app.23133

Brahmakumar M, Pavithran C, Pillai RM (2005) Coconut fibre reinforced polyethylene composites: effect of natural waxy surface layer of the fibre on fibre/matrix interfacial bonding and strength of composites. Compos Sci Technol 65:563–569. https://doi.org/10.1016/j.compscitech.2004.09.020

British Standards Institutions (2019) BSI standards publication plastics—determination of flexural properties

BS EN ISO (1998) Plastics—injection moulding of test specimens of thermoplastic materials—English

BS EN ISO (2008) Plastics—determination of water absorption. Management 3

BS EN ISO (2011) BSI standards publication plastics—determination of the melt mass-flow rate (MFR) and melt volume-flow rate (MVR) of thermoplastics. Part 2: method for materials sensitive to time-temperature history and/or moisture

BS EN ISO (2012) BSI standards publication plastics—determination of thermal conductivity and thermal diffusivity

BS EN ISO (2013) ISO 75-1:2013—Plastics—determination of temperature of deflection under load—part 1: general test method. Geneva 9

BS EN ISO (2019) BS EN ISO 1183-2:2019 BSI standards publication plastics—methods for determining the density of non-cellular plastics

BS EN ISO 180:2019 (2019) Plastic. Determination of Izod impact strength

Bugnicourt E, Cinelli P, Lazzeri A, Alvarez V (2014) Polyhydroxyalkanoate (PHA): review of synthesis, characteristics, processing and potential applications in packaging. Express Polym Lett 8:791–808. https://doi.org/10.3144/expresspolymlett.2014.82
Chan CM, Vandi LJ, Pratt S et al (2018) Composites of wood and biodegradable thermoplastics: a review. Polym Rev 58:444–494. https://doi.org/10.1080/15583724.2017.1380039
Chan CM, Vandi L-J, Pratt S et al (2019) Insights into the biodegradation of PHA/wood composites: micro- and macroscopic changes. Sustain Mater Technol 21: https://doi.org/10.1016/j.susmat.2019.e00099
Chang HM, Wang ZH, Luo HN et al (2014) Poly(3-hydroxybutyrate-co-3-hydroxyhexanoate)-based scaffolds for tissue engineering. Braz J Med Biol Res 47:533–539. https://doi.org/10.1590/1414-431X20143930
Cheung H, Ho M, Lau K et al (2009) Natural fibre-reinforced composites for bioengineering and environmental engineering applications. Compos Part B Eng 40:655–663. https://doi.org/10.1016/j.compositesb.2009.04.014
Chimeni DY, Toupe JL, Dubois C, Rodrigue D (2016) Effect of hemp surface modification on the morphological and tensile properties of linear medium density polyethylene (LMDPE) composites. Compos Interfaces 23:405–421. https://doi.org/10.1080/09276440.2016.1144163
Chuah JA, Yamada M, Taguchi S et al (2013) Biosynthesis and characterization of polyhydroxyalkanoate containing 5-hydroxyvalerate units: Effects of 5HV units on biodegradability, cytotoxicity, mechanical and thermal properties. Polym Degrad Stab 98:331–338. https://doi.org/10.1016/j.polymdegradstab.2012.09.008
Coskun S, Korkusuz F, Hasirci V (2005) Hydroxyapatite reinforced poly(3-hydroxybutyrate) and poly(3-hydroxybutyrate-co-3-hydroxyvalerate) based degradable composite bone plate. J Biomater Sci Polym Ed 16:1485–1502. https://doi.org/10.1163/156856205774576727
Council NR (1994) Manafacturing: materialas and Processing. Polymer science and engineering. The National Academies Press, Washington, DC, pp 76–78
Crosky A, Soatthiyanon N, Ruys D et al (2013) Thermoset matrix natural fibre-reinforced composites. Woodhead Publishing Limited
Da Silva ALN, Coutinho FMB (1996) Some properties of polymer blends based on EPDM/PP. Polym Test 15:45–52. https://doi.org/10.1016/0142-9418(95)00012-7
Das R, Saha NR, Pal A et al (2018) Comparative evaluation of physico-chemical characteristics of biopolyesters P(3HB) and P(3HB-co-3HV) produced by endophytic Bacillus cereus RCL 02. Front Biol (Beijing) 13:297–308. https://doi.org/10.1007/s11515-018-1509-z
Database PP (2015) Polydienes (diene elastomers, synthetic rubbers). https://polymerdatabase.com/polymerclasses/Polydienetype.html#Butylrubber. Accessed 26 Dec 2019
Dehghani A, Madadi Ardekani S, Al-Maadeed MA et al (2013) Mechanical and thermal properties of date palm leaf fiber reinforced recycled poly (ethylene terephthalate) composites. Mater Des 52:841–848. https://doi.org/10.1016/j.matdes.2013.06.022
Dhakal H, Bourmaud A, Berzin F et al (2018) Mechanical properties of leaf sheath date palm fibre waste biomass reinforced polycaprolactone (PCL) biocomposites. Ind Crops Prod 126:394–402. https://doi.org/10.1016/j.indcrop.2018.10.044
Dittenber DB, Gangarao HVS (2012) Critical review of recent publications on use of natural composites in infrastructure. Compos Part A Appl Sci Manuf 43:1419–1429. https://doi.org/10.1016/j.compositesa.2011.11.019
Elseify LA, Midani M, Shihata LA, El-Mously H (2019) Review on cellulosic fibers extracted from date palms (Phoenix Dactylifera L.) and their applications. Cellulose 26:2209–2232. https://doi.org/10.1007/s10570-019-02259-6
Eslami-Farsani R (2015) Effect of fiber treatment on the mechanical properties of date palm fiber reinforced PP/EPDM composites. Adv Compos Mater 24:27–40. https://doi.org/10.1080/09243046.2013.871177
Fan M, Fu F (2016) Advanced high strength natural fibre composites in construction. Woodhead Publishing Limited

Fan M, Naughton A, Bregulla J (2017) Fire performance of natural fibre composites in construction. In: Advanced high strength natural fibre composites in construction, pp 375–404

Farah S, Anderson DG, Langer R (2016) Physical and mechanical properties of PLA, and their functions in widespread applications—a comprehensive review. Adv Drug Deliv Rev 107:367–392. https://doi.org/10.1016/j.addr.2016.06.012

Fazeli N, Arabi H, Bolandi S (2006) Effect of branching characteristics of ethylene/1-butene copolymers on melt flow index. Polym Test 25:28–33. https://doi.org/10.1016/j.polymertesting.2005.09.008

Feldman D (2008) Polymer history. Des Monomers Polym 11:1–15. https://doi.org/10.1163/156855508X292383

Fiorelli J, Curtolo DD, Barrero NG et al (2012) Particulate composite based on coconut fiber and castor oil polyurethane adhesive: an eco-efficient product. Ind Crops Prod 40:69–75. https://doi.org/10.1016/j.indcrop.2012.02.033

García-Quiles L, Cuello ÁF, Castell P (2019) Sustainable materials with enhanced mechanical properties based on industrial polyhydroxyalkanoates reinforced with organomodified sepiolite and montmorillonite. Polymers (Basel) 11. https://doi.org/10.3390/polym11040696

Gent AN, Kawahara S, Zhao J (1998) Crystallization and strength of natural rubber and synthetic cis-1,4-polyisoprene. Rubber Chem Technol 71:668–678. https://doi.org/10.5254/1.3538496

Gheith MH, Aziz MA, Ghori W et al (2019) Flexural, thermal and dynamic mechanical properties of date palm fibres reinforced epoxy composites. J Mater Res Technol 8:853–860. https://doi.org/10.1016/j.jmrt.2018.06.013

Gierej A, Vagenende M, Filipkowski A et al (2019) Poly(D,L-Lactic Acid) (PDLLA) biodegradable and biocompatible polymer optical fiber. J Light Technol 37:1916–1923. https://doi.org/10.1109/JLT.2019.2895220

Goldberg D (1995) A review of the biodegradability and utility of poly(caprolactone). J Environ Polym Degrad 3:61–67. https://doi.org/10.1007/BF02067481

Grigore ME (2017) Methods of recycling, properties and applications of recycled thermoplastic polymers. Recycling 2

Halley PJ, Truss RW, Markotsis MG et al (2007) A review of biodegradable thermoplastic starch polymers, pp 287–300

Hammood AS (2015) Effect of erosion on water absorption and morphology for treated date palm fiber-reinforced polyester composites. Int J Mech Mechatron Eng 15

Hanhi K, Poikelispaa M, Tirila H-M (2007) Elastomeric materials. Tampere University of Technology. https://laroverket.com/wp-content/uploads/2015/03/Elastomeric_materials.pdf. Accessed 17 Jan 2020

Harandi MH, Alimoradi F, Rowshan G et al (2017) Morphological and mechanical properties of styrene butadiene rubber/nano copper nanocomposites. Results Phys 7:338–344. https://doi.org/10.1016/j.rinp.2016.11.022

Hossain KMZ, Parsons AJ, Rudd CD et al (2014) Mechanical, crystallisation and moisture absorption properties of melt drawn polylactic acid fibres. Eur Polym J 53:270–281. https://doi.org/10.1016/j.eurpolymj.2014.02.001

Huang C, Qian X, Yang R (2018) Thermal conductivity of polymers and polymer nanocomposites. Mater Sci Eng R Reports 132:1–22. https://doi.org/10.1016/j.mser.2018.06.002

Ibeh CC (1999) Amino and furan resins

Ibrahim H, Farag M, Megahed H, Mehanny S (2014) Characteristics of starch-based biodegradable composites reinforced with date palm and flax fibers. Carbohydr Polym 101:11–19. https://doi.org/10.1016/j.carbpol.2013.08.051

Ibrahim H, Mehanny S, Darwish L, Farag M (2017) A Comparative study on the mechanical and biodegradation characteristics of starch-based composites reinforced with different lignocellulosic fibers. J Polym Environ 26:2434–2447. https://doi.org/10.1007/s10924-017-1143-x

ISO 178 (2010) BSI standards publication plastics—determination of flexural properties

Jaillet F, Nouailhas H, Auvergne R et al (2014) Synthesis and characterization of novel vinylester prepolymers from cardanol. Eur J Lipid Sci Technol 116:928–939. https://doi.org/10.1002/ejlt.201300487

Janošová M, Lenfeld P (2015) A locally shrinkage of the biopolymers PLA and PLLA 4

Jiménez A, Fabra MJ, Talens P, Chiralt A (2012) Edible and biodegradable starch films: a review. Food Bioprocess Technol 5:2058–2076. https://doi.org/10.1007/s11947-012-0835-4

Joseph K, Thomas S, Pavithran C (1995) Effect of ageing on the physical and mechanical properties of sisal-fiber-reinforced polyethylene composites. Compos Sci Technol 53:99–110. https://doi.org/10.1016/0266-3538(94)00074-3

Kaddami H, Dufresne A, Khelifi B et al (2006) Short palm tree fibers—thermoset matrices composites. Compos Part A Appl Sci Manuf 37:1413–1422. https://doi.org/10.1016/j.compositesa.2005.06.020

Kim JS (2015) Production, separation and applications of phenolic-rich bio-oil—a review. Bioresour Technol 178:90–98. https://doi.org/10.1016/j.biortech.2014.08.121

Kim HS, Chae YS, Park BH et al (2008) Thermal and electrical conductivity of poly(l-lactide)/multiwalled carbon nanotube nanocomposites. Curr Appl Phys 8:803–806. https://doi.org/10.1016/j.cap.2007.04.032

Koo J (2016) Basics of polymer matrices and composites. In: Fundamentals, properties, and applications of polymer nanocomposites. Cambridge University Press, pp 109–129

Laoui T (2013) Mechanical and thermal properties of styrene butadiene rubber—functionalized carbon nanotubes nanocomposites. Fuller Nanotub Carbon Nanostructures 21:89–101. https://doi.org/10.1080/1536383X.2011.574324

Lule Z, Kim J (2019) Surface modification of aluminum nitride to fabricate thermally conductive poly (Butylene Succinate) nanocomposite. Polymers (Basel) 11. https://doi.org/10.3390/polym11010148

Mahdavi S, Kermanian H, Varshoei A (2010) Date palm properties

Mahdi E, Hernández DR, Eltai EO (2015) Effect of water absorption on the mechanical properties of long date palm leaf fiber reinforced epoxy composites. J Biobased Mater Bioenergy 9:173–181. https://doi.org/10.1166/jbmb.2015.1508

Mahmoudi N (2013) Use of date palm fibers as reinforcement for thermoplastic-based composites. Mech Ind 14:71–77. https://doi.org/10.1051/meca/2012043

Malik N, Shrivastava S, Ghosh SB (2018) Moisture absorption behaviour of biopolymer Polycapralactone (PCL)/Organo Modified Montmorillonite Clay (OMMT) biocomposite films. IOP Conf Ser Mater Sci Eng 346. https://doi.org/10.1088/1757-899X/346/1/012027

Masoodi R, Pillai KM (2012) A study on moisture absorption and swelling in bio-based jute-epoxy composites. J Reinf Plast Compos 31:285–294. https://doi.org/10.1177/0731684411434654

Masri T, Ounis H, Sedira L et al (2018) Characterization of new composite material based on date palm leaflets and expanded polystyrene wastes. Constr Build Mater 164:410–418. https://doi.org/10.1016/j.conbuildmat.2017.12.197

Meinander K, Niemi M, Hakola JS, Selin JF (1997) Polylactides—degradable polymers for fibres and films. Macromol Symp 123:147–153. https://doi.org/10.1002/masy.19971230115

Menon ARR, Pillai CKS, Nando GB (1996) Thermal degradation characteristics of natural rubber vulcanizates modified with phosphorylated cashew nut shell liquid. Polym Degrad Stab 52:265–271. https://doi.org/10.1016/0141-3910(96)00007-9

Milanese AC, Cioffi MOH, Jacobus H, Voorwald C (2011) Mechanical behavior of natural fiber composites. Procedia Eng 10:2022–2027. https://doi.org/10.1016/j.proeng.2011.04.335

Mirmehdi SM, Zeinaly F, Dabbagh F (2014) Date palm wood flour as filler of linear low-density polyethylene. Compos Part B Eng 56:137–141. https://doi.org/10.1016/j.compositesb.2013.08.008

Mizuno S, Katsumata S, Hiroe A, Tsuge T (2014) Biosynthesis and thermal characterization of polyhydroxyalkanoates bearing phenyl and phenylalkyl side groups. Polym Degrad Stab 109:379–384. https://doi.org/10.1016/j.polymdegradstab.2014.05.020

Mohan P (2013) A critical review: the modification, properties, and applications of epoxy resins. Polym Plast Technol Eng 52:107–125. https://doi.org/10.1080/03602559.2012.727057
Mohanty JR (2017) Investigation on solid particle erosion behavior of date palm leaf fiber- reinforced polyvinyl pyrrolidone composites. J Thermoplast Compos Mater 30:1003–1016. https://doi.org/10.1177/0892705715614079
Mohanty JR, Das SN, Das HC, Swain SK (2014) Effect of chemically modified date palm leaf fiber on mechanical, thermal and rheological properties of polyvinylpyrrolidone. Fibers Polym 15:1062–1070. https://doi.org/10.1007/s12221-014-1062-6
Mohd Ishak ZA, Mat Taib R (2015) Multifunctional polymer composites using natural fiber reinforcements. In: Multifunctionality of polymer composites. William Andrew, pp 71–101
Nafchi AM, Moradpour M, Saeidi M, Alias AK (2013) Thermoplastic starches: properties, challenges, and prospects. Starch/Staerke 65:61–72
Neher B, Bhuiyan MMR, Kabir H et al (2016) Thermal properties of palm fiber and palm fiber-reinforced ABS composite. J Therm Anal Calorim 124:1281–1289. https://doi.org/10.1007/s10973-016-5341-x
Njoku RE, Ofili I, Agbiogwu DO, Agu CV (2012) Effect of alkali treatment and fiber content variation on the tensile properties of coir fiber reinforced Cashew Nut Shell Liquid (CNSL) composite. Niger J Technol 31:108–110
Nofar M, Tabatabaei A, Sojoudiasli H et al (2017) Mechanical and bead foaming behavior of PLA-PBAT and PLA-PBSA blends with different morphologies. Eur Polym J 90:231–244. https://doi.org/10.1016/j.eurpolymj.2017.03.031
Noorunnisa Khanam P, AlMaadeed MA (2014) Improvement of ternary recycled polymer blend reinforced with date palm fibre. Mater Des 60:532–539. https://doi.org/10.1016/j.matdes.2014.04.033
Oksman K, Skrifvars M, Selin JF (2003) Natural fibres as reinforcement in polylactic acid (PLA) composites. Compos Sci Technol 63:1317–1324. https://doi.org/10.1016/S0266-3538(03)00103-9
Olabisi O, Adewale K (2016) Handbook of thermoplastics, 2nd edn. CRC Press
Paetau I, Chen CZ, Jane JL (1994) Biodegradable plastic made from soybean products. 1. Effect of preparation and processing on mechanical properties and water absorption. Ind Eng Chem Res 33:1821–1827. https://doi.org/10.1021/ie00031a023
Paluvai NR, Mohanty S, Nayak SK (2014) Synthesis and modifications of epoxy resins and their composites: a review. Polym Plast Technol Eng 53:1723–1758. https://doi.org/10.1080/03602559.2014.919658
Pivsa-Art W, Pavasupree S, O-Charoen N et al (2011) Preparation of polymer blends between poly (L-lactic acid), poly (butylene succinate-co-adipate) and poly (butylene adipate-co-terephthalate) for blow film industrial application. Energy Procedia 9:581–588. https://doi.org/10.1016/j.egypro.2011.09.068
Plummer CJG, Bourban P-E, Månson J-A (2016) Polymer matrix composites: matrices and processing. Ref Modul Mater Sci Mater Eng 1–9. https://doi.org/10.1016/b978-0-12-803581-8.02386-9
Ra I (2015) Effect of date palm seeds on the tribological behaviour of polyester composites under different testing conditions. J Mater Sci Eng 04. https://doi.org/10.4172/2169-0022.1000206
Raghavendra S, Lokesh GN (2019) Evaluation of mechanical properties in date palm fronds polymer composites. AIP Conf Proc 2057. https://doi.org/10.1063/1.5085592
Ramamoorthy SK, Di Q, Adekunle K, Skrifvars M (2012) Effect of water absorption on mechanical properties of soybean oil thermosets reinforced with natural fibers. J Reinf Plast Compos 31:1191–1200. https://doi.org/10.1177/0731684412455257
Ratna D (2009) Handbook of thermoset resins. Smithers Rapra, Shropshire, United Kingdom
Ray SS, Bandyopadhyay J, Bousmina M (2007) Thermal and thermomechanical properties of poly[(butylene succinate)-co-adipate] nanocomposite. Polym Degrad Stab 92:802–812. https://doi.org/10.1016/j.polymdegradstab.2007.02.002

Renner T, Pék L (2011) Comparing strength properties of natural and synthetic rubber mixtures 134–141
Rudnik E (2013) Compostable polymer properties and packaging applications. In: Plastic films in food packaging. Elsevier, pp 217–248
Rudnik E (2019) Properties and applications. In: Compostable polymer materials, 2nd edn. Elsevier, pp 49–98
Russo P, Cimino F, Acierno D et al (2014) Poly(butylene terephthalate) based composites containing alumina whiskers: influence of filler functionalization on dielectric properties. Int J Polym Sci 2014. https://doi.org/10.1155/2014/150589
Saba N, Alothman OY, Almutairi Z et al (2019) Date palm reinforced epoxy composites: tensile, impact and morphological properties. J Mater Res Technol 8:3959–3969. https://doi.org/10.1016/j.jmrt.2019.07.004
Saleh MA, Al Haron MH, Saleh AA, Farag M (2017) Fatigue behavior and life prediction of biodegradable composites of starch reinforced with date palm fibers. Int J Fatigue 103:216–222. https://doi.org/10.1016/j.ijfatigue.2017.06.005
Salih SI, Jasim AS, Hasan AM (2015) Investigation on mechanical properties of hybrid polymer composite reinforced by rice husks and date palm fibers as a construction material. J Al-Nahrain Univ 18:89–97. https://doi.org/10.22401/jnus.18.3.13
Sastri VR (2010) Materials used in medical devices. In: Plastics in medical devices. William Andrew, pp 21–32
Sbiai A, Maazouz A, Fleury E et al (2010) Short date palm tree fibers/polyepoxy composites prepared using RTM process: effect of tempo mediated oxidation of the fibers
Seggiani M, Gigante V, Cinelli P et al (2019) Processing and mechanical performances of Poly(Butylene Succinate–co–Adipate) (PBSA) and raw hydrolyzed collagen (HC) thermoplastic blends. Polym Test 77:105900. https://doi.org/10.1016/j.polymertesting.2019.105900
Setianto WB, Yoshikawa S, Smith RL et al (2009) Pressure profile separation of phenolic liquid compounds from cashew (Anacardium occidentale) shell with supercritical carbon dioxide and aspects of its phase equilibria. J Supercrit Fluids 48:203–210. https://doi.org/10.1016/j.supflu.2008.11.018
Shalwan A, Yousif BF (2014) Influence of date palm fibre and graphite filler on mechanical and wear characteristics of epoxy composites. Mater Des 59:264–273. https://doi.org/10.1016/j.matdes.2014.02.066
Sharath GP, Kishor BS, Vinod B, Naveen A (2016) Effect of fiber length on the mechanical properties of coir and wild date palm reinforced epoxy composites. IJSTE Int J Sci Technol Eng 2
Sharma M, Kumar V, Anek Pal G (2015) Thermoplastic Starch (TPS) based bio-disintegrable polymers—combination of modified potato starch with polyolefins. Int J Sci Res 4:403–407
Shenoy AV, Saini DR (1986) Melt flow index: more than just a quality control rheological parameter. Part I. Adv Polym Technol 6:1–58. https://doi.org/10.1002/adv.1986.060060101
Sim LN, Arof AK (2017) Elastomers and their potential as matrices in polymer electrolytes. In: Elastomers, p 21
Singh M, Kumar P, Ray S, Kalia VC (2015) Challenges and opportunities for customizing polyhydroxyalkanoates. Indian J Microbiol 55:235–249. https://doi.org/10.1007/s12088-015-0528-6
Sirin H, Tuna B, Ozkoc G (2014) The effects of thermomechanical cycles on the properties of PLA/TPS blends. Adv Polym Technol 33:1–9. https://doi.org/10.1002/adv.21458
Sreekumar PA, Thomas S (2008) Matrices for natural-fibre reinforced composites. Prop Perform Nat Compos 67–126. https://doi.org/10.1533/9781845694593.1.67
Sreekumar PA, Joseph K, Unnikrishnan G, Thomas S (2007) A comparative study on mechanical properties of sisal-leaf fibre-reinforced polyester composites prepared by resin transfer and compression moulding techniques. Compos Sci Technol 67:453–461. https://doi.org/10.1016/j.compscitech.2006.08.025
Suleiman BM (2018) Thermal conductivity and diffusivity of polymers. Int J Phys Study Res 1:54–58. https://doi.org/10.18689/ijpsr-1000107

Swain PTR, Das SN, Jena SP (2018) Manufacturing and study of thermo-mechanical behaviour of surface modified date palm leaf/glass fiber reinforced hybrid composite. Mater Today Proc 5:18332–18341

Teixeira E de M, Curvelo AAS, Corrêa AC et al (2012) Properties of thermoplastic starch from cassava bagasse and cassava starch and their blends with poly (lactic acid). Ind Crops Prod 37:61–68. https://doi.org/10.1016/j.indcrop.2011.11.036

Tonpheng B, Yu J, Andersson BM, Andersson O (2010) Tensile strength and Young's modulus of polyisoprene/single-wall carbon nanotube composites increased by high pressure cross-linking. Macromolecules 43:7680–7688. https://doi.org/10.1021/ma101484e

Tripathy S, Dehury J, Mishra D (2016) A study on the effect of surface treatment on the Physical and Mechanical properties of date-palm stem liber embedded epoxy composites. In: IOP conference series: materials science and engineering. Institute of Physics Publishing

Tschan MJ-L, Brule E, Haquette P, Thomas CM (2012) Synthesis of biodegradable polymers from renewable resources. Polym Chem 3:836–851. https://doi.org/10.1039/C2PY00452F

Vallejos ME, Curvelo AAS, Teixeira EM et al (2011) Composite materials of thermoplastic starch and fibers from the ethanol-water fractionation of bagasse. Ind Crops Prod 33:739–746. https://doi.org/10.1016/j.indcrop.2011.01.014

Van De Velde K, Kiekens P (2001) Thermoplastic polymers: overview of several properties and their consequences in flax fibre reinforced composites

Voirin C, Caillol S, Sadavarte NV et al (2014) Functionalization of cardanol: towards biobased polymers and additives. Polym Chem 5:3142–3162. https://doi.org/10.1039/C3PY01194A

Wang Y, Jiang X, Xu C et al (2013) Effects of partial replacement of silicone rubber with flurorubber on properties of dynamically cured poly(vinylidene fluoride)/silicone rubber/flurorubber ternary blends. Polym Test 32:1392–1399. https://doi.org/10.1016/j.polymertesting.2013.09.004

Wazzan AA (2005) Effect of fiber orientation on the mechanical properties and fracture characteristics of date palm fiber reinforced composites. Int J Polym Mater Polym Biomater 54:213–225. https://doi.org/10.1080/00914030390246379

Węcławski BT, Fan M, Hui D (2014) Compressive behaviour of natural fibre composite. Compos Part B Eng 67:183–191. https://doi.org/10.1016/j.compositesb.2014.07.014

Wei KK, Leng TP, Keat YC et al (2019) The potential of natural rubber (NR) in controlling morphology in two-matrix epoxy/NR/graphene nano-platelets (GNP) systems. Polym Test 77:105905. https://doi.org/10.1016/j.polymertesting.2019.105905

Wool RP, Sun XS (2011) Bio-based polymers and composites. Elsevier

Wu X, Liu H, Tang Z, Guo B (2016) Scalable fabrication of thermally conductive elastomer/boron nitride nanosheets composites by slurry compounding. Compos Sci Technol 123:179–186. https://doi.org/10.1016/j.compscitech.2015.12.010

Xu J, He Y (2015) Thermal analysis of NR composite with MWCNTs aligned in a magnetic field. Int J Polym Sci 2015. https://doi.org/10.1155/2015/305317

Xu Y, Zhang S, Wang P, Wang J (2018) Synthesis of Poly(butylene succinate) phosphorus-containing ionomers for versatile crystallization and improved thermal conductivity. Polymer (Guildf) 154:258–271. https://doi.org/10.1016/j.polymer.2018.09.025

Yeo JCC, Muiruri JK, Thitsartarn W et al (2018) Recent advances in the development of biodegradable PHB-based toughening materials: approaches, advantages and applications. Mater Sci Eng C 92:1092–1116. https://doi.org/10.1016/j.msec.2017.11.006

Zadeh KM, Inuwa IM, Arjmandi R et al (2017) Effects of date palm leaf fiber on the thermal and tensile properties of recycled ternary polyolefin blend composites. Fibers Polym 18:1330–1335. https://doi.org/10.1007/s12221-017-1106-9

Zhang Y, Rempel C, Liu Q (2014) Thermoplastic starch processing and characteristics—a review. Crit Rev Food Sci Nutr 54:1353–1370. https://doi.org/10.1080/10408398.2011.636156

Zhang X, Peng X, Zhang SW (2017) Synthetic biodegradable medical polymers: polymer blends. In: Science and principles of biodegradable and bioresorbable medical polymers. Woodhead Publishing Limited, pp 217–254

Zhao K, Deng Y, Chen JC, Chen GQ (2003) Polyhydroxyalkanoate (PHA) scaffolds with good mechanical properties and biocompatibility. Biomaterials 24:1041–1045. https://doi.org/10.1016/S0142-9612(02)00426-X

Zhou Y, Fan M, Chen L, Zhuang J (2015) Lignocellulosic fibre mediated rubber composites: an overview. Compos Part B Eng 76:180–191. https://doi.org/10.1016/j.compositesb.2015.02.028

Zhou Y, Fan M, Chen L (2016) Interface and bonding mechanisms of plant fibre composites: an overview. Compos Part B Eng 101:31–45. https://doi.org/10.1016/j.compositesb.2016.06.055

Zini E, Scandola M (2016) Polyvinyl alcohol-modified Pithecellobium clypearia Benth herbal residue fiberpolypropylene composites. Polym Compos 37:915–924. https://doi.org/10.1002/pc.23250

Production of plastics worldwide from 1950 to 2018 (in million metric tons). In: Statista. https://www.statista.com/statistics/282732/global-production-of-plastics-since-1950/. Accessed 15 Oct 2019

Polymer Science. www.polymerdatabase.com/index.html%0D. Accessed 9 Dec 2019

Plastic Material Melt And Mould Temperature. https://www.plastikcity.co.uk/useful-stuff/material-melt-mould-temperatures. Accessed 9 Dec 2019

PMMA 500—PMMA—Polykemi AB. https://www.materialdatacenter.com/ms/en/tradenames/Polyplex/Polykemi+AB/POLYplex+PMMA+500/c1a4fd40/1912. Accessed 10 Nov 2019

ASTM D3364 Flow Rates for Poly Vinyl Chloride. https://instron.us/testing-solutions/by-test-type/rheology/astm-d3364. Accessed 11 Nov 2019

AZO Materials. https://www.azom.com/

Density of Plastics: Technical Properties

Properties of PHA. https://www.ecobiomaterial.com/who-i-am/properties-of-pha/. Accessed 10 Dec 2019

Environmentally Friendly Polybutylene Succinate. https://www.succinicacidpowder.com/sale-10827793-environmentally-friendly-polybutylene-succinate-pbs-1-26g-cm-density.html. Accessed 9 Dec 2019

PBS Resin. https://www.pkbioplastic.com/products#1564033247047-7d16b4fc-8ce4. Accessed 9 Dec 2019

KANEKA Biodegradable Polymer PHBHTM

Date Palm Fiber Composite Fabrication Techniques

Ahmed H. Hassanin, Lobna A. Elseify, and Tamer Hamouda

Abstract Natural fibers are attractive for their biodegradability and exceptional properties. Date palm as a source of fibers has many advantages over other sources of natural fibers; since date palm fiber sources are regarded as agriculture waste and the tree itself is cultivated for the fruit. Natural fibers including date palm are used as reinforcement to thermoplastics or thermosets using different manufacturing technique. This chapter discusses different composite manufacturing techniques that can be used with natural fibers. The discussed techniques are hand layup, RTM, VARTM, filament winding, pultrusion, compression molding, extrusion, injection molding and 3D printing.

1 Introduction

The use of composite materials has seen a surge in many applications such as aerospace, automotive, marine, medical and sports equipment. The vast majority of the current commercial composites are made from synthetic polymers that are petroleum-based. Using such raw materials (petroleum-based) creates environmental problems and pollution. Meanwhile, there is a growing global demand for sustainable materials, which in turn increased the interest in natural fibers as reinforcement to composites. Natural fibers are not only environmentally friendly, but they also have high specific properties, due to their lightweight. The major challenge facing materials extracted from natural origin, is how to maximize the sustainable and profitable

A. H. Hassanin (✉)
Materials Science and Engineering, Egypt Japan University for Science and Technology (EJUST), New Borg El-Arab City, Alexandria, Egypt
e-mail: ahassan@ncsu.edu

Department of Textile Engineering, Alexandria University, Alexandria, Egypt

L. A. Elseify
Department of Materials Engineering, German University in Cairo, New Cairo City, Egypt

T. Hamouda
Textile Research Division, National Research Centre, 33 El Bohouth st., Dokki 12622, Egypt

M. Midani et al. (eds.), *Date Palm Fiber Composites*, Composites Science and Technology,
https://doi.org/10.1007/978-981-15-9339-0_5

use of these natural resources as a raw material for high value-added products. This requires highly innovative and environmentally friendly solutions (Mohareb et al. 2015).

This chapter provides an overview on Date Palm fiber manufacturing methods. It will give insights into the different fabrication and processing technique by which date palm fiber composites can be prepared. Including but not limited to, manual hand layup, Spray layup, VARTM, RTM, filament winding, Pultrusion, compression moulding, injection moulding and 3D printing. Advantages and limitations of each method will be reviewed as well.

2 Manufacturing Techniques

After constructing the fiber preform, it will go through the final stage or component production stage. Final properties of composite structure are not only dependent only on the material type, but also strongly related to the way in which the composites are manufactured. Manufacturing date Palm fiber and natural fiber composites can mostly be manufactured using the conventional composite manufacturing processes such as hand layup, resin transfer moulding RTM, filaments winding and compression moulding. In this chapter major techniques used in Date Palm fiber composites, will be reviewed.

2.1 Hand Layup

In hand layup, the preforms which could be in the form of woven, knitted, nonwoven… etc. are impregnated by hand with resin or matrix material. This is usually carried out by rollers or brushes, with the use of nip-roller type impregnators for pressing resin into the fabrics. Laminates are left to cure under standard atmospheric conditions or could be cured inside ovens or autoclaves. The main advantages of this method are the simplicity and low tooling cost (Netcomposites 2019). A schematic diagram showing the hand layup process is shown in Fig. 1.

Many natural fiber reinforced composites were manufactured using this technique. For instance, sisal in the form of short chopped fibers, were used as reinforcement to epoxy resin (Yuvaraj et al. 2017). Moreover, plain woven sisal, kenaf and alovera fibers were used as reinforcement to epoxy resin using this technique (Palani Kumar et al. 2017). A bath tub was made from unsaturated polyester resin reinforced with jute fibers using hand layup method (Xiao et al. 2015). Flax fiber reinforced composites have been manufactured using this technique several times in woven and nonwoven forms. Woven flax preforms were used to reinforce epoxy (Muralidhar et al. 2012; Muralidhar 2013). Moreover, unidirectional and nonwoven flax fibers were used with epoxy resin (Santulli 2000; Di Bella et al. 2010; Bertomeu et al. 2012; Dhakal et al. 2016). Finally, date palm mesh fibers were used with epoxy and graphite; hand

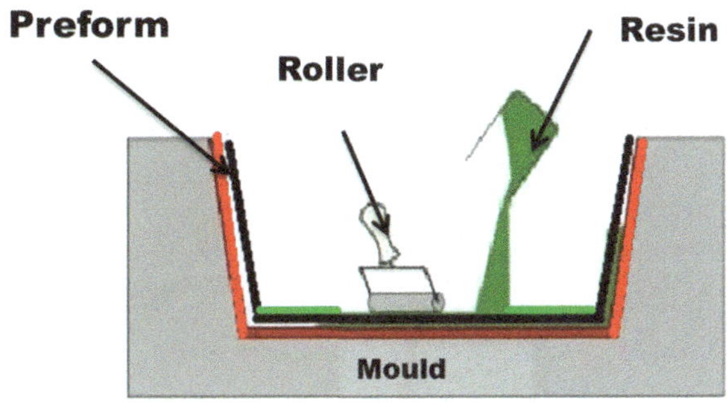

Fig. 1 Hand layup technique

layup technique was used then the composite was put in vacuum oven for curing (Abdal-hay et al. 2012).

Main Advantages (Strong 2008; Netcomposites 2019)

- Widely used and simple principles to teach.
- Low cost tooling, especially if room-temperature cure resins are used.
- Wide choice of suppliers and material types.
- Higher fiber contents, and longer fibers than with spray lay-up.

Main Disadvantages (Strong 2008; Netcomposites 2019)

- The whole process is very dependent on the skills of labor.
- It is difficult to insure complete impregnation.
- Low resin content laminates cannot usually be achieved without the incorporation of excessive quantities of voids.
- Health and safety considerations of resins because the process is carried out in open moulds.
- Low viscosity resins must be used to be workable by hand.

2.2 Spray Layup

Fibers are chopped in a hand-held gun and fed into a spray of catalyzed resin directed at the mould. Afterwards, the deposited materials are left to cure at room temperature. A schematic diagram of the process is shown in Fig. 2.

This technique was rarely used with natural fibers. The fibers are supposed to be supplied in continuous filaments. However, natural fibers do not exist in a form of continues filament. This may hinder the usage of this technique with natural fibers. In order to be able to use this technique, the authors suggest that the cut or chopped fibers be supplied from a reservoir then be sprayed with liquid resin.

Main Advantages (Strong 2008; Netcomposites 2019)

- Widely used and simple to teach.
- Low cost of tooling

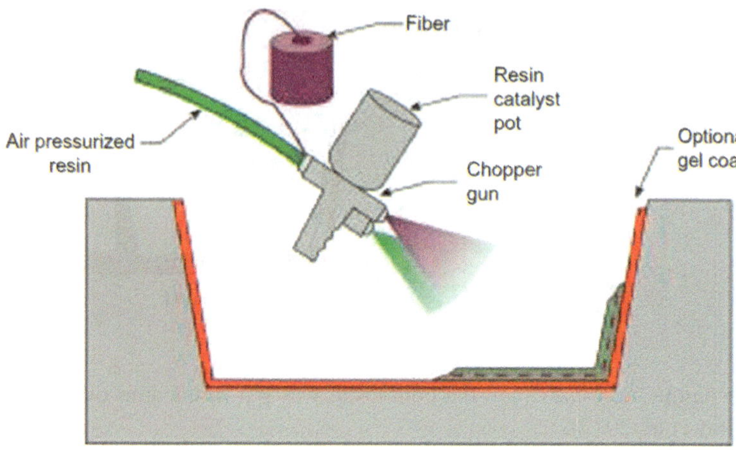

Fig. 2 Spray layup technique schematic diagram of the process (Balasubramanian et al. 2018)

- Suitable for large size pieces
- Less labor compares to hand layup and easy to be automated

Main Disadvantages (Strong 2008; Netcomposites 2019)

- Low fiber volume fraction can be obtained.
- Only short fibers are incorporated and only random fiber orientation can be achieved which significantly limits the mechanical properties of the produced composites.
- Small and complex parts are not suitable for spray layup
- Limited to low viscosity resins to be sprayable.
- Health and safety considerations of resins because the process is carried out in open moulds

2.3 Resin Transfer Moulding (RTM)

In this process RTM, preforms are laid up in form of single or multi stacks. These stacks of preforms are sometimes pre-pressed to the mould shape and held together by means of a binder. Then these preforms are stacked into the mould tool. The second mould tool is then clamped over the first mould. Resin is injected into the cavity mainly by pressure and sometimes vacuum can also be applied to assist resin in being drawn into the fabrics as shown in Fig. 3. Once all the stack of preforms is wet out, the resin inlets are closed, and the laminate is allowed to cure. This process can take place at either room temperature or elevated temperature. Generally, this method is used with low viscosity thermoset resins such as unsaturated polyester, vinyl ester and epoxy.

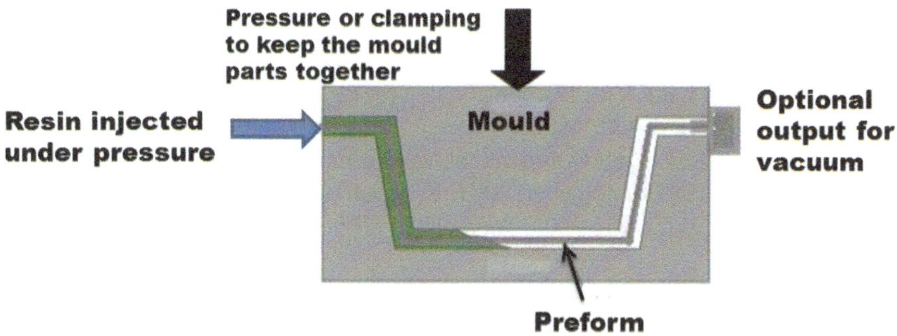

Fig. 3 Schematic diagram of Resin Transfer Moulding (RTM) technique

Date palm leaflet fibers were used as a reinforcement to epoxy in RTM (Sbiai et al. 2010). Similarly, other natural fibers, in unidirectional form, like sisal and flax fibers were used to reinforce epoxy resin and unsaturated polyester (Goutianos et al. 2007; Li et al. 2015).

Main Advantages (Strong 2008; Advani and Hsiao 2012; Netcomposites 2019)

- High fiber volume can be obtained (55–65%) with low void contents, can be reach to 1%
- Excellent surface quality can be achieved
- Parts with complex geometries can be made
- Good health and safety, and environmental control due to enclosure of resin.
- Less labor cost

Main disadvantages (Strong 2008; Advani and Hsiao 2012; Netcomposites 2019)

- Matched tooling is expensive and heavy in order to withstand pressures.
- Generally limited to smaller components.
- Pressure can cause fibers to move in the mould, this is called fiber wash.

2.4 Resin Infusion or Vacuum Assisted Resin Transfer Moulding

Resin infusion or Vacuum Assisted Resin Transfer Moulding (VARTM) is a complicated technique for manufacturing composite structures, and usually it creates void free or very low voids composites even in large or complicated moulds. In this process, the preform is laid into the mould in a dry form without any resin, and then enclosed by a specific stack of bagging materials (such as peel ply, infusion mesh and bagging film) before being subjected to vacuum pressure using a vacuum pump. Once all the air has been removed or vacuumed from the bag and the preform has been fully compressed, resin is introduced to the preform through a pipe known as spiral tube, then resin infuses through the preform under the vacuum pressure. After the resin has completely infused through the preform, the supply of resin is closed, and the composite is left to cure under vacuum pressure. Figure 4 shows a schematic diagram of VARTM and VARTM consumable tools.

The use of VARTM with natural fibers is not very popular. However, flax and hemp fibers were used in nonwoven and woven forms as reinforcement to epoxy and soybean oil resins (O'Donnell et al. 2004; Rayyaan et al. 2019).

Main Advantages (Strong 2008; Advani and Hsiao 2012; Netcomposites 2019)

- Much lower tooling cost compare to RTM, due to one side of the tool being a vacuum bag, and traditional metal mould in the other side
- Fabrication of large parts
- Flexibility, where standard wet lay-up tools may be able to be modified for this process.
- Cored structures can be produced in one operation.

Main disadvantages (Strong 2008; Advani and Hsiao 2012; Netcomposites 2019)

- Surface quality is not as good as RTM
- Complex compared to other methods
- Low viscosity resins must be used which significantly affect mechanical properties
- Dry spot areas can be formed
- Some elements of this process are covered by patents (SCRIMP).

2.5 Filament Winding

This process is mainly used for creating tubular components that are hollow and have circular or oval cross section, such as pipes and tanks. Yarns are passed through a resin bath then the impregnated yarns are wound onto a mandrel as shown in Fig. 5. Yarns orientations can be controlled by the traverse feeding mechanism, in conjunction with the speed of rotation of the mandrel. Filament winding is mostly used for producing pipes especially for petroleum industry and pressure vessels.

The use of filament winding technique with natural fibers was very limited in the literature. However, some researchers investigated the potential of using natural fibers in composite fabrication using this technique. Lehtiniemi et al. (2011) used flax yarn as reinforcement in composites; however, the flax fibers lacked adhesion compared to tubes reinforced with glass fibers. Another attempt was made by Mahdi et al. (2002). They used epoxy resin reinforced with oil palm frond fibers. Also, kenaf fibers was used with unsaturated polyester (Misri et al. 2016). Overall, it's important to optimize the process parameters in filament winding for making natural fiber reinforced composites for use in light weight tubular applications (Salit et al. 2015).

Main Advantages (Strong 2008; Advani and Hsiao 2012; Netcomposites 2019)

- The process is very fast and can be automated easily, therefore it is economical technique

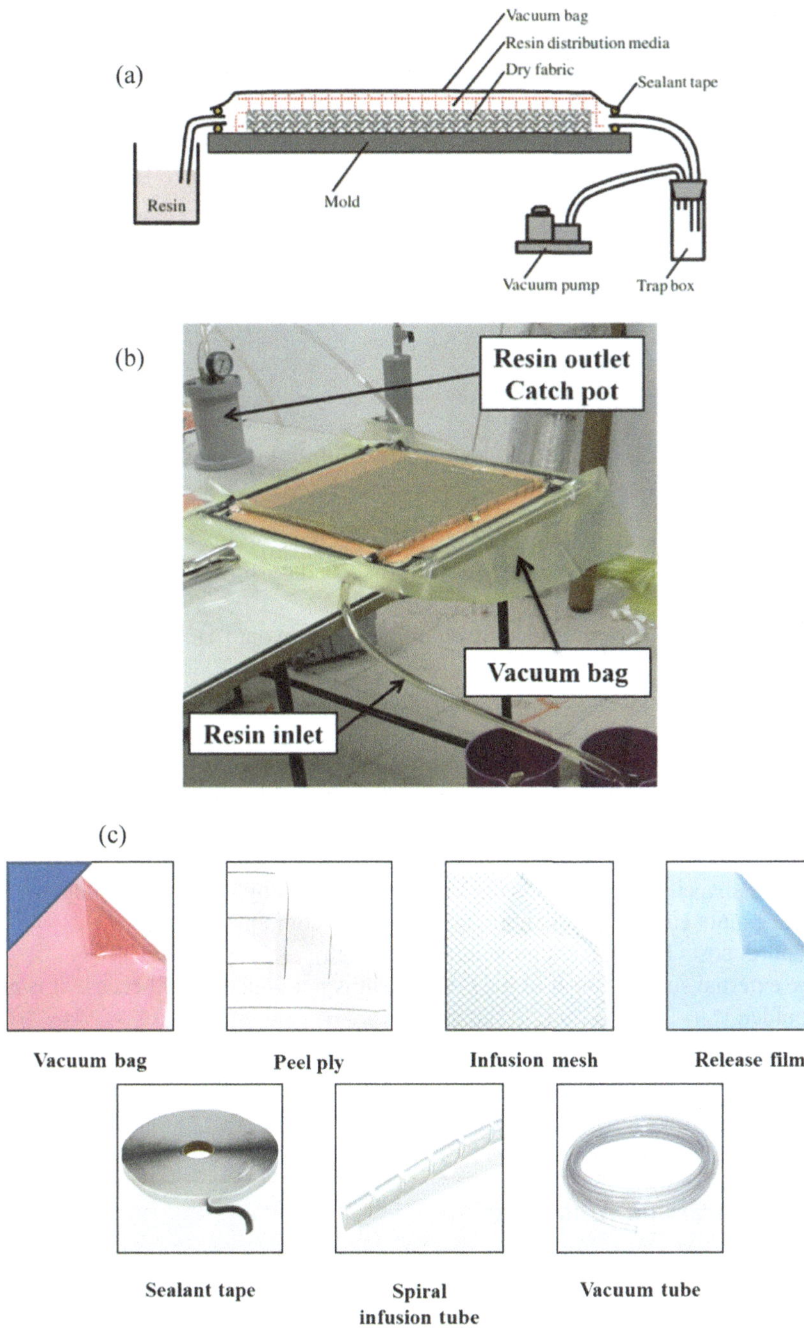

Fig. 4 **a** Schematic diagram of resin infusion technique (Matsuzaki et al. 2011). **b** VARTM process setup (Hassanin et al. 2016). **c** VARTM consumables and tools

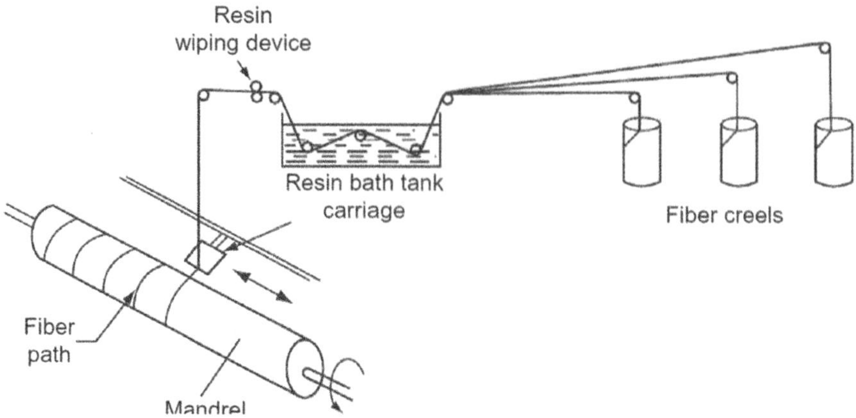

Fig. 5 Filaments winding technique (Balasubramanian et al. 2018)

- Labor intensive, usually one labor per each machine
- Fiber cost is minimized since there is no secondary process to manufacturing preforms
- Resin content can be controlled by metering the resin onto each fiber tow through nips or dies
- Void content can be very low, depending on the tension on the filament and speed of winding
- Structural properties of produced composites are very good since straight fibers can be laid in a complex pattern to match the applied loads.

Main disadvantages (Strong 2008; Advani and Hsiao 2012; Netcomposites 2019)

- Convex shaped components cannot be manufactured by filament winding
- Fiber cannot easily be laid along the length of a component
- Mandrel costs for large components can be high
- The external surface finish of the component is not high quality because it is not moulded
- Low viscosity resins should be used which affect the mechanical properties
- Health and safety considerations of resins because the process is carried out in open moulds.

2.6 Compression Moulding

Compression moulding is the most popular technique used in fabrication of polymeric composites. There are two types of compression moulding; cold compression moulding and hot compression moulding. In compression moulding process, two matched metal moulds (flat or shaped) are used to fabricate composite products as shown in Fig. 6. Compression moulding could be further classified into two cate-

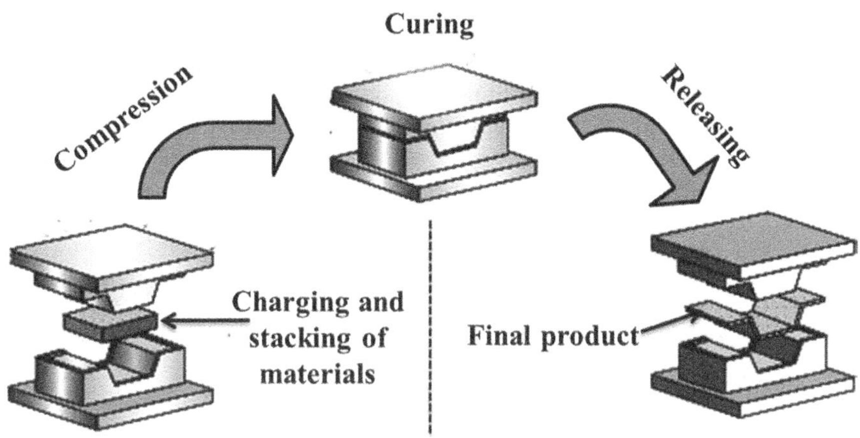

Fig. 6 Compression moulding process

gories; Sheet Moulding Compound (SMC) and Bulk Moulding Compound (BMC). In BMC the resin is mixed with the reinforcement. The BMC material has a dough-like appearance. On the other hand, in SMC the resin and fibers are not mixed together, but rather supplied in the form of sheets. In SMC the composite has a sandwich structure with an alternative layers of reinforcements and resin (Strong 2008). In compression moulding, one plate is stationary while the other plate is movable. Preform and matrix are placed in between the two moulds then heat and pressure is applied as required for composite for a specific period of time. In case of thermoplastic matrix heat will be required to melt the matrix and let it flow through the preform. For thermoset matrix curing of the composite may take place either at room temperature or at some elevated temperature (Advani and Hsiao 2012).

As mentioned earlier, compression moulding is one of the most extensively used technique in the literature. Therefore, there is a large amount of research made on the possibility of using natural fibers with compression moulding. Date palm leaflet fibers were used as reinforcement to epoxy resin using compression moulding (Sbiai et al. 2008). Also, mesh fibers were mixed with a starch based resin then compressed to form the final composite (Ali and Alabdulkarem 2017). Midrib fibers were also used as reinforcement to thermoplastic matrices such as HDPE and LDPE (Mahdavi et al. 2010; Mirmehdi et al. 2014). Sisal fiber composites and thermoset or thermoplastic matrix have been manufactured using compression moulding. Sisal was used in woven and the nonwoven forms. The most commonly used matrix materials with sisal fibers were epoxy and unsaturated polyester (Lu et al. 2003; Jiang et al. 2009; Kejariwal and Keerthi Gowda 2017; Sathishkumar et al. 2017; Senthilkumar et al. 2017; Yorseng et al. 2020). Bamboo and coir fiber composites were manufactured using compression moulding with starch based resins or epoxy resin (Ochi 2006; Mulinari et al. 2011; Romli et al. 2012; Keerthi Gowda et al. 2016; Abdullah et al. 2020). Moreover, coir fibers were used with polypropylene as a matrix (Lai et al. 2005; Mir et al. 2013; Arya et al. 2015). As for jute fibers, they were used in plain woven and unidirectional forms with PLA and starch based resin respectively (Ochi 2006; Papa et al. 2017). Additionally flax fiber composites were manufactured using epoxy resin, tannin resin, Bioplast films, and polypropylene in woven and nonwoven fiber forms (Zhu et al. 2013; Duquesne et al. 2015; Berges et al. 2016; Bar et al. 2018). Finally, limited research was performed on composites reinforced with abaca and ramie fiber manufactured using compression moulding (Ochi 2006; Banowati et al. 2016).

Main Advantages (Advani and Hsiao 2012)

- Excellent reproducibility of parts
- The fiber content and type can be easily controlled
- Finished interior and exterior surfaces
- Complex shapes can be fabricated easily
- High production rates and low labors needed
- Minimum scrap materials are generated.

Main Advantages (Advani and Hsiao 2012)

- Expensive equipment is needed compare to other techniques such as hand layup
- Surface imperfections can be occurred such as pitting and waviness
- Chopped or short fibers are only used in compression moulding techniques
- Sharp corners should be avoided since they are not only stress concertation locations, but also, they are places where the fibers may not flow during the moulding process.
- Very fine and thin section is very difficult to be performed
- Large structure needs a huge presser to perform compression moulding process.

2.7 Pultrusion

Pultrusion is a composite manufacturing technique where continuous yarns are completely impregnated with the thermoset polymeric matrix then pulled through a heated die to form the composite. Pultrusion process is accomplished by pull rather than push as in case of extrusion. The main advantage of the pultrusion process is the ability of forming complex shapes such as I beam, C section, tubes, rods…etc. Pultrusion process is a continuous production process. A schematic diagram of pultrusion process is shown in Fig. 7.

However, the pultrusion process depends on pulling the fibers; hence, the tensile strength of fibers must be large enough to withstand the pulling force. Limited research was conducted on the possibility of manufacturing natural fiber reinforced composites using pultrusion (Salit et al. 2015). Jute and kenaf fibers were used as reinforcement to unsaturated polyester (Safiee et al. 2011; Zamri et al. 2014). Moreover, jute and kenaf fibers were used in hybrid composites with glass fibers as a reinforcement to polyester (Akil et al. 2010; Zamri et al. 2012). Van De Velde and Kiekens (2001) developed a special machine to be used with thermoplastics. Also, Nguyen-Chung et al. (2007) and Angelov et al. (2007) researched the idea of

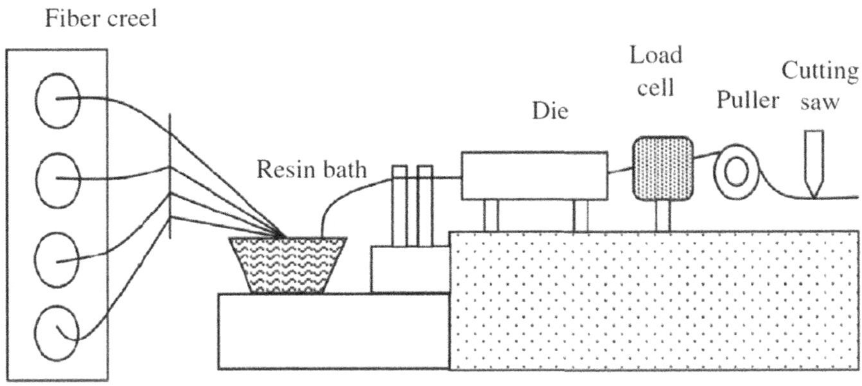

Fig. 7 Pultrusion process (Balasubramanian et al. 2018)

pultruding flax using PP; it was concluded that as the pulling speed was increased the flexural modulus of the composite was decreased.

Main advantages (Strong 2008; Advani and Hsiao 2012; Netcomposites 2019)

- Length can be customized due to the continuous nature of the pultrusion process, so any length can be produced ranging from few centimeters to kilometers
- Fiber volume fraction can be accurately controlled and high fiber content can be achieved easily
- Fiber cost is minimized since there in no intermediate process to manufacture preforms
- Good structural properties of laminates due to the aligned fibers and high fiber volume fractions
- More automation and less human interferences which allow high quality consistence
- Good health, safety and environmental control since the resin impregnation area can be enclosed which limiting volatile emissions.

Main disadvantages (Strong 2008; Advani and Hsiao 2012; Netcomposites 2019)

- Used with constant cross-section components, so Tapered and complex shapes are difficult or impossible to be produced
- The accuracy of part dimensions and their tolerances are not very high compare to other manufacturing methods
- Thin parts are very challengeable to be produced via pultrusion.
- Heated die costs can be high.

2.8 3D Printing

3D printing of natural fiber reinforced composites in the literature was performed using fused deposition modeling. In fused deposition modeling a thermoplastic in the form of filaments is used. The 3D printer depends on melting the thermoplastic filament to create or form a structure. Continuous fibers are also supplied at the nozzle along with the melted thermoplastic filament (Lee et al. 2019). A schematic 3D printer using the fused deposition modeling is shown in Fig. 8.

Matsuzaki et al. (2016) developed a method to fabricate composites from natural fibers using 3D printing following the fused deposition modeling. Thermoplastic PLA filament and continuous jute fibers were supplied separately to the 3D printer. The fibers were heated inside the nozzle of the printer. Jute composites were compared to composites reinforced with carbon fiber. Obviously, the mechanical properties of carbon composites were superior to jute composites. However, the mechanical properties of jute composites were better than neat PLA. A 3D printed dog bone tensile specimen is shown in Fig. 9.

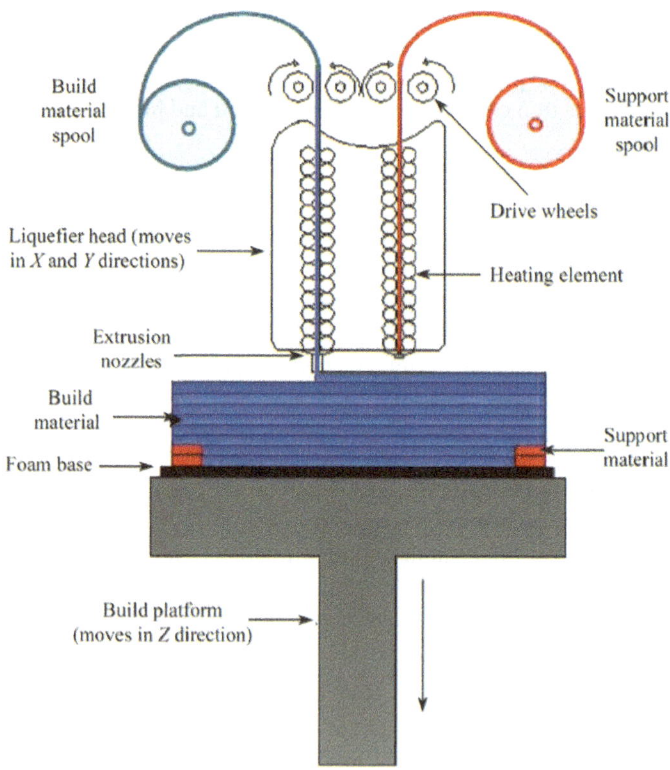

Fig. 8 Schematic diagram of fused deposition modeling 3D printer (Mohamed et al. 2015)

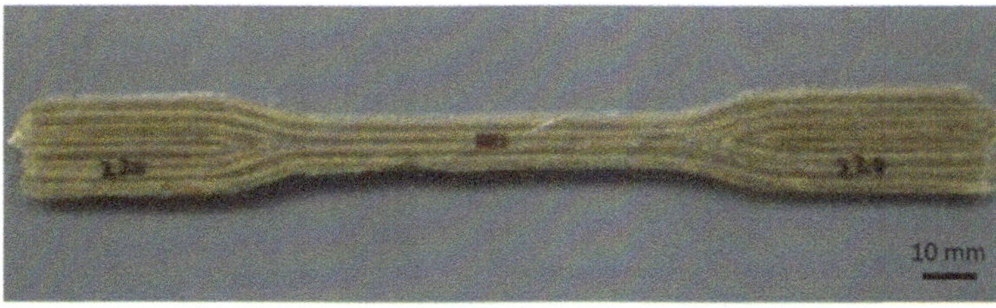

Fig. 9 3D printed tensile specimen (Matsuzaki et al. 2016)

Another study by Le Duigou et al. (2019), investigated the potential of 3D printing composites reinforced with flax fibers using fused deposition modeling. Flax composites were printed longitudinally and transversally and their properties were investigated. The mechanical properties of the printed composites were comparable to continuous glass fiber/polyamide.

2.9 Extrusion

Extrusion is a continuous process during this process; thermoplastic raw materials are melted and shaped into continuous length of polymeric products with fixed cross section profile. The products can be cut into the desired length by certain post-die equipment. The extrusion products cover many applications such as textile fibers, pipes, tubes, sheets, and insulated wire. Also profile shapes such as window frames and doors can be manufactured. The major extrusion equipment is the extruder, which consists of three major components: feeding section, melting section, and finally head box which comprising a die to form the required profile (Yang et al. 2016).

2.10 Injection Moulding

Injection moulding is a technique used to shape polymers into various structures. Injection moulding process can be divided into three main stages: filling, packing-holding, and cooling. Process starts with filling which starts after complete closing of mould. During the filling stage, injection screw pushes the molten polymer into the mould cavity until the mould is completely or nearly completely filled. Then the packing-holding stage begins, during this stage additional material is "packed" into the mould cavity under pressure to compensate any the shrinkage can be associated with cooling and solidification. The packing-holding stage continues until the material at the mould gate, a narrow entrance to the mould, is frozen and the material inside the mould is no longer influenced by that at the injection nozzle. The cooling stage begins, and the material is cooled down inside the mould until it is rigid enough

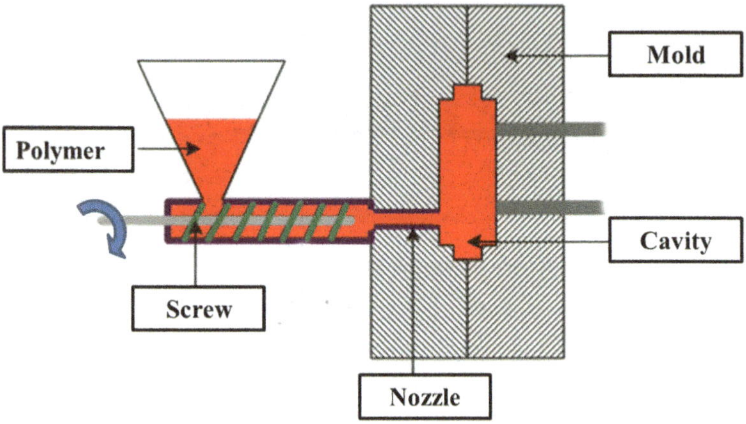

Fig. 10 Injection moulding process

to be ejected. Simultaneously with the material solidification in the cooling stage, plastication takes place inside the barrel, in this process polymer melt is conveyed to the screw tip by the screw rotation. The screw rotation stops after a sufficient amount of melt is accumulated in front of the screw. When the part in the mould becomes rigid enough, the mould is opened, and the part is ejected (Yang et al. 2016) (Fig. 10).

Composites reinforced with date palm fibers have been made using extrusion followed by injection moulding or compression moulding using thermosets and thermoplastics (Elseify et al. 2019). Date palm midrib fiber composites have been extruded with HDPE and LLDPE then hot pressed (Mahdavi et al. 2010; Mirmehdi et al. 2014). Moreover, some researchers used PP as a matrix with leaflet, spadix stems, and mesh fibers; the fibers were first extruded into pellets then used in injection moulding (Abu-Sharkh and Hamid 2004; Alawar et al. 2009; Zadeh et al. 2017; Boukettaya et al. 2018). Additionally, date palm leaflet fibers were used with recycled matrices like PETr, RLDPE, and RHDPE; the manufacturing techniques were extrusion followed by injection moulding (Dehghani et al. 2013; Zadeh et al. 2017). Other natural fibers like coir and bamboo fibers were also manufactured tusing extrusion followed by either injection moulding or compression moulding. Bamboo fibers in emulsion form were used with PLA in composite fabrication using extrusion followed by compression moulding (Li et al. 2019). Also, coir fibers were used with PP using extrusion then injection moulding (Haque et al. 2010).

3 Manufacturing of Date Palm Fiber-Based Composites

In a previous study, biodegradable polymer was used as a matrix reinforced with date palm fibers. Starch based biodegradable polymer was used with a fiber to matrix weigh fraction equals to 20:80. Prior to the compression process, date palm fibers and flax fibers were chemically treated using 5% NaOH. Date palm fibers were cut into length ranged between 20 and 30 mm and flax fibers' length ranged between 15 and 30 mm. The positive mould was coated with steric acid as release agent to facilitate taking out the sample off the mould. Hybrid fibers of date palm and flax were distributed in the mould and then the prepared thermo plastic starch (TPS) poured over the fibers. TPS was mixed with water in ratios of 1:1, 1:2, 1:3, 1:4, and 1:8 for 0, 20, 40, 50, 60, and 80 wt% respectively and then poured over the hybrid fibers. After that, mould was closed and hot pressed in two stages. First, mould with pressed and preheated at 140 ± 3 °C for 30 min followed by hot pressing under 5 MPa at 160 ± 3 °C for 30 min. Then mould was cooled down (Ibrahim et al. 2014).

In another study, date palm fiber reinforced recycled plastic was investigated as an environmentally friendly artificial wood. Three different types of plastic were collected namely; Polycarbonate (PC), Polystyrene (PS), and Polyvinyl chloride (PVC). Collected plastics were soaked in diluted 5% Clorox solution for 60 min and then washed with detergent and left to air dry. Washed and cleaned plastics were shredded into length ranged between 1 and 4 mm. Collected date palm fibers left to dry and cut into length ranged between 0.63 and 4 mm. Then, recycled plastics were mixed with date palm fibers at ratios of 1:1. Each mixture was pretreated with coupling agent to enhance the interfacial adhesion between date palm fibers and matrix. Table 1 illustrates the mixture and the used coupling agent. Pre-treated plastics and date palm fibers were mixed in a mixer at speed of 40–50 rpm for 3–4 min and then air dried, this process was repeated two times. After mixing process, mixed pellets of each sample were extruded in single extruder machine equipped with single screw 25 mm. Three different heating profile were used based on plastic type. For the PC-mix, the extruded heating profile ranged between 120 and 240 °C, for the PVC-mix the extruded heating profile ranged between 100 and 220 °C, and for PS-mix the extruded heating profile ranged between 120 and 180 °C. Extruded mixture was then pressed using hydraulic press of 119 tons. During the compression moulding, the mould temperature for the PC and PVC-mix kept at 185 and 160 °C for the PS-mix (Binhussain and El-Tonsy 2013).

Another study was made to evaluate the effect of chemical treatment using HCL, NaOH, and CH_3COOH for their effect on the mechanical properties of date palm fiber reinforced composites (Abdellah et al. 2019). Compression moulding was used to fabricate the composite samples. Date palm fibers were washed and cleaned from dust and left to air dry. Fibers were chopped using electrical mixer for 15 min. Following the chopping process, grinding date palm fibers are chemical treated using the three different chemicals. Mixture of date palm fibers and epoxy resin (2:1 resin

Table 1 Material mixture with coupling agent (Binhussain and El-Tonsy 2013)

Mixture	Coupling agent for pre-treatment
PC-mix	Saturated solution of polycarbonate in methylene chloride
PVC-mix	Pre-treatment with mixture of acetone + carbon disulfide + PVC polymer
PS-mix	Saturated solution of polystyrene in chloroform

to hardener) was poured into 300 mm × 200 mm × 10 mm mould. Aluminum foil was used on the top and bottom of the open mould to prevent the sticking of the mixture to the mould surface. Mould was then placed under 3-ton manual hydraulic press. Samples were pressed for 48 h until the full curing is occurred.

In a study by Al-Kaabi et al. (2005) compression moulding was used to study the effect of date palm fiber length on flexural properties. Date palm fiber length 0.5, 1, 2, and 3 cm were used after treatments with detergent, 5% NaOH, and bleaching with dioxin solution. Composite date palm fibers produced by using aluminum mould. Polyester resin with 1% methyl ethyl ketone peroxide was used as a matrix. Date palm fibers' mixture was placed in vacuum desiccator to get all the air bubbles out. Mixture was then poured into the mould and placed under a hydraulic press for 24 h for curing. Cured samples placed in oven under 100 °C for 2 h for post-curing.

Composite reinforced with date palm stem fibers and epoxy matrix was prepared with different fiber loading 40, 50, and 60 wt% (Saba et al. 2019). Acquired date palm fibers were washed in distilled water to remove any impurities such as dust and sands. Washed fibers were then ground into 0.8–1 mm length and placed in oven for air drying to acquire 6–8% moisture content. Mixture of epoxy and hardener was prepared with ratio 2:1 and date palm fibers were add to the prepared epoxy resin. Mechanical stirrer was used for 20 min to ensure the homogeneity of the mixture, then the mixture was then poured into stainless steel mould and left to cure for 24 h.

Hybrid laminates of date palm fiber, polyester and carbon fiber mats with epoxy resin as a matrix were fabricated (Raghavendra 2018). Date palm fibers with length of 5 mm were extracted from date palm fronds using crushing machine. Date palm fibers were chemically treated with 10% NaOH followed by washing and drying. In this study, vacuum bagging technique was used to fabricate the composite samples. Laminates of DPF + Polyester mat (NP), DPF + polyester + carbon mats (NPC), and DPF + carbon mats (NC) were wet laid with epoxy resin and sealed by plastic film. Vacuum pump was used to vacuum all the air under the plastic film using negative pressure value of 14 psi. Samples left to be cure and fully consolidate for testing and evaluation.

A researcher has used the hand layup technique to study the effect of fiber length of the reinforced natural fibers (coir and date palm) on the composite mechanical properties (Sharath et al. 2016). Reinforcement fibers, with different length 3, 6, 9, 12, and 15 mm, were placed in 200 mm × 150 mm × 4 mm granite mould. The laminates were prepared by pouring the resin over the reinforcement for fibers wetting and roller was used to spread the resin over the fibers.

Additionally, hand layup was integrated with other methods such as compression moulding (Alshammari et al. 2019; Gheith et al. 2019). In this study, different date palm fillers were used as reinforcement. Date palm fibers were cut into length ranged between 0.8 and 1 mm and placed in steel mould, then hand layup with epoxy resin. Fiber loading was maintained at 50 wt%. After the hand layup, mould was placed under hydraulic press for 10 min at 110 °C and then cold pressed for 5 min.

Another study by Dehghani et al. (2013) investigated the mechanical properties of date palm leaf fibers (DPLF) with recycled polyethylene terephthalate (PET) fabricated by injection moulding technique as mentioned earlier. DPLF were treated with 5% NaOH for 1 h at 100 °C followed by washing and drying. For best interfacial adherence between recycled polyester and DPLF, maleic anhydride composed of 70 wt% of ethylene-butylene block and 30 wt% of styrene blocks was used as a coupling agent. Prior the injection moulding process, pellets of PET, DPLF, and maleic anhydride (MA) were prepared. Firstly, PET and MA were dried at 60 °C for 24 h. and then premixed according to the designed ratios. Counter-rotating twin-screw extruder was used for compounding the mixture. Two zone temperature compounding was used, as the feeding temperature kept at 225 °C and the die temperature kept at 245 °C. Produced pellets of PET, MA, and DPLF was dried at 60 °C for 24 h. Dried pellets were fed into injection moulding machine to produce standard mechanical test specimens. Injection moulding was performed at 190–240 °C for 25 s.

Some of the recently studies on the date palm reinforced composites with different fabrication methods are presented in Table 2.

Table 2 Literature on date palm reinforced polymer composites

Reinforcement	Matrix	Fabrication	References
DPF	Polyester	RTM	Wazzan (2006)
DPF, glass fiber	Epoxy	Hand layup	Tripathy et al. (2016)
Date palm wood fronds	Polyester	Hand layup	Sadik et al. (2017)
DPF	Polypropylene	Injection mould	Alawar et al. (2008)
DPF	High density polyethylene	Compression moulding	Mahdavi et al. (2010)
Short DPF	Poly-epoxy thermoset	RTM	Sbiai et al. (2010)
Aligned date palm frond fibers	Low Density Polyethylene	Compression moulding	Alzebdeh et al. (2017)
Date palm frond fibers	Polyester	Hand lay-up	Anyakora (2013)
DPF and jute fibers	PP/EPDM	Injection moulding	Asadzadeh et al. (2012)
DPF, glass fiber	Recycled high density polyethylene	Compression moulding	Aldousiri et al. (2013)
DPF	Polypropylene	Compression moulding	Mahmoudi (2013)
DPF and flax fibers	Starch	Compression moulding	Strong (2008)
DPF	PVA and starch	Injection moulding	Mohanty et al. (2013)
DPF	Thermoplastic starch	Compression moulding	Saleh et al. (2017)
DPF	Phenolic, Bisphenol	Hand layup, Vacuum bagging	Al-Sulaiman (2002)
DPF	Polyester	Hand layup	Hammood (2015)

4 Conclusions

Date palm fiber composites could be manufactured using various techniques such as hand layup, RTM, VARTM, filament winding, compression moulding, extrusion, and injection moulding. Date palm fibers have high potential to be used as a replacement to already existing natural fibers. Each manufacturing technique has its own advantages and limitations. Therefore, choosing the suitable technique is a critical step that interfere with the final composite properties. Optimizing the manufacturing technique by controlling the processing parameters like temperature and pressure, will render a composite with high properties that could be used in high-end applications like the automotive industry.

References

Abdal-hay A, Suardana NPG, Jung DY et al (2012) Effect of diameters and alkali treatment on the tensile properties of date palm fiber reinforced epoxy composites. Int J Precis Eng Manuf 13:1199–1206. https://doi.org/10.1007/s12541-012-0159-3

Abdellah MY, Seleem AHA, Marzok WW, Hashem AM (2019) Tensile and fracture properties of chemically treatment date palm tree fibre reinforced epoxy

Abdullah M, Alharbi H, Hirai S et al (2020) Effects of chemical composition, mild alkaline pretreatment and particle size on mechanical, thermal, and structural properties of binderless lignocellulosic biopolymers prepared by hot-pressing raw microfibrillated Phoenix dactylifera and Cocos nucifera. Polym Test 106384. https://doi.org/10.1016/j.polymertesting.2020.106384

Abu-Sharkh BF, Hamid H (2004) Degradation study of date palm fibre/polypropylene composites in natural and artificial weathering: mechanical and thermal analysis. Polym Degrad Stab 85:967–973. https://doi.org/10.1016/j.polymdegradstab.2003.10.022

Advani SG, Hsiao KT (2012) Manufacturing techniques for polymer matrix composites (PMCs)

Akil HM, De Rosa IM, Santulli C, Sarasini F (2010) Flexural behaviour of pultruded jute/glass and kenaf/glass hybrid composites monitored using acoustic emission. Mater Sci Eng A 527:2942–2950. https://doi.org/10.1016/j.msea.2010.01.028

Al-Kaabi K, Al-Khanbashi A, Hammami A (2005) Date palm fibers as polymeric matrix reinforcement: DPF/polyester composite properties. Polym Compos 26:604–613. https://doi.org/10.1002/pc.20130

Al-Sulaiman FA (2002) Mechanical properties of date palm fiber reinforced composites. Appl Compos Mater 9:369–377. https://doi.org/10.1023/A:1020216906846

Alawar A, Hamed AM, Al-Kaabi K (2009) Characterization of treated date palm tree fiber as composite reinforcement. Compos Part B Eng 40:601–606. https://doi.org/10.1016/j.compositesb.2009.04.018

Alawar A, Hamed AM, Al-Kaabi K (2008) Date palm tree fiber as polymeric matrix reinforcement, DPF-polypropylene composite characterization. Adv Mater Res 47–50:193–196. https://doi.org/10.4028/www.scientific.net/AMR.47-50.193

Aldousiri B, Alajmi M, Shalwan A (2013) Mechanical properties of palm fibre reinforced recycled HDPE. Adv Mater Sci Eng 2013. https://doi.org/10.1155/2013/508179

Ali ME, Alabdulkarem A (2017) On thermal characteristics and microstructure of a new insulation material extracted from date palm trees surface fibers. Constr Build Mater 138:276–284. https://doi.org/10.1016/j.conbuildmat.2017.02.012

Alshammari BA, Saba N, Alotaibi MD et al (2019) Evaluation of mechanical, physical, and morphological properties of epoxy composites reinforced. Materials (Basel) 12:17

Alzebdeh KI, Nassar MM, Al-Hadhrami MA et al (2017) Characterization of mechanical properties of aligned date palm frond fiber-reinforced low density polyethylene. J Eng Res 14:115–123. https://doi.org/10.24200/tjer.vol.14iss2pp115-123

Angelov I, Wiedmer S, Evstatiev M et al (2007) Pultrusion of a flax/polypropylene yarn. Compos Part A Appl Sci Manuf 38:1431–1438. https://doi.org/10.1016/j.compositesa.2006.01.024

Anyakora AN (2013) Investigation of impact strength properties of oil and date palm frond fiber reinforced polyester composites

Arya A, Tomlal JE, Gejo G, Kuruvilla J (2015) Commingled composites of polypropylene/coir-sisal yarn: effect of chemical treatments on thermal and tensile properties. E-Polymers 15:169–177. https://doi.org/10.1515/epoly-2014-0186

Asadzadeh M, Khalili SMR, EslamiFarsani R, Rafizadeh S (2012) Bending properties of date palm fiber and jute fiber reinforced polymeric composite. Int J Adv Des Manuf Technol 5:59–63

Balasubramanian K, Sultan MTH, Rajeswari N (2018) Manufacturing techniques of composites for aerospace applications. Elsevier Ltd.

Banowati L, Hadi BK, Suratman R, Faza A (2016) Tensile strength of ramie yarn (spinning by machine)/HDPE thermoplastic matrix composites. AIP Conf Proc 1717:0–4. https://doi.org/10.1063/1.4943456

Bar M, Das A, Alagirusamy R (2018) Effect of interface on composites made from DREF spun hybrid yarn with low twisted core flax yarn. Compos Part A Appl Sci Manuf 107:260–270. https://doi.org/10.1016/j.compositesa.2018.01.003

Berges M, Léger R, Placet V et al (2016) Influence of moisture uptake on the static, cyclic and dynamic behaviour of unidirectional flax fibre-reinforced epoxy laminates. Compos Part A Appl Sci Manuf 88:165–177. https://doi.org/10.1016/j.compositesa.2016.05.029

Bertomeu D, García-Sanoguera D, Fenollar O et al (2012) Use of eco-friendly epoxy resins from renewable resources as potential substitutes of petrochemical epoxy resins for ambient cured composites with flax reinforcements. Polym Compos 33:683–692. https://doi.org/10.1002/pc.22192

Binhussain MA, El-Tonsy MM (2013) Palm leave and plastic waste wood composite for outdoor structures. Constr Build Mater 47:1431–1435. https://doi.org/10.1016/j.conbuildmat.2013.06.031

Boukettaya S, Alawar A, Almaskari F et al (2018) Modeling of water diffusion mechanism in polypropylene/date palm fiber composite materials. J Compos Mater. https://doi.org/10.1177/0021998317752228

Dehghani A, Madadi Ardekani S, Al-Maadeed MA et al (2013) Mechanical and thermal properties of date palm leaf fiber reinforced recycled poly (ethylene terephthalate) composites. Mater Des 52:841–848. https://doi.org/10.1016/j.matdes.2013.06.022

Dhakal HN, MacMullen J, Zhang ZY (2016) Moisture measurement and effects on properties of marine composites. Elsevier Ltd.

Di Bella G, Fiore V, Valenza A (2010) Effect of areal weight and chemical treatment on the mechanical properties of bidirectional flax fabrics reinforced composites. Mater Des 31:4098–4103. https://doi.org/10.1016/j.matdes.2010.04.050

Duquesne S, Samyn F, Ouagne P, Bourbigot S (2015) Flame retardancy and mechanical properties of flax reinforced woven for composite applications. J Ind Text 44:665–681. https://doi.org/10.1177/1528083713505633

Elseify LA, Midani M, Shihata LA, El-Mously H (2019) Review on cellulosic fibers extracted from date palms (Phoenix Dactylifera L.) and their applications. Cellulose 26. https://doi.org/10.1007/s10570-019-02259-6

Gheith MH, Aziz MA, Ghori W et al (2019) Flexural, thermal and dynamic mechanical properties of date palm fibres reinforced epoxy composites. J Mater Res Technol 8:853–860. https://doi.org/10.1016/j.jmrt.2018.06.013

Goutianos S, Peijs T, Nystrom B, Skrifvars M (2007) Textile reinforcements based on aligned flax fibres for structural composites. Compos Innov 1–13

Hammood AS (2015) Effect of erosion on water absorption and morphology for treated date palm fiber-reinforced polyester composites. Int J Mech Mechatron Eng 15:108–114

Haque M, Islam N, Huque M et al (2010) Coir fiber reinforced polypropylene composites: physical and mechanical properties. Adv Compos Mater 19:91–106. https://doi.org/10.1163/092430409X12530067339325

Hassanin AH, Hamouda T, Candan Z et al (2016) Developing high-performance hybrid green composites. Compos Part B Eng 92:384–394. https://doi.org/10.1016/j.compositesb.2016.02.051

Ibrahim H, Farag M, Megahed H, Mehanny S (2014) Characteristics of starch-based biodegradable composites reinforced with date palm and flax fibers. Carbohydr Polym 101:11–19. https://doi.org/10.1016/j.carbpol.2013.08.051

Jiang X, Rui Y, Chen G (2009) Unsaturated polyester-toughened epoxy composites: effect of sisal fiber on thermal and dynamic mechanical properties of sisal fiber on thermal and dynamic mechanical properties. J Vinyl Addit Technol 21:129–133. https://doi.org/10.1002/vnl

Keerthi Gowda BS, Easwara Prasad GL, Velmurgan R (2016) Probabilistic study of tensile properties of coir fiber reinforced polymer matrix composite probabilistic study of tensile properties of coir fiber reinforced polymer matrix composite. Int J Adv Mater Sci 06:7–17

Kejariwal RK, Keerthi Gowda BS (2017) Flammability and moisture absorption behavior of sisal-polyester composites. Mater Today Proc 4:8040–8044. https://doi.org/10.1016/j.matpr.2017.07.142

Lai CY, Sapuan SM, Ahmad M et al (2005) Mechanical and electrical properties of coconut coir fiber-reinforced polypropylene composites. Polym Plast Technol Eng 44:619–632. https://doi.org/10.1081/PTE-200057787

Le Duigou A, Barbé A, Guillou E, Castro M (2019) 3D printing of continuous flax fibre reinforced biocomposites for structural applications. Mater Des 180:107884. https://doi.org/10.1016/j.matdes.2019.107884

Lee J, Unnithan AR, Park CH, Kim CS (2019) 3D bioprinting for active drug delivery. Elsevier Inc.

Lehtiniemi P, Dufva K, Berg T et al (2011) Natural fiber-based reinforcements in epoxy composites processed by filament winding. J Reinf Plast Compos 30:1947–1955. https://doi.org/10.1177/0731684411431019

Li W, He X, Zuo Y et al (2019) Study on the compatible interface of bamboo fiber/polylactic acid composites by in-situ solid phase grafting. Int J Biol Macromol 141:325–332. https://doi.org/10.1016/j.ijbiomac.2019.09.005

Li Y, Ma H, Shen Y et al (2015) Effects of resin inside fiber lumen on the mechanical properties of sisal fiber reinforced composites. Compos Sci Technol 108:32–40. https://doi.org/10.1016/j.compscitech.2015.01.003

Lu X, Zhang MQ, Rong MZ et al (2003) Self-reinforced melt processable composites of sisal. Compos Sci Technol 63:177–186. https://doi.org/10.1016/S0266-3538(02)00204-X

Mahdavi S, Kermanian H, Varshoei A (2010) Comparison of mechanical properties of date palm fiber-polyethylene composite. BioResources 5:2391–2403. https://doi.org/10.15376/biores.5.4.2391-2403

Mahdi E, Hamouda AMS, Sahari BB, Khalid YA (2002) Effect of material and geometry on crushing behaviour of laminated conical composite shells. Appl Compos Mater 9:265–290. https://doi.org/10.1023/A:1019695903534

Mahmoudi N (2013) Use of date palm fibers as reinforcement for thermoplastic-based composites. Mech Ind 14:71–77. https://doi.org/10.1051/meca/2012043

Matsuzaki R, Kobayashi S, Todoroki A, Mizutani Y (2011) Full-field monitoring of resin flow using an area-sensor array in a VaRTM process. Compos Part A Appl Sci Manuf 42:550–559. https://doi.org/10.1016/j.compositesa.2011.01.014

Matsuzaki R, Ueda M, Namiki M et al (2016) Three-dimensional printing of continuous-fiber composites by in-nozzle impregnation. Sci Rep 6:1–7. https://doi.org/10.1038/srep23058

Mir SS, Nafsin N, Hasan M et al (2013) Improvement of physico-mechanical properties of coir-polypropylene biocomposites by fiber chemical treatment. Mater Des 52:251–257. https://doi.org/10.1016/j.matdes.2013.05.062

Mirmehdi SM, Zeinaly F, Dabbagh F (2014) Date palm wood flour as filler of linear low-density polyethylene. Compos Part B Eng 56:137–141. https://doi.org/10.1016/j.compositesb.2013.08.008

Misri S, Ishak MR, Sapuan SM, Leman Z (2016) Split-disk properties of kenaf yarn fibre-reinforced unsaturated polyester composites using filament winding method. Pertanika J Sci Technol 24:475–482

Mohamed OA, Masood SH, Bhowmik JL (2015) Optimization of fused deposition modeling process parameters: a review of current research and future prospects. Adv Manuf 3:42–53. https://doi.org/10.1007/s40436-014-0097-7

Mohanty JR, Das SN, Das HC, Swain SK (2013) Effective mechanical properties of polyvinylalcohol biocomposites with reinforcement of date palm leaf fibers. Polym Compos 34:959–966. https://doi.org/10.1002/pc.22502

Mohareb ASO, Hassanin AH, Badr AA et al (2015) Novel composite sandwich structure from green materials: mechanical, physical, and biological evaluation. J Appl Polym Sci 132:4–11. https://doi.org/10.1002/app.42253

Mulinari DR, Baptista CARP, Souza JVC, Voorwald HJC (2011) Mechanical properties of coconut fibers reinforced polyester composites. Procedia Eng 10:2074–2079. https://doi.org/10.1016/j.proeng.2011.04.343

Muralidhar BA (2013) Tensile and compressive properties of flax-plain weave preform reinforced epoxy composites. J Reinf Plast Compos 32:207–213. https://doi.org/10.1177/0731684412469136

Muralidhar BA, Giridev VR, Raghunathan K (2012) Flexural and impact properties of flax woven, knitted and sequentially stacked knitted/woven preform reinforced epoxy composites. J Reinf Plast Compos 31:379–388. https://doi.org/10.1177/0731684412437987

Netcomposites (2019) Manufacturing. https://netcomposites.com/guide/manufacturing/. Accessed 7 Apr 2020

Nguyen-Chung T, Friedrich K, Mennig G (2007) Processability of pultrusion using natural fiber and thermoplastic matrix. Res Lett Mater Sci 2007:1–5. https://doi.org/10.1155/2007/37123

O'Donnell A, Dweib MA, Wool RP (2004) Natural fiber composites with plant oil-based resin. Compos Sci Technol 64:1135–1145. https://doi.org/10.1016/j.compscitech.2003.09.024

Ochi S (2006) Development of high strength biodegradable composites using Manila hemp fiber and starch-based biodegradable resin. Compos Part A Appl Sci Manuf 37:1879–1883. https://doi.org/10.1016/j.compositesa.2005.12.019

Palani Kumar K, Shadrach Jeya Sekaran A, Pitchandi K (2017) Investigation on mechanical properties of woven alovera/sisal/kenaf fibres and their hybrid composites. Bull Mater Sci 40:117–128. https://doi.org/10.1007/s12034-016-1343-3

Papa I, Lopresto V, Simeoli G et al (2017) Ultrasonic damage investigation on woven jute/poly (lactic acid) composites subjected to low velocity impact. Compos Part B Eng 115:282–288. https://doi.org/10.1016/j.compositesb.2016.09.076

Raghavendra S (2018) Mechanical properties of hybrid composites with date palm fibre reinforcement. Adv Mater 7:78. https://doi.org/10.11648/j.am.20180703.14

Rayyaan R, Kennon WR, Potluri P, Akonda M (2019) Fibre architecture modification to improve the tensile properties of flax-reinforced composites. J Compos Mater 002199831986315. https://doi.org/10.1177/0021998319863156

Romli FI, Alias AN, Rafie ASM, Majid DLAA (2012) Factorial study on the tensile strength of a coir fiber-reinforced epoxy composite. AASRI Procedia 3:242–247. https://doi.org/10.1016/j.aasri.2012.11.040

Saba N, Alothman OY, Almutairi Z et al (2019) Date palm reinforced epoxy composites: tensile, impact and morphological properties. J Mater Res Technol 8:3959–3969. https://doi.org/10.1016/j.jmrt.2019.07.004

Sadik T, Sivaram NM, Senthil P (2017) Evaluation of mechanical properties in date palm fronds polymer composites. Int J ChemTech Res 10:558–564. https://doi.org/10.1063/1.5085592

Safiee S, Md Akil H, Mazuki AAM et al (2011) Properties of pultruded jute fiber reinforced unsaturated polyester composites. Adv Compos Mater 20:231–244. https://doi.org/10.1163/092430410X547047

Saleh AA, Saleh MA, Al Haron MH, Farag M (2017) Insights into the effect of moisture absorption and fiber content on the mechanical behavior of starch–date-palm fiber composites. Starch/Staerke 69:1–8. https://doi.org/10.1002/star.201600254

Salit MS, Jawaid M, Bin N et al (2015) Manufacturing of natural fibre reinforced polymer composites

Santulli C (2000) Mechanical and impact properties of untreated jute fabric reinforced polyester laminates compared with different E-glass fibre reinforced laminates. Sci Eng Compos Mater 9:177–188. https://doi.org/10.1515/secm.2000.9.4.177

Sathishkumar TP, Naveen J, Navaneethakrishnan P et al (2017) Characterization of sisal/cotton fibre woven mat reinforced polymer hybrid composites. J Ind Text 47:429–452. https://doi.org/10.1177/1528083716648764

Sbiai A, Kaddami H, Fleury E et al (2008) Effect of the fiber size on the physicochemical and mechanical properties of composites of epoxy and date palm tree fibers. Macromol Mater Eng 293:684–691. https://doi.org/10.1002/mame.200800087

Sbiai A, Maazouz A, Fleury E et al (2010) Short date palm tree fibers/polyepoxy composites prepared using RTM process: effect of tempo mediated oxydation of the fibers. BioResources 5:672–689

Senthilkumar K, Siva I, Sultan MTH et al (2017) Static and dynamic properties of sisal fiber polyester composites—effect of interlaminar fiber orientation. BioResources 12:7819–7833. https://doi.org/10.15376/biores.12.4.7819-7833

Sharath GP, Kishor B, Vinod B, Naveen A (2016) Effect of fiber length on the mechanical properties of coir and wild date palm reinforced epoxy composites. Int J Sci Technol Eng 2:555–559

Strong A (2008) Fundamentals of composites manufacturing: materials, methods and applications, 2nd edn. Society of Manufacturing Engineers

Tripathy S, Dehury J, Mishra D (2016) A study on the effect of surface treatment on the physical and mechanical properties of date-palm stem liber embedded epoxy composites. IOP Conf Ser Mater Sci Eng 115:012036. https://doi.org/10.1088/1757-899X/115/1/012036

Van De Velde K, Kiekens P (2001) Thermoplastic pultrusion of natural fibre reinforced composites. Compos Struct 54:355–360. https://doi.org/10.1016/S0263-8223(01)00110-6

Wazzan AA (2006) Effect of fiber orientation on the mechanical properties and fracture characteristics of date palm fiber reinforced composites 213–225

Xiao B, Yang Y, Wu X et al (2015) Hybrid laminated composites molded by spray lay-up process. Fibers Polym 16:1759–1765. https://doi.org/10.1007/s12221-015-5298-6

Yang Y, Chen X, Lu N, Gao F (2016) Injection molding process control, monitoring, and optimization. Hanser Publisher

Yorseng K, Rangappa SM, Pulikkalparambil H et al (2020) Accelerated weathering studies of kenaf/sisal fiber fabric reinforced fully biobased hybrid bioepoxy composites for semi-structural applications: morphology, thermo-mechanical, water absorption behavior and surface hydrophobicity. Constr Build Mater 235:117464. https://doi.org/10.1016/j.conbuildmat.2019.117464

Yuvaraj G, Kumar H, Saravanan G (2017) An experimentation of chemical and mechanical behaviour of epoxy-sisal reinforced composites. Polym Polym Compos 25:221–224. https://doi.org/10.1177/096739111702500307

Zadeh KM, Inuwa IM, Arjmandi R et al (2017) Effects of date palm leaf fiber on the thermal and tensile properties of recycled ternary polyolefin blend composites. Fibers Polym 18:1330–1335. https://doi.org/10.1007/s12221-017-1106-9

Zamri MH, Akil HM, Bakar AA et al (2012) Effect of water absorption on pultruded jute/glass fiber-reinforced unsaturated polyester hybrid composites. J Compos Mater 46:51–61. https://doi.org/10.1177/0021998311410488

Zamri MH, Akil HM, Ishak ZAM, Abu Bakar A (2014) Effect of different fiber loadings and sizes on pultruded kenaf fiber reinforced unsaturated polyester composites. Polym Compos 36:1224–1229. https://doi.org/10.1002/pc.23025

Zhu J, Zhu H, Njuguna J, Abhyankar H (2013) Recent development of flax fibres and their reinforced composites based on different polymeric matrices. Materials (Basel) 6:5171–5198. https://doi.org/10.3390/ma6115171

Date Palm Nanofibres and Composites

Ramzi Khiari and Mohamed Naceur Belgacem

Abstract Over the last decades, the development of renewable resources of fibres was intensified, which amplified the market request of such a raw material. This is due to three potential advantages associated with these natural substances, namely: very cheap, bio-renewable and their availability in large quantities with various varieties forms. This chapter is devoted to the discussion of the potential valorization of the date palm waste to prepare fibres and nanofibres. This agricultural waste has lately noticed considerable attention, as an important source to produce cellulosic fibres, especially in forest-poor regions. The morphologies features, the chemical composition of date palm and their comparison with different lignocellulosic fibres sources are described. Then, the delignification methods and characterization of fibres from date palm are also presented. Finally, the preparation of nanocellulose (Cellulose nanocrystals (CNCs) and cellulose nanofibres (CNF)) and their use as nanofibre-reinforced nanocomposite materials will be detailed and discussed.

Keywords Date palm · Cellulose fibre · Natural fibre · Nanomaterials · Nanocellulose · CNC · CNF

1 Introduction

Lignocellulosic biomass is a bio-based material well known by its renewability and biodegradability. This outstanding property makes it the most abundant polymer on this earth. Hence, this biomaterial is characterized by the large production thanks to photosynthesis reaction, which is estimated to be give between 10^{11} and 10^{12} tons per year (Klemm et al. 2011; Belgacem and Gandini 2008). Cellulose, in general, and that

R. Khiari (✉)
Department of Textile, Higher Institute of Technological Studies of Ksar Hellal, Ksar Hellal, Tunisia
e-mail: khiari_ramzi2000@yahoo.fr

Research Unity of Applied Chemistry & Environment, University of Monastir, Monastir, Tunisia

R. Khiari · M. N. Belgacem
Université Grenoble Alpes, CNRS, Grenoble INP, LGP2, 38000 Grenoble, France

M. Midani et al. (eds.), *Date Palm Fiber Composites*, Composites Science and Technology,
https://doi.org/10.1007/978-981-15-9339-0_6

produced from agricultural and food wastes been exploited in very old applications, such as in papermaking fields (Khiari et al. 2011a, b, c; Ferhi et al. 2014a, b; Jonoobi et al. 2015), glue (Khiari et al. 2011a, b, c), textile (Botdrof and Soap 1962; Mansouri et al. 2015; Elseify et al. 2019), cosmetics or pharmaceuticals (Olaru et al. 1998). Recently a new window were focused more deeply in developing new high-tech materials based in the application of knowledge especially in nanotechnology such as the production of nanocellulose (Bettaieb et al. 2015a, b, c; Dufresne 2012).

Nowadays, the production of nanocellulose has quickly interesting many fields and one of the fastest budding study fields of applications (CNPs). Cellulose nanocrystals (CNCs) and cellulose nanofibres (CNF) can be extracted from several cellulosic biomasses. The valorisation of nanocellulose in nanocomposites has involved large attention due to their significantly excellent performances, in particular, what concerns mechanical properties, when compared with neat polymers or conventional polymer composites (Dufresne 2012; Ibrahim et al. 2010; Hussain et al. 2006). In fact, availability, excellent mechanical properties (strength and stiffness), weak density and biodegradability are some important characteristic for the nano-scale cellulose fibre products which make them potential as nano-filler for bio-nanocomposites (Siró and Plackett 2008, 2010). In addition, a large number of researches were devoted to the extraction, preparation and characterization of novel forms of cellulose. In all the case, the extracted cellulosic materials with one dimension in the nanometer range are referred to generically as nanocelluloses.

Nanocellulose and cellulose fibres can be prepared from large sources of biomass lignocellulosic biomass residues. The last one has received more attention in the last decades especially where the wood quantity biomass are very low. Furthermore, the agricultural residues and/or annual plants concluded rational and advanced methods of exploitation and valorisation, which seems to be of potential interest in such countries. Thus, these products can be considered as new sources to produce cellulosic fibres. That is why the tendency is currently practical to valorise various annual crops available for example in Egypt (Abouzeid et al. 2015; Midani et al. 2018; Hassan et al. 2012), India (Dutt et al. 2005, 2008), Malaysia (Abdul Khalil et al. 2014; Alotaibi et al. 2019), Tunisia (Bettaieb et al. 2015a, b, c; Khiari et al. 2010, 2011a, b, c, Khiari and Belgacem 2017; Khiari 2007; Naili et al. 2017) or Morocco (Benhamou et al. 2015; Bendahou et al. 2008, 2009, 2010), to give only limited number of examples.

In this context, date palms agricultural residues have attracted a large attention. Date palms are widely grown in the Middle East and North Africa and their estimated population in the region is almost 85 million palms. It can be found with large quantities in different country such as Tunisia (Khiari et al. 2010, 2011a; b, c), Egypt (El Morsy 1980; El Morsy et al. 1981; Elseify et al. 2019), Morocco (Benhamou et al. 2015; Bendahou et al. 2008, 2009, 2010), Sudan (Khristova et al. 2005). It is viewed as renewable sources to produce cellulosic fibres and their valorisation in the field of composites materials (Khiari et al. 2011a, b, c; Mansouri et al. 2015; Taha et al. 2007; Bendahou et al. 2008; Kriker et al. 2005, 2008). On the opposite, few studies where focused to study the potential use of date palm by-products for nanocellulose production. Besides, most of works are devoted to the extraction fibre from date palm (Ezzat 1974; El Morsy et al. 1981; Khristova et al. 2005; Khiari et al. 2010, 2011a, b, c). Date palm (*Phoenix dactylifera*) is one of the most cultivated plants in the arid and semi-arid regions of the world. *Phoenix Dactylifera L.* is sometimes referred to as the true date palm; it presents one of the richest sources of cellulosic fibres (Khiari et al. 2010, 2011a, b, c). Fibres from date palm could be isolated from various parts namely; rachis, spadix stems, leaves, and mesh (Hakeem et al. 2014; Midani et al. 2018).

This chapter is devoted to the characterization of this raw material in terms of morphology and chemical composition. Then, pulps isolated from date palm are firstly discussed. Finally, the preparation of nanocellulose and their use as nanofibre-reinforced nanocomposite materials will be detailed. The illustrated results are detailed and a comparison with different sources such as wood, non-wood, or annual plant crops was discussed.

2 Date Palm Characterisation

2.1 Morphology of Date Palm

The general morphology, using the scanning electron microscopy (SEM) at various amplifications, of date palm was examined by several researches from different countries (Khiari et al. 2010; 2011a, b, c; Midani et al. 2018; Benhamou et al. 2015; Bendahou et al. 2008, 2009, 2010; El Morsy et al. 1981; Khristova et al. 2005; Elseify et al. 2019). The surface and longitudinal cross section observations of date palm fragments has a very porous structure (Fig. 1) compared to other annual plants, taking as an example, of *posidonia oceanica* balls and/or alfa, bagasse, rice straw, vine stem, etc. However, the raw materials present in all case the same structure and it is mainly composed of parenchyma, vascular tissue, epidermis, sclereid cells (or fibres) with a diameter and a wall thickness of around 25 μm and 10–15 μm, respectively. Bendahou et al. (2009) and Elseify et al. (2019) reported that a disposable difference in the composition in terms of can be observed between the leaves and rachis (called midrib) fragments isolated from date palm. It can be noticed that the leaves present a regular structure of slots which is different to compare with the rachis counterpart.

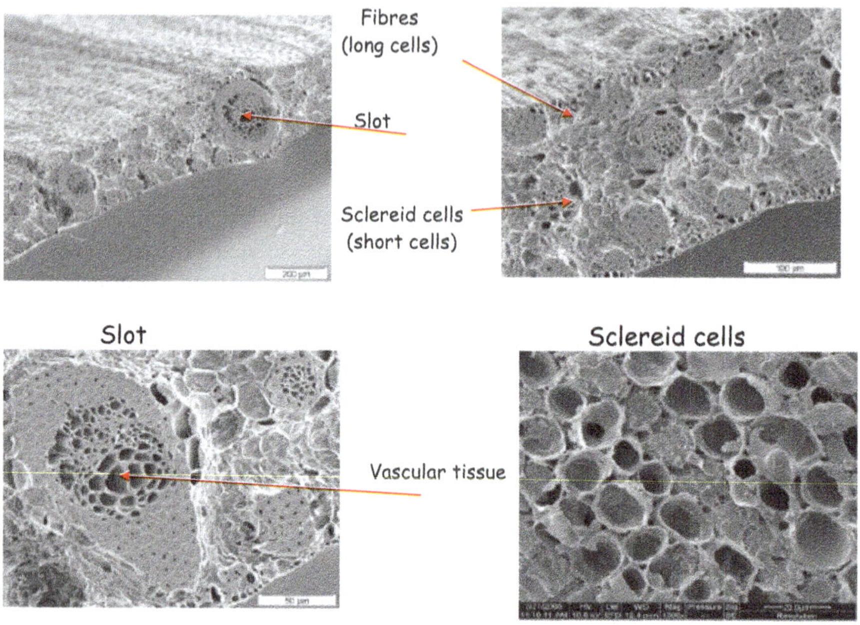

Fig. 1 Electron scanning microscopy of date palm (Bendahou 2009)

2.2 *Chemical Composition*

As reported by several researches, the date palm chemical composition was studied and evaluated according the TAPPI standard methods and the results are illustrated in Table 1.

It can be noticed that the chemical composition of date palm (leaves and or rachis…) strongly depends on the climate conditions and soil chemical composition. In fact, the amount of holocellulose or lignin seems to be different. In addition, it can be mentioned that the rachis or midrib part present a high amount of cellulose and holocellulose than the leaves element. Such results encourage the research to valorise the rachis element in the context of production of fibre for papermaking applications, for composites and/or as a raw material for cellulose derivatives.

Table 2 summarizes the data for several lignocellulosic materials such as wood (hardwood and softwood sources), agricultural residues, like olive pruning, wheat straw, sunflower stalks, sorghum stalks, rice straw, sugarcane bagasse and/or marine biomass, as well as other non-wood sources (Jimenez and Lopez 1990; Jimenez et al. 1993, 2008; Eugenio et al. 2006; Copur et al. 2008; Antunes et al. 2000; Cordeiro et al. 2004; Schott 2000; Eugenio et al. 2006; Fiserova et al. 2006, Gominho et al. 2001; Manfred 1993; Khristova et al. 2005; Dutt et al. 2005).

From these results (Tables 1 and 2), it can be noticed that the quantity of extractive substances (in cold water, hot water and in organic reagent) are different in the leaves and those in the rachis. This amount is comparable to that associated to annual plants

Table 1 Date chemical composition (Reprinted from Ref. Khiari et al. (2010), with permission from Elsevier 2020)

Amounts in % (w/w with respect to oven dried raw material)	Date palm rachis				Date palm leaves		
	Khiari et al. (2010)	Khristova et al. (2005)	Bendahou et al. (2009)	El Morsy et al. (1980)	Khristova et al. (2005)	Bendahou et al. (2009)	Ezzat (1974)
Cold water extractives (%)	5.0	n.d	n.d	n.d	n.d	n.d	n.d
Hot water extractives (%)	8.1	8.7	n.d	n.d	10.8	n.d	n.d
1% NaOH extractives (%)	20.8	25.6	n.d	n.d	29.9	n.d	n.d
Ethanol-toluene extractives (%)	6.3	12.8[a]	4		11.7[a]	3	5.9[b]
Ash (%)	5	5.6	2.5	3.4	9.6	6.5	3.9
Lignin (%)	27.2[c]	23.8[d]	14	25.8	31.2[d]	27	n.d
Holocellulose (%)	74.8	n.d	72	n.d	n.d	59.5	n.d
Cellulose (%)	45	43.1	44	n.d	30.3	33.5	n.d

[a]in ethanol-cyclohexane; [b]in alcohol-benzene; [c]Klason lignin; [d]residual lignin (Klason and soluble lignin)

which does not exceed 10%, w/w. However, this amount is generally lower than that noticed in the case of wood and non-wood biomass. Regarding, the amount of soluble compound in 1% NaOH (i.e. 20% for date palm rachis) seems to be easy for delignification process and this result can be explained by the greater porosity and subsequently date palm present higher accessibility to the reagent or solvent. In all case, this amount is comparable to that of non-wood and wood sources which already does not overdo 20%. The amount of ash contained in date palm rachis does not exceed 5%, w/w. This value seems to be slightly higher than that found in wood and comparable to that of annual plant (Table 3). In addition, these amounts are lower than those of hardwood and softwood for which the ash content is less that 2%. As reported by Khiari et al. (2010), the ashes chemical composition are mostly composed by calcium (Ca), chlorine (Cl), potassium (K), magnesium (Mg) and Silicon (Si). The silicon quantity found in the leaves and/or rachis date palm is negligible. This is also true for Banana pseudo–stems (Cordeiro et al. 2004) and amaranth (Fiserova et al. 2006).

The content of polysaccharides in terms of holocellulose and α-cellulose is comparable to that of wood and some non-wood raw material and higher than some known for annual plants. This significant amount makes the date palm as a rational and potential source of lignocellulosic source fibres for many applications such as: paper-making applications and fibre-reinforced materials. Cellulose derivatives could also

Table 2 Comparison of chemical composition between some lignocellulosic biomass (Reprinted from Ref. Khiari et al. (2010), with permission from Elsevier 2020)

	References	C.W	H.W	A.B	1% NaOH	Ash	Hol	Lign	Hemi	Cell
Wood										
Brutia pine	Copur et al. (2008)	2.2	2.8	1.94	16.1	0.4	75.5	26.1	28.5	47
Pine pinaster **	Jimenez and Lopez (1990), Jimenez et al. (2008)	n.d	2	1–2.6	7.9–10.3	0.3–0.5	69–67	26–28	13. 7	56
*Eucalyptus globulus***	Jimenez and Lopez (1990), Jimenez et al. (2008)	n.d	2.8	1.15	12.42	0.6	80.5	19.9	27.7	53
Olive wood*	Jimenez and Lopez (1990),Jimenez et al. 1992, 2008)	15.5	17	10.4	30.0	1.4	65.83	15.64	24.33	41.5
Holm Oak (*Quercus Ilex*)	Eugenio al. (2006)	n.d	n.d	n.d	n.d	n.d	71.2	16.3	28.3	43
Non wood										
Prosopis alba	Jimenez et al. (2008)	n.d	4.7	4.65	20.8	n.d	63.6	19.3	22.0	42
Chamaecytisus	Jimenez et al. (2008)	n.d	3	3.43	16.1	n.d	75.3	14.8	31.7	44
Phragmites	Jimenez et al. (2008)	n.d	5.4	6.36	34.7	n.d	64.1	23.6	24.4	40
Retama monosperma	Jimenez et al. (2008)	n.d	3.8	5.03	16.9	n.d	71.7	21.5	29.0	43
Arundo donax	Shatalov et al. (2001)	n.d	6.7	9.2	n.d	4.8	61.2	20.9	32.1	29.2
Banana pseudo -stems	Cordeiro et al. (2004)	n.d	5.4	2.7	n.d	14	65.2	12.7	25.2	40

(continued)

Table 2 (continued)

	References	C.W	H.W	A.B	1% NaOH	Ash	Hol	Lign	Hemi	Cell
Paulownia fortuna	Jimenez et al. (1993,2008)	n.d	9.6	5.5	31.5	n.d	70.7	22.4	33.3	37
Annual and perennial plants										
Wheat straw	Schott (2000)	5.8–11	14	4.6–9.2	41–42.8	4–9	n.d	11–21	21–28.5	33–45.5
Rice dishes	Alcaide et al. (1990)	10.6	13	4.6–5.7	49.1	13–20	n.d	11–13.5	13–26.2	42–49.8
Barley fodder	Alcaide et al. (1990)	16	16	4.7	47	4.9–7	n.d	7–18	24.5	34–48
Ryestraw	Alcaide et al. (1990)	8.4	9.4	3.2–5.2	37.4	4- 4.3	n.d	18.5–19	23–30.5	55
Oat straw	Alcaide et al. (1990)	13.2	15	4.4	41.8	7–7.5	n.d	11–19.6	16–27	37–53.6
Sorghum stalks	Jimenez et al. (1993)	n.d	21.7	7.99	41.6	4.85	71.7	13.4	29.3	42
Amaranth	Fiserova et al. (2006)	23.5	28	2.51	46.8	12	58.4	13.2	26.1	32
Orache	Fiserova et al. (2006)	4.6	6.5	1.87	27.5	2	74.9	19.5	38.8	36
Jerusalem artichoke	Fiserova et al. (2006)	26.6	31	2.86	48.5	3	51.6	14.7	23.1	29
Cynara Cardunculus L*	Antunes et al. (2000), Gominho et al. (2001)	n.d	10	6	n.d	8	64	20	26	38
Miscanthus sinensis	Barba et al. (2002)	n.d	9.1	3.1	n.d	0.7	72.5	19.9	30.3	42
Kenaf (*Hibiscus cannabinus*)	Manfred (1993)	n.d	n.d	n.d	n.d	1.7–5	n.d	14.5–18.7	n.d	31–39

C.W. Cold water solubility; *H.W.* Hot water solubility; *A.B* solubility in various organic solvents; *1%NaOH* 1% sodium hydroxide solubility; *Hol.* Holocellulose; *Lign* Klason lignin (%); *Hemi* hemicellulose; *Cell.* Cellulose; *Average of 2 or 3 varieties; **Average of 11 varieties

Table 3 The comparison of chemical composition ash between some lignocellulosic biomass and date palm (Reprinted from Ref. Khiari et al. (2010), with permission from Elsevier 2020)

%	*Posidonia oceanica* balls	Date palm rachis		Date palm leaves	Amaranth	Banana pseudo—stems
	This work Khiari et al. (2010)	Khristova et al. (2005)		Khristova et al. (2005)	Fiserova et al. (2006)	Cordeiro et al, (2004)
Si	17.7	2.8	1.1	5.8	0.25	2.7
Ca	9.12	21.5	n.d	n.d	4.17	7.5
Mg	3.89	3.53	n.d	n.d	0.035	4.3
Fe	3.78	240 ppm	n.d	n.d	n.d	n.d
Cu	<100 ppm	360 ppm	n.d	n.d	0.01	n.d
K	2.04	10.2	n.d	n.d	36.67	33.4
P	0.12	0.7	n.d	n.d	n.d	2.2
S	1.92	1.69	n.d	n.d	n.d	n.d
C	<0.3	1.5	n.d	n.d	n.d	n.d
Cl	0.72	18.6	n.d	n.d	n.d	n.d
Na	2.49	6.79	n.d	n.d	n.d	n.d
Absolute silicon contents in raw materials						
Si	2.13	0.14	0.06	0.56	0.03	0.38

be a target of application, if one submits the obtained fibres to further steps of purifications. Nevertheless, the lignin quantities are close to those known for non-wood as well as the typical amounts encountered in annual plants. Nevertheless, they are lower than those established for wood, i.e., close to 20%.

3 Extraction and Delignification of Fibres

Several works were focused to evaluate the best conditions of delignification process of the different parts taken from date palm. Many processes were adapted from existing technologies and tested using the mechanical and/or chemical method. Table 4, summarizes the used experimental conditions and the composition of the obtained fibres suspension, as collected from previously published studies dealing with pulping of date palm.

From this table, whatever the starting raw material, a decrease in terms of pulping yield, ethanol-toluene extractives, amount of lignin and the DP of cellulose was observed when the temperature of pulping increases. It can be clearly observed that the rachis date palm gives better pulps than those obtained from leaves. In all case, the pulping yield appears to be higher than that noticed for agricultural biomass, usually about 35% (Alcaide et al. 1990; Jimenez and Lopez 1990, Jimenez et al. 1993, 2008;

Table 4 Chemical composition and delignification condition of unbleached fibres from date palm (Reprinted from Ref. Khiari et al. (2010), with permission from Elsevier 2020)

	Date palm rachis							Date palm leaves			
	Khiari et al. (2010)			Khristova et al. (2005)		El Morsy (1980)	El Morsy (1981)	Khristova et al. (2005)		El Morsy (1981)	Ezzat (1974)
Total akali charge expressed in NaOH, %[a]	20			20	23	12	15	13	20	12	18[c]
Anthraquinone concentration	0.1			0.1		0	0	0.1		0	0
Time at constant temperature, min	120			120		120	20	120		20	300
Temperature (°C)	150	160	170	165–170		100	150	165		150	150
Cooking yield (%)[a]	46.8	44.8	41.8	44.2	42.3	78.5	58.5	39.7	29.5	22.6	38.5
Screening yield (%)[a]	94			43.1	42	n.d	n.d	39.7	28.2	n.d	n.d
Ethanol-toluene extractives (%)[b]	n.d	1.81	0.91	n.d	n.d	n.d	n.d	n.d	n.d	n.d	n.d

(continued)

Table 4 (continued)

	Date palm rachis							Date palm leaves			
	Khiari et al. (2010)			Khristova et al. (2005)		El Morsy (1980)	El Morsy (1981)	Khristova et al. (2005)		El Morsy (1981)	Ezzat (1974)
Ash (%)[b]	4			n.d	n.d	2.2	1.8	n.d	n.d	3.6	6.15
Holocellulose (%)[b]	n.d	69.3	75.2	n.d	n.d	n.d	n.d	n.d	n.d	n.d	n.d
Lignin (%)[b]	n.d	5.2	3.4	n.d	n.d	22.2	6.3	n.d	n.d	10.8	2.2
Kappa-number	59	54	47	25.5	20.7	n.d	n.d	50	20.9	n.d	n.d
Pulp viscosity (mPa.s) or (mL.g^{-1})	n.d	15.7[d]	15.3[d]	937[e]	845[e]	n.d	n.d	618[e]	780[e]	n.d	n.d
DP (pulp)	n.d	1203	1188	1403	1252	n.d	880	886	1146	510	n.d

[a]w/w with respect to oven dried raw material; [b]w/w with respect to oven dried pulps; [c]Sulfidity 25%; [d]Pulp viscosity in mPa.s; [e]Pulp viscosity in mL g^{-1}

Schott 2000; Fiserova et al. 2006). From this table, it can be noticed also that the total quantities of all the constituents mostly ranged 90%. This trend can be explain by the use of standard methods considered for wood without any modifications for the chemical composition evaluation of annual plants and non-wood, as reported by Khiari et al. (2010). Concerning, the obtained DP, it seems to be comparable to unbleached pulp which it is generally around 1300 and 1500.

Khiari et al. (2011a; b, c, 2010) and Khristova et al. (2005) applied soda-anthraquinone, as pulping method, under the following conditions: 160 °C, a total alkali charge of 20% expressed in NaOH, an anthraquinone concentration of (0.1%) and a cooking time of 120 min. Using this protocol, the average length (mm), width (μm) and the amount in fine elements of fibres obtained from rachis date palm are ($\bar{l}_A$ = 0.69 mm – $\bar{l}_W$ = 0.89 mm), 22 μm and 31%, respectively. This morphological behavior seems to be the same as that observed to common annual plants (Fiserova et al. 2006; Eugenio et al. 2006; Dutt et al. 2008; Antunes et al. 2000; Abrantes et al. 2007). However, the fibre length obtained from date palm whatever the starting constituent (leaves, rachis, trunk…) is lower than that obtained from wood and non-wood fibres.

4 Preparation and Characterization of Nanocellulose and Their Applications

Before starting to describe the processes of preparation and extraction of cellulose nanocrystals or cellulose nanofibrils, it is important to establish the terminology relating to cellulosic nanomaterials or nanocellulose. The latter term is used to describe cellulosic materials isolated from lignocellulosic materials with a dimension on the nanometer scale. In the literature, the designation has only been standardized recently, which justifies the existence of several names to describe the same types of cellulose nanoparticles. Thus, recently TAPPI (Technical Association of the Pulp and Paper Industry) proposed to standardize the terminology (standardization of terms as well as the definition for cellulose nanomaterials WI 3021). The nomenclature, the abbreviation, and the dimensions applicable to each sub-group are reported in Fig. 2.

In the last decades, nanocellulose seems to gain a large attention in nanotechnology industries. The research, development and innovation of using nanomaterial, in general, and that of cellulose nanofibres (CNF) and/or cellulose nanocrystals (CNC), especially, has involved important attention due to their interesting properties such as high surface area to volume ratio, high Young's modulus, high tensile strength and low thermal expansion coefficient (Siró and Plackett 2008; Habibi et al. 2009; Eichhorn et al. 2010). The general way of making cellulose nanocrystals is the hydrolysis treatment with acids compounds such as HCl and H_2SO_4. To prepare an aqueous suspension of cellulose nanofibres, a mechanical disintegration is necessary with and/or without a chemical treatment stage. The nanomaterials used as bio

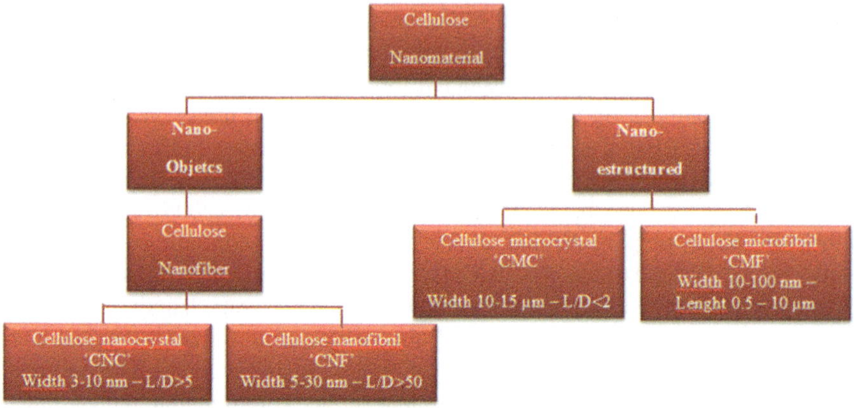

Fig. 2 Standard terms for cellulose nanomaterials (TAPPIWI3021)

and/or nanocomposites presents a high efficiency and fulfil a large demand for green nanotechnology. That is why a strong tendency to exploit the preparation of nanoscale material from different sources is observed, including for *date palm.*

4.1 CNCs: Preparation, Characterization and Nanocomposites Application

Several researches (Bendahou et al. 2009, 2010; Benhamou et al. 2015) were reported the preparation and the characterization of CNC from date palm fragments (rachis and leaves) using hydrolysis reaction (H_2SO_4 with ca. 65%). The reaction yield was estimated to 15%. As reported by Bendahou et al. (2009, 2010), this amount seems to be interesting and more profitable in the case the rachis to compare with that obtained for leaflets. However, these yields are very low comparable to annul plants such as *Posidonia oceanica* leaves and balls, which were established at 24.6 and 20.1%, respectively (Bettaieb et al. 2015a, b, c).

Figure 3 illustrates the AFM image taken as an example from Benhamou's et al. (2015) study. The dimensions of cellulose nanocrystal from date palm (rachis and leaves) were calculated thanks to Image J software, using the analysis of TEM and AFM micrographs. This evaluation concerns the analysis of 200 rod-like nanomaterials. The average length and width of CNC from date palm were about 260 nm and 6.1 nm, respectively, giving consequently to an aspect ratio of 43. Concerning, the CNC from leaves date palm, the average length and width were 180 nm and 6.1 nm, respectively, which gives an aspect ratio around 30. It can be noticed that whatever the part of date palm, the obtained aspect ratios seems to be elevate than those of CNCs prepared from others sources for example ramie (L/d = 28 (Alloin et al. 2011)), bagasse pulp (L/d = 13 (Dufresne 2012)), rice husk (L/d = 10 – 15,

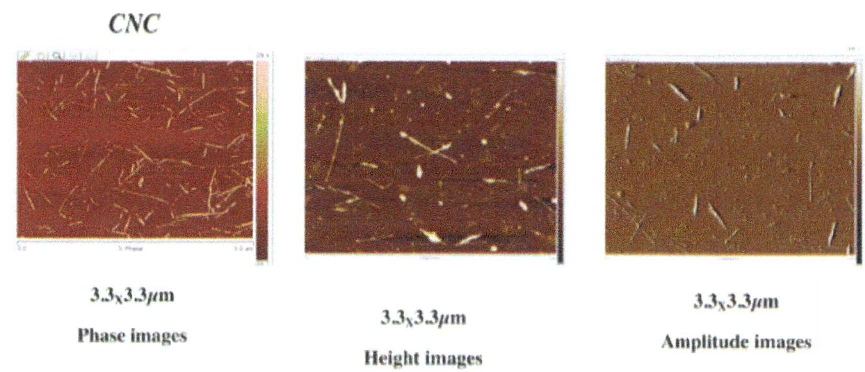

Fig. 3 AFM observation of CNC from the rachis of the date palm (Reprinted from Ref. Benhamou et al. (2015), with permission from Elsevier 2020)

(Johar et al. 2012)), kenaf bast fibres (L/d = 13, (Kargarzadeh et al. 2012)), alfa (L/d = 20, (Ben Mabrouk et al. 2009)) and eucalyptus wood pulp (L/d = 24, (Beck-Candanedo et al. 2005)). However, the aspect ratio of date palm has a similar to those reported for pea hull (L/d = 34, (Chen et al. 2009)), sisal (L/d = 43, (Siqueira et al. 2010)), wheat Straw (L/d = 45, Johar et al. 2012) and bleached softwood (L/d = 33 – 47, (Orts et al. 1998)). It can be concluded that the CNC from date palm present interesting aspect ratio properties and therefore it could be considered as a promising potential candidate as reinforcing nanoparticles in polymer matrices. Different papers were devoted to the study of the effect of use CNC from date palm (rachis and/or leaves) in nanocomposites. Bendahou et al. (2009) prepared various nanocomposites films from CNC obtained from rachis and/or leaves and natural rubber prepared by casting and evaporation method. The efficiency of the reinforcing effect was studied by successive tensile tests and DMA.

Bendahou et al. (2010) was reported that the mechanical properties of the nanocomposite films reinforced by CNC in terms of the elongation at break decreases with increasing the nanocrystal content. The tensile modulus and the strength increased considerably with the quantity of added cellulose nanocrystals. Thus, the modulus increases from 0.5 MPa for the neat matrix up to 187 MPa for the nanocomposite film reinforced with 15 wt% nanofibres. Concerning the glass transition temperature of the natural rubber-based composites, it remains approximately constant but tends to marginally increase with the added amounts of CNCs. A high reinforcing effect was evidenced from both DMA and tensile tests, in the region above the Tg of the matrix (see also Table 5 and Fig. 4).

However, the reinforcing effect is lower than that predicted from the percolation formalism, suggesting strong filler matrix interactions. On the other hand, high strain mechanical experiments show a radical change of the mechanical behavior with the cellulose nanocrystals content. In fact, a ductile behavior for lower nanofiller content and a quasi-brittle behavior for composites reinforced with high nanofiller content were observed. These findings demonstrated clearly a higher efficiency in terms of

Table 5 Mechanical properties of prepared nanocomposites reinforced with cellulose nanocrystals (Reprinted from Ref. Bendahou et al. (2010), with permission from Elsevier 2020)

Sample	E (MPa)	σ_R (MPa)	ε_R (%)
NR	0.5 ± 0.15	0.56 ± 0.12	575 ± 35
NR-CNC 1%	1.7 ± 0.5	0.86 ± 0.06	408 ± 49
NR-CNC 2.5%	2.8 ± 0.4	1.17 ± 0.24	358 ± 22
NR-CNC 5%	8.4 ± 1.1	2.71 ± 0.10	231 ± 53
NR-CNC 10%	118 ± 6	8.93 ± 1.23	16 ± 3
NR-CNC 15%	187 ± 0.5	12.15 ± 1.48	14 ± 1
NR-CNF 1%	1.27 ± 0	0.70 ± 0.13	209 ± 29
NR-CNF 2.5%	10.52 ± 0.66	0.80 ± 0.26	14.6 ± 4.5
NR-CNF 5%	35.46 ± 5.79	2.17 ± 0.38	13.8 ± 2.3
NR-CNF 7.5%	121.2 ± 8.8	4.15 ± 0.71	8.18 ± 1.65
NR-CNF 10%	172 ± 62	5.99 ± 2.56	6.65 ± 2.08
NR-CNF 15%	233 ± 57	6.26 ± 2.70	3.95 ± 1.14

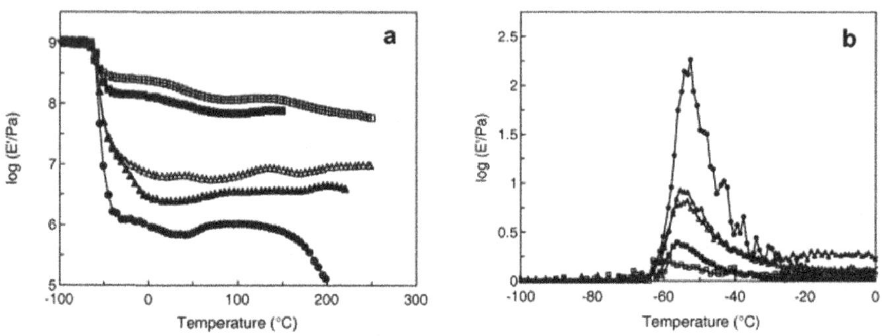

Fig. 4 Dynamic analysis mechanical (DMA) for nanocomposites based on CNC (Reprinted from Ref. Bendahou et al. (2010), with permission from Elsevier 2020)

reinforcing can be observed and it already attribute to the effect of CNC percolation and their high aspect ratio. As mentioned above, the aspect ratio of the isolated CNCs from the date palm rachis and leaves is 43 and 30, respectively.

4.2 CNFs: Preparation, Characterization and Nanocomposites Application

Turbak et al. (1983) have used a homogenizer and reported for the first time the cellulose nanofibres. This pioneering nanomaterial involved properties and important characteristics which makes this research in the top level. As described by Rol et al. (2019), a large attention was focused on production and characterization of cellulose nanofibrils (CNF). Since 2008, an exponential in scientific papers number devote the preparation of cellulose nanofibrils was observed. This nanomaterial is of high importance due to much higher yield of the nanoparticles produced. That is why a large number of books, reviews and article appeared recently. They are mostly devoted to the promising properties of such a nano-material (Klemm et al. 2011; Dufresne 2012; Nechyporchuk et al. 2014; Rol et al. 2017, 2018, 2019). In fact, cellulose nanofibrils were investigated and their efficiency in many research areas were pointed out. These include (i) reinforcing element to produce nanofilms or nanocomposites (Abdul Khalil et al. 2014; Bettaieb et al. 2015a, b, c; Belgacem and Gandini 2008; Iwamoto et al. 2005, 2008), (ii) retention and adsorbent nano-material for waste treatment, as illustrated by the work reported on the oil and/or dye adsorption, (iii) the squeezing of heavy metal, such as the removal of silver ions (Ag^{+}) from contaminated water, (iv) removing of radioactive species from water ($UO2^{2+}$), and organic matter etc. (Ma et al. 2011; Pei et al. 2013; Abouzeid et al. 2018; Liu et al. 2014; Sehaqui et al. 2014; Gebald et al. 2011). The CNFs were also used in the production of nanopapers for food packaging (Kang et al. 2017), printed electronics and conductive composites using 3D printing (Agate et al. 2018; Hoeng et al. 2016; De France et al. 2017; Sehaqui et al. 2011).

Recently, the CNFs were used in medicine applications (Carlsson et al. 2014; Gopi et al. 2018; Jorfi and Foster 2015). In order to prepare the CNF, different mechanical devices are available, such as a high-pressure homogeniser (Bettaieb et al. 2015a, b, c; Alila et al. 2013; Andresen et al. 2006; Andresen and Stenius 2007; Djafari et al. 2014; Erkisen et al. 2008; Nakagaito and Yano 2004; Rezayati et al. 2013; Stenstad et al. 2008; Winuprasith and Suphantharika 2013; Zhang et al. 2012; Syverud and Stenius 2009), micro-fluidiser (Bendahou et al. 2010), ultrafine friction grinder (Subramanian et al. 2008; Nechyporchuk et al. 2014; Iwamoto et al. 2005, 2008; Hassan et al. 2012; Abe et al. 2007, Abe and Yong 2009), cryo-crushing (Janardhnan and Sain 2006; Chakraborty et al. 2005), or ultrasonic methods (Zhao et al. 2007), extruded (Rol et al. 2019),…

The first CNF from date palm was prepared using the Micro Fluidizer. Then different other devices were reported to efficient in such a context. Figure 5 shows the morphological features of the CNF obtained after two treatments. The first one was chemical (TEMPO-mediated oxidation), whereas the second one consisted on mechanical defibrillation (a high intensity homogenizer). The TEM and AFM images demonstrated clearly that nanoscale fibrils can be prepared from date palm rachis and their diameter varies from 5 to 29 nm. However, it is impossible to calculate the aspect ratio due the large length distribution. This result was confirmed by using optical microscopy images which can be prove the presence of some residual microscopic fibres, as some big size fragments were detected.

Numerous reports were also devoted to the addition of CNC or CNF nanoparticles from date palm, in order to enhance the physical performances. Table 5 illustrates an example the mechanical properties of the nanocomposites film reinforced by CNF using natural-rubber as matrix. As reported by Bedanhou et al. (2010), the elongation at break decreased and the tensile modulus and the strength increased significantly when adding the nanofillers content. Moreover, the DMA results confirmed that high mechanical performances can be reached. Figure 6 shows a presentation of the

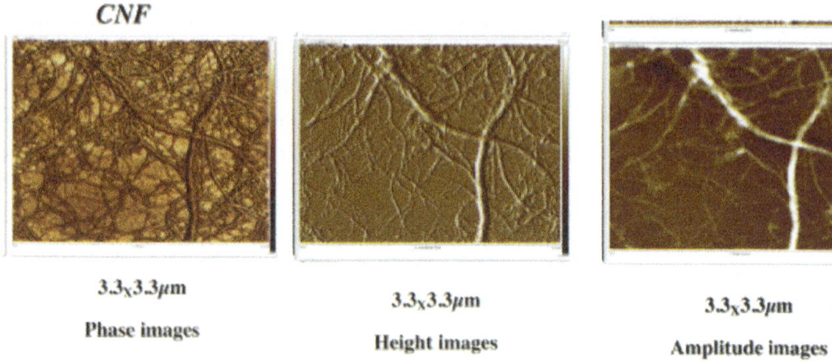

Fig. 5 AFM images of date palm CNF (Reprinted from Ref. Benhamou et al. (2015), with permission from Elsevier 2020)

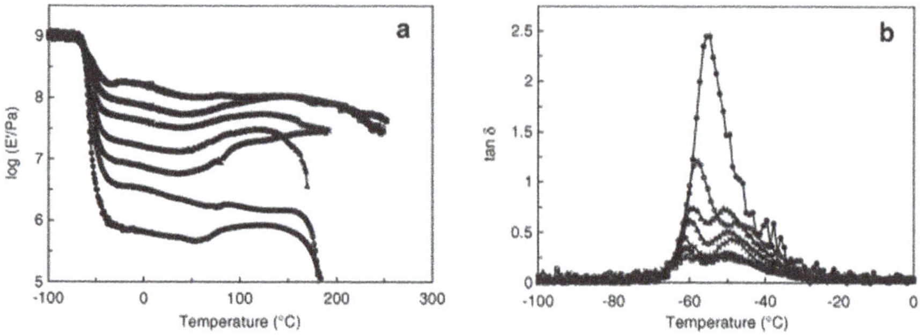

Fig. 6 Dynamic analysis mechanical (DMA) for nanocomposites based on CNFs (Reprinted from Ref. Bendahou et al. (2010), with permission from Elsevier 2020)

storage modulus E' evolution for various CNF amounts, as a function of temperature. Obviously, adding the cellulose nanofibrils into neat matrix (chosen for example natural rubber as matrix) has different effects depending on the fixed temperature. Before the Tg temperature, there is no significant evolution in the storage modulus upon the adding of cellulose nanofibrils from date palm. However, it can be noticed that for temperatures higher than the Tg of the polymer, the storage modulus is more sensitive to the presence of CNF. Thus, the modulus increased significantly with CNF addition at the rubbery plateau. This phenomenon was already observed for other systems, including well-known cellulose-reinforced materials (Dufresne 2010). This results on limiting the polymer chains movement beyond the Tg. This it can be observed a higher strong modulus reinforcing from Tg temperature which a rigid entanglement can be making occur through hydrogen interaction.

Figure 6 presents the dynamic analysis mechanical (DMA) results illustrate by the evolution of the storage modulus (E') in a rubbery state (estimated at −60 °C) with different CNF additions. The obtained results illustrate clearly a significant increase in the modulus with adding of CNFs, which is consistent with their strong reinforcing effect. At 25 °C, as illustration, the nanocomposite filled with 5 and 10 wt.% CNF displays a storage modulus of 2.48 and 106 MPa, respectively. This it can be represent a 5 and 210-fold enlargement over the matrix (0.507 MPa). In addition, it is important to mention that when the nanocomposite filled with a less 5% CNF, the storage modulus is closely comparable to that of nanocomposite materials prepared from wood-based CNF and/or non-wood base CNF. With high nanofibril cellulose adding, high modulus storage can be reached. It can be nevertheless concluded that CNFs prepared from date palm demonstrate an excellent reinforcing potential when added into an elastic polymer.

5 Conclusions

Phoenix Dactylifera L. is abundant sources of cellulosic fibres which are cultivated worldwide, but they are mostly concentrated in the Middle East and North Africa with an estimated population of 85 million palms. The valorization of their large quantities wastes constitutes an important role in terms of environmental and economical issues, especially in the country where the sources of cellulosic fibres are limited. This biomass contains high percentage of cellulose i.e. more than 45% which justifies the huge amounts of studies focused on the production of fibres, cellulose derivatives, composites etc. The present chapter is devoted to the description of the potential use of lignocellulosic materials from date palm, in order to produce fibres then nanocellulose filaments, whatever the starting part (leaves or rachis). Nanocellulose namely nanocrystal cellulose "CNC" and nanofibril cellulose "CNF" constitutes new high value-added products. The nanocellulose from date palm presents an interesting aspect ratio when compared to common wood and non-wood plants. The interesting properties encourage the researchers to investigate their efficiency as reinforcement in nanocomposites. Finally, the utilization of CNC and CNF from date palm constitutes a new challenge to produce high added value material and it opens the door for their use in new application such as nanopapers destined to food packing, functional and advanced materials, etc.

References

Abdul Khalil HPS, Davoudpour Y, Islam MN, Mustapha A, Sudesh K, Dungani R, Jawaid M (2014) Production and modification of nanofibrillated cellulose using various mechanical processes: A review. Carbohyd Polym. https://doi.org/10.1016/j.carbpol.2013.08.069

Abe K, Yano H (2009) Comparison of the characteristics of cellulose microfibril aggregates of wood rice straw and potato tuber. Cellulose 16(6):1017–1023

Abe K, Iwamoto S, Yano H (2007) Obtaining cellulose nanofibers with a uniform width of 15 nm from wood. Biomacromol 8(10):3276–3278

Abouzeid RE, Hassan EA, Bettaieb F, Khiari R, Hassan ML (2015) Use of cellulose and oxidized cellulose nanocrystals from olive stones in chitosan bionanocomposites. J Nanomat 2015. https://doi.org/10.1155/2015/687490

Abouzeid RE, Khiari R, Beneventi D, Dufresne A (2018) Biomimetic mineralization of three-dimensional printed alginate/TEMPO-oxidized cellulose nanofibril scaffolds for bone tissue engineering. Biomacromol 19(11):4442–4452

Abrantes S, Amaral ME, Costa AP, Duarte AP (2007) Cynara cardunculus L. alkaline pulps: alternative fibres for paper and paperboard production. Biores Technol 98:2873–2878

Agate S, Joyce M, Lucian L, Pal L (2018) Cellulose and nanocellulose-based flexible-hybrid printed electronics and conductive composites—a review. Carbohyd Polym 198:249–260

Alcaide LJ, Parra IS, Baldovin F (1990) Characterization of Spanish agricultural residues with a view to obtaining cellulose pulp. Tappi J 73(5):173–176

Alila S, Besbes I, Vilar MR, Mutjé P, Boufi S (2013) Non-woody plants as raw materials for production of microfibrillated cellulose (NFC): a comparative study. Ind Crops Prod 41(1):250–259

Alloin F, D'Aprea A, Dufresne A, El Kissi N, Bossard F (2011) Poly(oxyethylene) and ramie whiskers based nanocomposites: influence of processing: extrusion and casting/evaporation. Cellulose 18:957–973

Alotaibi MD, Alshammari BA, Saba N, Alothman OY, Sanjay MR, Almutairi Z, Jawaid M (2019) Characterization of natural fibre obtained from different parts of date palm tree (Phoenix dactylifera L.). Int J Biol Macromol 135:69–76

Andresen M, Stenius P (2007) Water-in-oil emulsions stabilized by hydrophobized nanofibrillar cellulose. J Dispersion Sci Technol 28:837–844

Andresen M, Johansson LS, Tanem BS, Stenius P (2006) Properties and characterization of hydrophobized nanofibrillar cellulose. Cellulose 13:665–677

Antunes A, Amaral E, Belgacem MN (2000) Cynara cardunculus L.: chemical composition and soda-anthraquinone cooking. Ind Crops Prod 12:85–91

Beck-Candanedo S, Roman M, Gray DG (2005) Effect of reaction conditions on the properties and behavior of wood cellulose nanocrystal suspensions. Biomacromol 6(2):1048–1054

Belgacem MN, Gandini A (2008) Monomers polymers and composites from renewable resources. Elsevier. ISBN: 978-0-08-045316-3

Ben Mabrouk A, Thielemans W, Dufresne A, Boufi S (2009) Preparation of poly(styrene-co-hexylacrylate)/cellulose whiskers nanocomposites via miniemulsion polymerization. J Appl Polym Sci 114(5):2946–2955

Bendahou A (2009) Nouveaux matériaux nanocomposites à base de monocristaux de cellulose et de polymère: relation structure – propriétés, thèse. Institut Polytechnique de Grenoble, France

Bendahou A, Kaddami H, Sautereau H, Raihane M, Erchiqui F, Dufresne A (2008) Short palm tree fibres polyolefin composites: effect of filler content and coupling agent on physical properties. Macromol Mater Eng 293(2):140–148

Bendahou A, Habibi Y, Kaddami H, Dufresne A (2009) Physico-chemical characterization of palm from Phoenix Dactylifera – L, preparation of cellulose whiskers and natural rubber–based nanocomposites. Journal of Biobased Materials and Bioenergy 3(1):81–90

Bendahou A, Kaddami H, Dufresne A (2010) Investigation on the effect ofcellulosic nanoparticles morphology on the properties of natural rubber based nanocomposites. Eur Polymer J 46:609–620

Benhamou K, Kaddami H, Magnin A, Dufresne A, Ahmad A (2015) Bio-based polyurethane reinforced with cellulose nanofibres: a comprehensive investigation on the effect of interface. Carbohyd Polym 122:202–211

Bettaieb F, Khiari R, Hassan ML, Belgacem MN, Bras J, Dufresne A, Mhenni MF (2015a) Preparation and characterization of new cellulose nanocrystals from marine biomass Posidonia oceanica. Ind Crop Prod 72:175–182

Bettaieb F, Khiari R, Dufresne A, Mhenni MF, Belgacem MN (2015b) Mechanical and thermal properties of *Posidonia oceanica* cellulose nanocrystal reinforced polymer. Carbohyd Polym 123:99–104

Bettaieb F, Khiari R, Dufresne A, Mhenni MF, Putaux JL, Boufi S (2015c) Nanofibrillar cellulose from *Posidonia oceanica*: properties and morphological features. Ind Crop Prod 72:97–106
Botdrof J, Soap B (1962) CMC in liquid detergents. Chem Spec 38:55–58
Carlsson DO, Hua K, Forsgren J, Mihranyan A (2014) Int J Pharm 461:74–81
Chakraborty A, Sain M, Kortschot M (2005) Cellulose microfibrals: a novel method of preparation using high shear refining and cryocrushing. Holzforschung 59:102–107
Chen Y, Liu C, Chang P, Cao X, Anderson D (2009) Bionanocomposites based on pea starch and cellulose nanowhiskers hydrolyzed from pea hull fibre: Effect of hydrolysis time. Carbohyd Polym 4:607–615
Cordeiro N, Belgacem MN, Torres IC, Mourad JCVP (2004) Chemical composition and pulping of banana pseudo-stems. Ind Crops Prod 19:147–154
De France KJ, Hoare T, Cranston ED (2017) Review of hydrogels and aerogels containing nanocellulose. Chem Mater 29:4609–4631
Djafari Petroudy SR, Syverud K, Chinga-Carrasco G, Ghasemain A, Resalati H (2014) Effects of bagasse microfibrillated cellulose and cationic polyacrylamide on key properties of bagasse paper. Carbohydrate Polymer 99:311–318
Dufresne A (2010) Processing of polymer nanocomposites reinforced with polysaccharide nanocrystals. Mol J 15:4111–4128
Dufresne A (2012) Nanocellulose from nature to high performance tailored materials. Walter de Gruyter.ISBN: 978-3110254563
Dutt D, Upadhyaya JS, Malik RS, Tyagi CH (2005) Studies on the pulp and papermaking characteristics of some Indian non-woody fibrous raw materials. Cellul Chem Technol 39(1–2):115–128
Dutt D, Upadhyaya JS, Tyagi CH, Kumar A, Lal M (2008) Studies on Ipomeacarnea and Cannabis sativa as an alternative pulp blend for softwood: an optimization of kraft delignification process. Ind Crops Prod 28:128–136
Eichhorn SJ, Dufresne A, Aranguren M, Marcovich NE, Capadona JR, Rowan SJ, Weder C, Thielemans W, Roman M, Renneckar S, Gindl W, Veigel S, Keckes J, Yano H, Abe K, Nogi M, Nakagaito AN, Mangalam A, Simonsen J, Benight AS, Bismarck A, Berglund LA, Peijs T (2010) Review: current international research into cellulose nanofibres and nanocomposites. J Mater Sci 45(1):1–33
El Morsy MMS (1980) Studies on the rachis of Egyptian date palm leaves for hardboard production. Fibre Sci Technol 13(4):317–321
El Morsy MMS, Riad BY, Mohamed MAS (1981) Pulp and paper from the Egyptian date palm leaves. Fibre Sci Technol 14(2):157–161
Elseify LA, Midani M, Shihata LA (2019) Review on cellulosic fibres extracted from date palms (*Phoenix Dactylifera L.*) and their applications. Cellulose 6. https://doi.org/10.1007/s10570-019-02259-6
Erkisen O, Syverud K, Gregersen O (2008) The use of microfibrillated cellulose produced from kraft pulp as strength enhancer in TMP paper. Nordic Pulp Paper Res J 23:299–304
Eugenio ME, Alaejos J, Diaz MJ, Lopez F, Vidal T (2006) Evaluation of Holm oak (Quercus Ilex) wood as alternative source for cellulose pulp. Cellul Chem Technol 40(1–2):53–61
Ezzat S (1974) Leaves of date palm tree (Phoenix dactylifera) as a technical feasible source of raw material for paper production. Cellul Chem Technol 8:627–634
Ferhi F, Das S, Elaloui E, Moussaoui Y, Yanez JG (2014a) Chemical characterisation and suitability for papermaking applications studied on four species naturally growing in Tunisia. Ind Crops Prod 61:180–185
Ferhi F, Das S, Elaloui E, Moussaoui Y, Yanez JG (2014b) Chemical characterisation and suitability for papermaking applications studied on four species naturally growing in Tunisia. Ind Crop Prod 61:180–185
Fiserova M, Gigac J, Majtnerova A, Szeiffova G (2006) Evaluation of annual plants (Amaranthus caudatus L., Atriplex hortensis L., Helianthus tuberosus L.) for pulp production. Cellul Chem Technol 40(6):405–412

Gebald C, Wurzbacher JA, Tingaut P, Zimmermann T, Steinfeld A (2011) Amine-based nanofibrillated cellulose as adsorbent for CO2 capture from air. Environ Sci Technol 45:9101–9108
Gominho J, Fernandez J, Pereira H (2001) Cynara cardunculus L.: a new fibre crop for pulp and paper production. Ind Crops Prod 13:1–10
Gopi S, Balakrishnan P, Geethamma VG, Pius A, Thomas S (2018) Applications of cellulose nanocomposite materials in drug delivery, pp 75–95
Habibi Y, Mahrouz M, Vignon MR (2009) Nanofibrillar cellulose from the peel of prickly pear fruits. Food Chem 115:423–429
Hakeem KR, Jawaid M, Rashid U (2014) Biomass and bioenergy processing and properties. Springer. https://doi.org/10.1007/978-3-319-07641-6
Hassan ML, Mathew AP, Hassan EA, El-Wakil NA, Oksman K (2012) Nanofibers from bagasse and rice straw process optimization and properties. Wood Sci Technol 46:193–205
Hoeng F, Denneulin A, Bras J (2016) Use of nanocellulose in printed electronics: a review. Nanoscale 8(27):13131–13154
Hussain F, Hojjati M, Okamoto M, Gorga RE (2006) Review article: polymer–matrix nanocomposites, processing, manufacturing, and application: an overview. J Compos Mater 40:1511–1575
Ibrahim MM, El-Zawawy WK, Nassar MA (2010) Synthesis and characterization of polyvinyl alcohol/nanospherical cellulose particle films. Carbohyd Polym 79:694–699
Iwamoto S, Nakagaito AN, Yano H, Nogi M (2005) Optically transparent composites reinforced with plant fibers-based nanofibers. Appl Phys A Mater 81:1109–1112
Iwamoto S, Abe K, Yano H (2008) The effect of hemicelluloses on wood pulp nanofibrillation and nanofiber network characteristics. Biomacromol 9:1022–1026
Janardhnan S, Sain M (2006) Isolation of cellulose microfibrils–an enzymatic approach. BioResources 1:176–188
Jiménez L, Lopez F, Martınez C (1993) Paper from sorghum stalks. Holzforschung 47:529–533
Jimenez L, Lopez F (1990) Characterization of Spanish agricultural residues with a view to obtaining cellulose pulp. Tappi J 73(8):173–176
Jimenez L, Rodriguez A, Perez A, Moral A, Serrano L (2008) Alternative raw materials and pulping process using clean technologies. Ind Crops Prod 28(1):11–16
Johar N, Ahmad I, Dufresne A (2012) Extraction, preparation and characterization of cellulose fibres and nanocrystals from rice husk. Ind Crops Prod 37(1):93–99
Jonoobi M, Oladi R, Davoudpour Y, Oksman K, Dufresne A, Hamzeh Y, Davoodi R (2015) Different preparation methods and properties of nanostructured cellulose from various natural resources and residues: a review. Cellulose 22:935–969
Jorfi M, Foster EJ (2015) Recent advances in nanocellulose for biomedical applications. J Appl Polym Sci 132(14):41719 (1–19)
Kang X, Sun P, Kuga S, Wang C, Zhao Y, Wu M, Huang Y (2017) Thin cellulose nanofibre from corncob cellulose and its performance in transparent nanopaper. ACS Sustain Chem Eng 5:2529–2534
Kargarzadeh H, Ahmad I, Abdullah SY, Dufresne A (2012) Effect of hydrolysis conditions on the morphology, crystallinity and thermal stability of cellulose nanocrystals extracted from kenaf bast fibres. Cellulose 19(3):855–866
Khiari R (2017) Valorization of agricultural residues for cellulose nanofibrils production and their use in nanocomposite manufacturing. Int J Polym Sci. https://doi.org/10.1155/2017/6361245
Khiari R, Belgacem MN (2017) Potential for using multiscale posidonia oceanica waste: current status and prospects in material science. https://doi.org/10.1016/B978-0-08-100959-8.00021-4
Khiari R, Mhenni MF, Belgacem MN, Mauret E (2010) Chemical composition and pulping of date palm rachis and Posidonia oceanica—a comparison with other wood and non-wood fibre sources. Biores Technol 101:775–780
Khiari R, Mhenni F, Belgacem N, Mauret E (2011a) Valorisation of vegetal wastes as a source of cellulose and cellulose derivatives. J Polym Environ 19:80–89

Khiari R, Marrakchi Z, Belgacem MN, Mauret E, Mhenni MF (2011b) New lignocellulosic fibres-reinforced composite materials: a step forward in the valorisation of the Posidonia oceanica balls. Compos Sci Technol 71:1867–1872

Khiari R, Mauret E, Belgacem MN, Mhenni MF (2011c) Tunisian date palm rachis used as an alternative source of fibres for papermaking applications. BioResources 6:265–281

Khristova P, Kordsachia O, Khider T (2005) Alkaline pulping with additives of date palm rachis and leaves form Sudan. Biores Technol 96:79–85

Klemm D, Kramer F, Moritz S, Lindström T, Ankerfors M, Gray D, Dorris A (2011) Nanocelluloses: A New Family Of Nature-Based Materials. Angew Chem Ed 50:5438–5466

Kriker A, Debicki G, Bali A, Khenfer MM, Chabannet M (2005) Mechanical properties of date palm fibres and concrete reinforced with date palm fibres in hot-dry climate. Cem Concr Compos 27:554–564. https://doi.org/10.1016/j.cemconcomp.4192004.09.015

Kriker A, Bali A, Debicki G, Bouziane M, Chabannet M (2008) Durability of date palm fibres and their use as reinforcement in hot dry climates. Cem Concr Compos 30(639–648):416. https://doi.org/10.1016/j.cemconcomp.2007.11.006

Liu P, Sehaqui H, Tingaut P, Wichser A, Oksman K, Mathew AP (2014) Cellulose and Chitin nanomaterials for capturing silver ions (Ag+) from water via surface adsorption. Cellulose 21:449–461

Ma H, Hsiao BS, Chu B (2011) Ultrafine Cellulose Nanofibres as Efficient Adsorbents for Removal of UO22+ in water. ACS Macro Lett 1:213–216

Mansouri S, Khiari R, Bettaieb F, El-Gendy AA, Mhenni F (2015) Synthesis and characterization of carboxymethyl cellulose from tunisian vine stem: study of water absorption and retention capacities. J Polym Environ 23(2):190–198

Midani M, Elseify LA, Hassanin A, Hamouda T (2018) Optimizing the extraction process of a novel long fibrillated fibres from the midribs of date palm (Phoenix Dactylifera L.). In: 1st world conference on by-products of palm trees and their application

Naili H, Jelidi A, Limam O, Khiari R (2017) Extraction process optimization of Juncus plant fibres for its use in a green composite. Ind Crops Prod 107:172–183

Nakagaito AN, Yano H (2004) The effect of morphological changes from pulp fibre towards nano-scale fibrillated cellulose on the mechanical properties of high-strength plant fibre based composites. Appl Phys A Mater Sci Process 78:547–552

Nechyporchuk O, Belgacem MN, Pignon F (2014) Rheological properties of micro-/nanofibrillated cellulose suspensions: wall-slip and shear banding phenomena. Carbohydr Polym 112:432–439

Olaru N, Olaru L, Stoleriu A, Timpu D (1998) Carboxymethylcellulose synthesis in organic media containing ethanol and/or acetone. J Appl Polym Sci 67:481–486

Orts T, Godbout L, Marchessault RH, Revol JF (1998) Enhanced ordering of liquid crystalline suspensions of cellulose microfibrils: a small angle neutron scattering study. Macromolecules 31:5717–5727

Pei A, Butchosa N, Berglund LA, Zhou Q (2013) Surface quaternized cellulose nanofibrils with high water absorbency and adsorption capacity for anionic dyes†. Soft Matter 9:2047–2055

Rezayati Charani P, Dehghani-Firouzabadi M, Afra E, Blademo Å, Naderi A, Lindström T (2013) Production of microfibrillated cellulose from unbleached kraft pulp of kenaf and Scotch pine and its effect on the properties of hardwood kraft microfibrillated cellulose paper. Cellulose 20:2559–2567

Rol F, Karakashov B, Nechyporchuk O, Terrien M, Meyer V, Dufresne A, Belgacem MN, Bras J (2017) Pilot-scale twin screw extrusion and chemical pretreatment as an energy-efficient method for the production of nanofibrillated cellulose at high solid content. ACS Sustain Chem Eng 5:6524–6531

Rol F, Banvillet G, Meyer V, Petit-conil M, Bras J (2018) Combinaison of Twin-screw extruder and homogenizer to produce high quality nanofibrillated cellulose with low energy consumption. J Mater Sci 53(17):12604–12615

Rol F, Belgacem MN, Gandini A, Bras J (2019) Recent advances in surface-modified cellulose nanofibrils. Prog Polym Sci 88:241–264

Schott S (2000) Valorisation de la pâte de paille de blé obtenue par le procédé d'explosion en phase vapeur, pour une application en papeterie, thèse. Institut Polytechnique de Grenoble, France

Sehaqui H, De larraya UP, Liu P, Pfenninger N, Mathew AP, Zimmermann T, Tingaut P (2014) Enhancing adsorption of heavy metal ions onto biobased nanofibres from waste pulp residues for application in waste water treatment. Cellulose 21(4):2831–2844

Sehaqui H, Zhou Q, Berglund LA (2011) High-porosity aerogels of high specific surface area prepared from nanofibrillated cellulose (NFC). Compos Sci Technol 71:1593–1599

Siqueira G, Tapin-Lingua S, Bras J, Da Silva Perez D, Dufresne A (2010) Morphological investigation of nano-particles obtained from combined mechanical shearing and enzymatic and acid hydrolysis of sisal fibres. Cellulose 17(6):1147–1158

Siró I, Plackett D (2008) Characterization of nanofibrillar cellulose (NFC) films made of different types of raw material. In: Nordic Polymer Days, 11–13 June, Stockholm, Sweden.

Siró I, Plackett D (2010) Microfibrillated cellulose and new nanocomposite materials: a review. Cellulose 17:459–494

Stenstad P, Andresen M, Tanem BS, Stenius P (2008) Chemical surface modifications of nanofibrillar cellulose. Cellulose 15:35–45

Subramanian R, Knononov A, Kang T, Paltakari J, Paulapura H (2008) Nat Cellul Fibrils Bioresour 3:192–203

Syverud K, Stenius P (2009) Strench and barrier properties of NFC films. Cellulose 16:75–85

Taha I, Steuernagel L, Ziegmann G (2007) Optimization of the alkali treatment process of date palm. Compos Interfaces 14:669–684

Turbak AF, Snyder FW, Sandberg KR (1983) Nanofibrillar cellulose a new cellulose product: properties, uses, and commercial potential. J Appl Polym Sci 37:815–827

Winuprasith T, Suphantharika M (2013) Microfibrillated cellulose from mangosteen (Garcinia Mangostana L,) rind: preparation characterization and evaluation as an emulsion stabilizer. Food Hydrocolloids 32:383–394

Zhang J, Song H, Lin L, Zhuang J, Pang C, Liu S (2012) Microfibrillated cellulose from bamboo pulp and its properties. Biomass Bioenergy 39:78–83

Zhao HP, Feng XQ, Gao H (2007) Ultrasonic technique for extracting nanofibres from nature materials. Appl Phys Lett 90:073–112

Design and Modeling of Date Palm Fiber Composites

Shafaat Ahmed Salahudeen

Abstract In this chapter, recent trends in design and modeling of various numerical methods used in Date Palm fiber composites (DPFC) are discussed. The microstructures of DPFC are complex and quality of fiber is affected by weather conditions, crop variety and climate. It is a challenging task to analyze how the complicated microstructures of DPFC affect the macroscopic behaviors, and then predict the macroscopic properties of the targeted materials. Compared to experimental methods, numerical approximation method is predictable and more convenient to characterize the properties in three dimensional networks. Among the various approximate methods, Finite element model (FEM) is an engineer's choice due to any structure with complex shape, material and boundary conditions could be easily solved and represent the results in graphical user interface environment. Mesh convergence techniques are discussed in details. FEM is used to predict the thermal conductivity, mechanical properties and water absorption or diffusion coefficient of DPFC. Through Representative volume element (RVE) models, FEM could predict the relationships between micro-structures and macro-behaviors of DPFC and helps to select the proper DPFC for suitable applications.

1 Introduction

The term "fiber reinforced composite" is defined as a combination of two or more synergic constituents which differs in physical form or chemical composition. Fiber reinforced composite has three phases, matrix, reinforcement constituent as fiber and interface as shown in Fig. 1.

The matrix such as polymer, concrete or metal is the adhesive medium for the fibers. The interface is the contact space region between fiber and matrix, where it facilitates stress, heat and mass transfer from the matrix to the fiber.

S. A. Salahudeen (✉)
Department of Polymer Engineering, B.S.Abdur Rahman Crescent Institute of Science and Technology, Chennai, Tamilnadu, India
e-mail: shafaatht@gmail.com

M. Midani et al. (eds.), *Date Palm Fiber Composites*, Composites Science and Technology,
https://doi.org/10.1007/978-981-15-9339-0_7

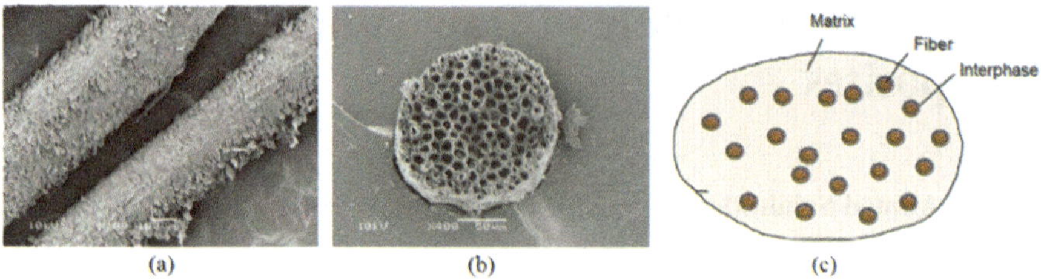

Fig. 1 Scanning electron microscope (SEM) image of **a** DPF and **b** sectional view (Hakeem et al. 2014); **c** components of DPFC

Based on reinforcement fiber, composites are classified as natural fiber and synthetic fiber composite. Glass fibers and carbon fibers composites are examples of synthetic fiber used in wind turbine, automobile and aircraft application. Coir, hemp, jute and date palm fiber composites are examples of natural fiber.

Based on the length, natural fibers are classified as short and long fibers. Long fibers are used as reinforcement in matrix to improve the strength and stiffness for load bearing applications. Whereas, matrices reinforced with short fibers are used primarily in non-load-bearing applications. Schematic representations of short and long fiber are shown in Fig. 2.

Among the many natural short fibers based composites, Date Palm Fiber (DPF) is gaining much popularity in many industrial applications due to its availability, biodegradability and low material cost. Most reported Industrial applications of DPFC are automotive parts and building material (Alawar et al. 2009; Hakeem et al. 2014; Noorunnisa Khanam and AlMaadeed 2014; Boukhattem et al. 2017; Ghori et al. 2018; Alshammari et al. 2019; Gheith et al. 2019). Mokhtari introduced the concept of DPF as filler for mortar (Mokhtari et al. 2015) and found that replacing sand with DPF in mortar increased the overall flexural strength of the material.

The properties such as water absorption, thermal conductivity and mechanical properties are predominant factors to determine the adaptability of the product. As DPFC found many industrial applications, there is a need to improve the compatibility of the composites at complex geometries. DPFC properties can be determined approximately in complex geometries by many methods such as finite difference method, Galerkin's method, RayleigheRitz method, finite element method and finite volume method.

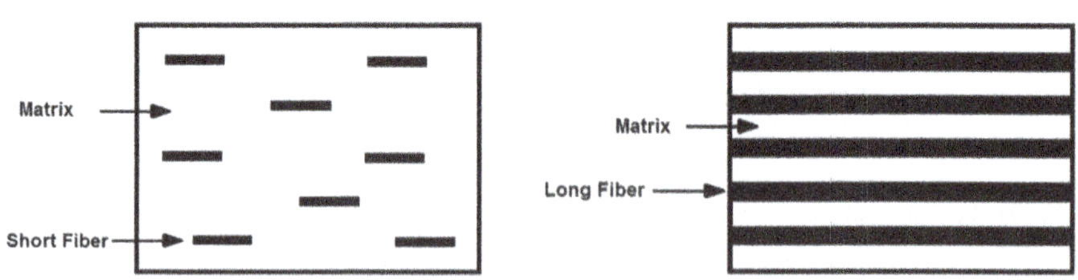

Fig. 2 Schematic Representation of short and long fiber composite

Due to recent advancement in computation power of modern computer, Finite element method (FEM) is the engineer's choice. FEM is a user interface through graphical environment where, user can access 1D, 2D, and 3D portfolio of products with the real time virtual experience. Users can perform many iterations using FEM to optimize design, decrease down time, decrease prototype product cost and simulate the extreme environments. Today, FEM technology is able to analyze the failure of layered fiber composite structures such as tensile fiber fracture, layer delamination and compressive fiber breakage with layer delamination.

2 How FEM Works

In general, engineering problems can be solved based on the known parameters. For complex problems, many parameters are unknown for examples, flow parameter around blades of turbo jets, magnetic field around moving coils, temperature profile between the composites, water absorption kinetics around the fibers and so on. The infinite unknown parameters can be predicted by divide the domain into many finite small elements. Each finite small element transfers the field variables between the neighboring elements through specified points called as nodes. At any point of time and frame, field variable can be calculated using interpolation functions.

3 How to Perform FEM

FEM is the three step serial hierarchy, Pre-processer, solver and post-processor as shown in Fig. 3.

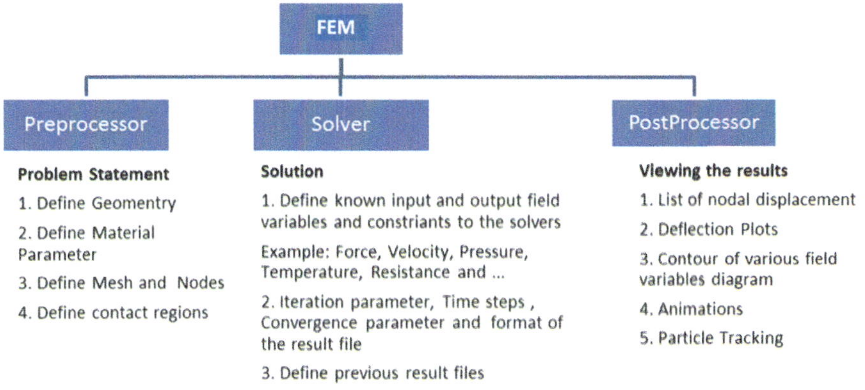

Fig. 3 Overview of Finite Element Method

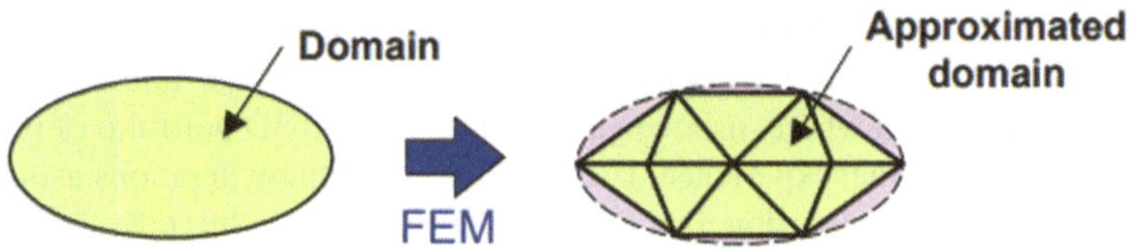

Fig. 4 Representation of Domain in FEM framework

3.1 *Preprocessing*

Preprocessing is the initial startup process to prepare the FEM to execute the solving of the engineering problem. At preprocessing stage, different variables of FEM such as geometry mesh model, material properties and boundary conditions should be set.

Preprocessing, also known as model preparation is often the most working step of FEM. The process of breaking the physical model into small pieces (finite elements) is called meshing as shown in Fig. 4. There are many different types of elements are used in FEM. Commonly, line shapes represent beam or springs used in one dimensional model. Quadrilateral element represents triangles and squares in two dimensional frame model. Hexahedron or tetrahedron elements represent 3D solid model. Hexahedron models are the preferred mesh for simple geometries due to less simulation time. For curved surfaces, tetrahedron meshes are the preferred over Hexahedron for accurate results. Choosing a suitable type of mesh and number of mesh elements determines the quality of the simulation results is termed as mesh convergence (will be discussed in Sect. 4).

After choosing a proper mesh, material property of the model has to be defined. In general, FEM has two types of material properties, solid and fluid. For accurate results, defining the material properties is the critical parameter. Actual material properties of the solid or fluid should be determined using experimental method. Material properties such as tensile strength, flexural modulus, density, Poisson ratio, thermal conductivity and soon are applied for solid. Properties such as viscosity, diffusion coefficient, thermal and other relevant parameters are applied for fluids.

To facilitate suitable FEM environment, boundary condition are the control parameters to define the problems. Boundary conditions set the limits for the governing equation and provide accurate results in FEM. There are many boundary conditions based on the nature of material domain such as flow inlet, flow outlet, wall slip, heat conduction, insulation, normal vectors, tangential vectors and leakage.

3.2 *Solver*

Solver is the second step in FEM process where actual numerical calculation performed based on the input provided at the preprocessing stage. Solver uses topological information of the finite element model and solves the differential equations according to the algorithm set by the respective software provider. Solver step

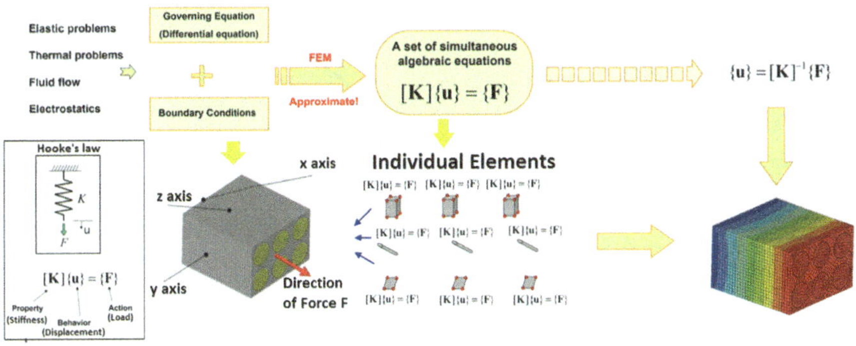

Fig. 5 Working principle of FEM Solver

runs the calculation without any interference from the user or software designer. Basic finite element equation is shown in Eq. 1. This equation is the combination of governing equation (differential) and boundary conditions as shown in Fig. 5. For each element, Eq. 1 is framed based on the material properties (K) and its behavior (U). The combination and collection of all the individual element equation is termed as set of simultaneous algebraic equations. The behavior of the elements is the end result, calculated based on the action (F) performed on the elements as shown in Eq. 2.

$$[\mathrm{k}]\{U\} = \{F\} \tag{1}$$

$$\{U\} = [K]^{(-1)}\{F\} \tag{2}$$

3.3 *Post Processing*

Post processing is the final step of viewing the result of the particular FEM simulation. The results are depicted in graphs, contour, tables and animation. Post processer shows the result parameter of the FEM model in pressure, velocity, stress–strain relationship, temperature, deflection, load, flow pattern and soon. Results are viewed in graphical interface as shown in Fig. 5 and viewed with all color patterns as technically known as contour. The 3D FE model can be viewed in section planes and customized the view in any location of the model. In advance post processer, particle tracking technology is available which can help to identify the movement of each element in the model at transient phase.

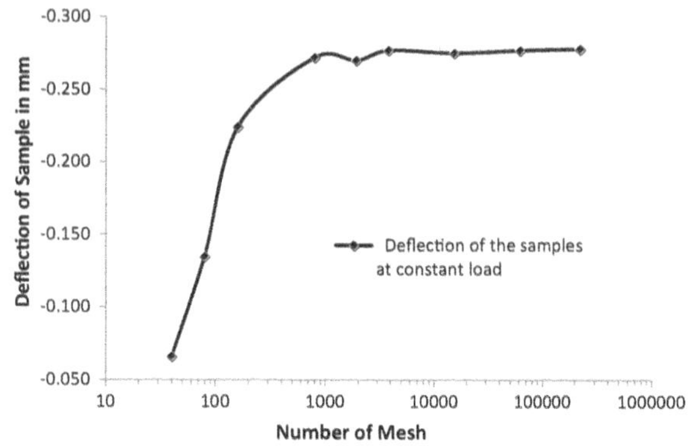

Fig. 6 FEM results of deflected samples verses different number of mesh at constant load

4 Convergence in Finite Element Analysis

It is a good practice to carry out convergence evaluation for every finite element computation. Convergence is the art of choosing a suitable number of nodes for the given problem. As a thump rule, finer the mesh exhibits more accurate results in the FEM. Albeit fine meshes exhibits good results, the simulation time increases exponentially with magnitude of increase in number of meshes. Technically, Mesh convergence determines the minimum number of meshes requires performing the simulation without altering the end results of the simulation. Mesh can be categorized into four types, course mesh, normal mesh, fine mesh and ultra fine mesh. Figure 6 shows the schematic view of various types of mesh in the flexural modules test specimen.

To understand the convergence phenomenon, a case study of plastic test specimen of dimension 100 mm * 10 mm * 2 mm (ASTM D790) is analyzed in FEM. To measure the minimum number of meshes required to generate reliable results, different number of mesh elements were used in the case study. The constant load is applied on the specimen as per ASTM D790 flexural modulus measurement and results such as displacement of sample and solve time is recorded as shown in Fig. 6 and Table 1. The results show that coarse mesh exhibits unreliable displacement values, whereas reliable values found in normal, fine and ultra fine mesh. On contrary, solve time increased from 5 s to 3 h as number of mesh increases.

5 Finite Element Analysis in Date Palm Fiber Composites

Finite element analysis of date palm fiber composite follows the same three steps as the regular FEM as discussed in previous section. Yet, date palm fiber is a short natural fiber, reinforced in a matrix in discontinuous phase. In regular structural analysis, structural model is considered as a single part, i.e. made up of one material and input material properties for FEM preprocessor such as tensile strength and

Table 1 FEM mesh characterization of deflected samples verses different number of mesh at constant load

Convergence and mesh independence											
FEM mesh characteristics	Case study		1	2	3	4	5	6	7	8	9
	Number of divisions	Width	2	2	2	2	4	4	8	8	8
		Depth	2	4	4	8	8	8	16	32	32
		Span	10	10	20	50	60	120	120	240	480
	Number of elements		40	80	160	800	1920	3840	15,360	61,440	222,880
			Coarse Mesh			Normal Mesh			Fine and Ultra Fine Mesh		
FEM Results	Maximum Deflection (mm)		−0.066	−0.135	−0.224	−0.272	−0.270	−0.277	−0.275	−0.277	−0.278
	Solve Time (Sec)		5	7	10	23	92	181	1367	4738	11,154

flexural modulus are identified using experimential practices. In contrast, DPFC has two different kinds of materials, matrix and fiber. Matrix is the continuous phase and short fibers are the discontinuous phase. In some cases, compatibility between matrix and fibers plays a critical role and interface between matrix and fiber also considered as a discontinuous phase.

In classical FEM model, DPFC are considered as a single domain and assumes that fiber and matrix are one phase as shown in Fig. 7. Classical model can be called as one phase model. A classical FEM result showed great deviation from the actual experimental results for all the composites. So, researchers developed a micro-mechanic inspired, homogenization based multi scale computational technique, called as Representative volume element method (RVE). In RVE, composite factors such as fiber orientation, fiber aspect ratio, volume fraction and cross sectional geometry of the composite are considered. RVE method analyzes the model with micro structure material properties at local and global level. The local level analysis is used to calculate the effective elastic properties of the homogenous microstructure and relationship between average RVE strain to local strain within the RVE. The global analysis is used to calculate the average stress–strain relationships of all collective RVE equivalents to real composite structure. RVE model can also be termed as multiphase model.

In general, RVE model is constructed based on the microscopic morphology of the DPFC composite. Date palm Fibers are distributed as small threads in the matrix and fiber distribution morphology of DPFC are shown in Fig. 9a. The number of fibers, aspect ratio of fiber, volume fraction is calculated from DPFC morphology. Based on the application, fibers are arranged in unidirectional and multi-directional or random directional. For compression molding, fibers are multidirectional and fibers are distributed randomly in multidirectional in the FEM model as shown in Fig. 9b. For injection molding, fibers are aligned unidirectional due to injection pressure and FEM model resembles similar to Fig. 10.

RVE can be constructed as 1D, 2D or 3D and the recommended mesh type is shown in Fig. 10 for better mesh convergence of DPFC. For mechanical or structural

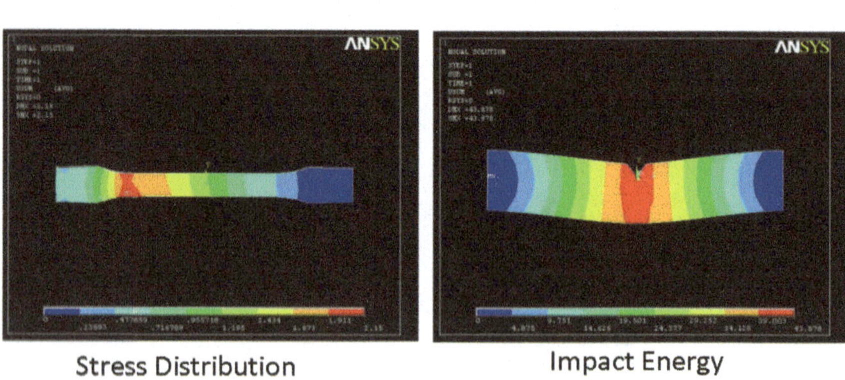

Fig. 7 Stress distribution and impact energy results of classical FEM model

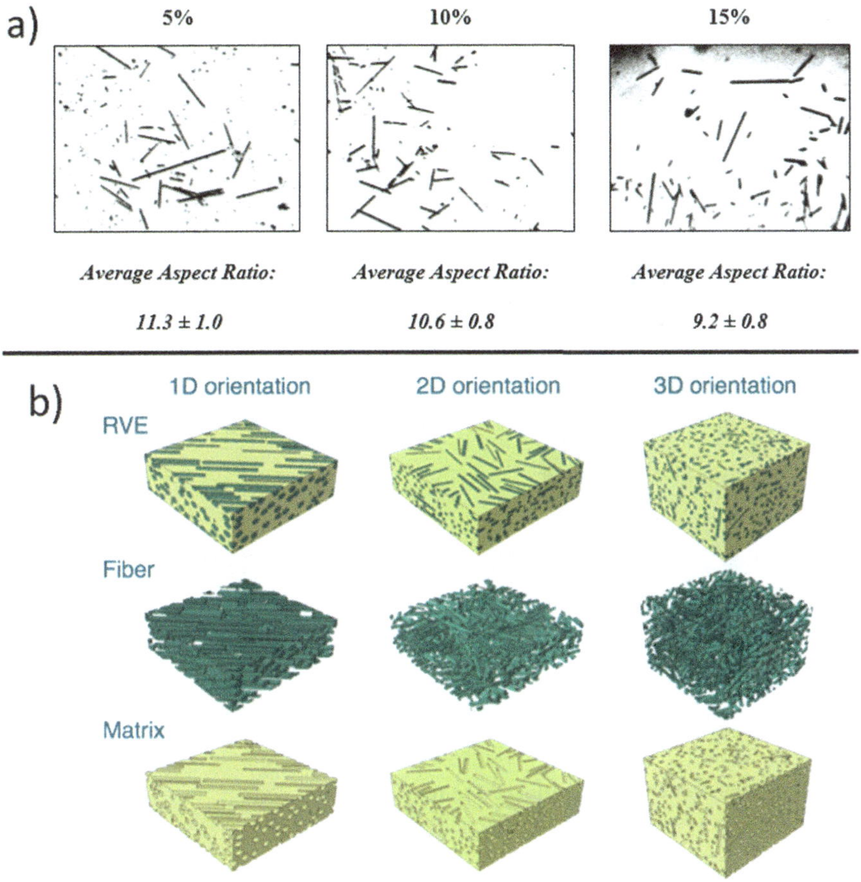

Fig. 9 RVE model with random distribution of fibers in the matrix a) SEM image of DPFC b) 1D, 2D & 3D view of fiber reinforced matrix

analysis, number of fiber and interaction between the fibers at RVE, determines the accuracy of the post processing results such as crack propagation and crazing effect between fibers. As for water diffusion or heat conduction analysis, single fiber with matrix is more sufficient to conduct the analysis as shown in Fig. 11. To decrease the computation time, one quarter or one eight of the RVE can be used in FE model to analyze the results due to line of symmetry.

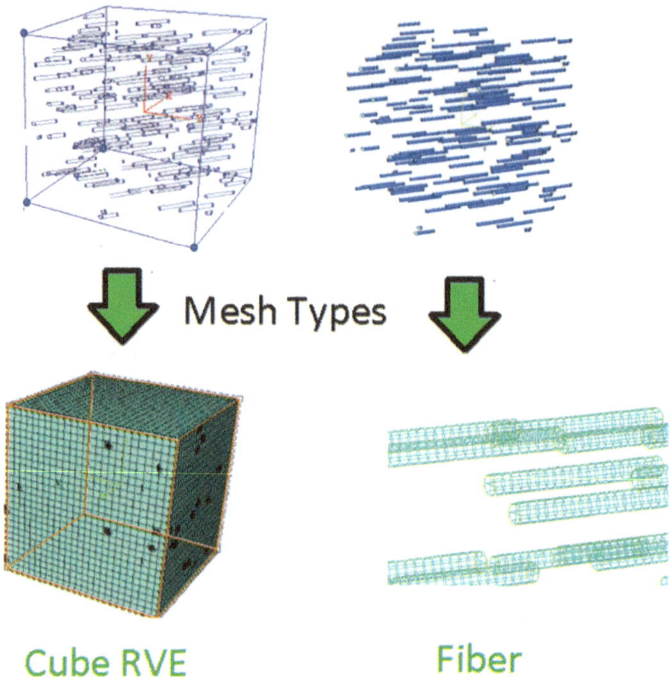

Fig.10 RVE model with unidirectional distribution of fibers in the matrix

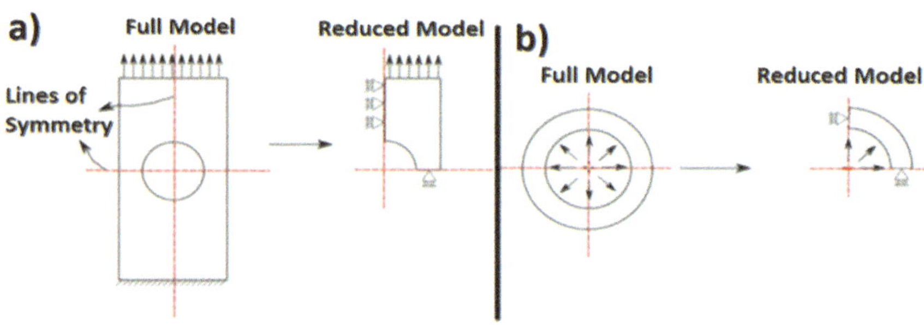

Fig. 11 RVE model for heat conduction and diffusion analysis

6 Finite Element Thermal Analysis of Date Palm Fiber Composites

Construction industry is one of the largest consumers of materials and energy, which eventually generates bulk of waste polluted materials. Eco-friendly material at minimum cost is the requirement of the construction sector and Date palm fiber composite is proven to be a promising candidate. Date fiber-based composites

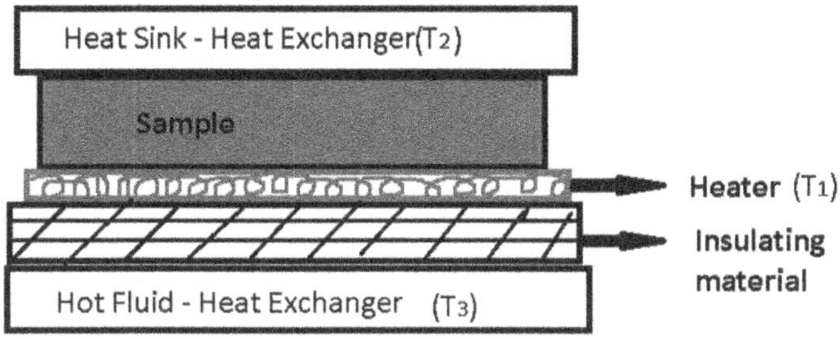

Fig. 12 Basic experimental setup of thermal conductivity

are widely used in many industries due to their light weight, eco-friendliness, low cost, and good mechanical properties. For construction sector, thermal conductivity is one of the critical parameter to select the thermal building insulation material for practical application. Researchers studied the thermal conductivity of DPFC based on fiber length and fiber orientation angle in simple rectangular structures like beam and column (Agoudjil et al. 2011; Wattanakul et al. 2011). They found that thermal conductivity of composite increase with fiber length and decrease with fiber orientation. As for complex structures, finite element analysis is the interesting way to calculate the thermal conductivity with respect to size, shape, volume fraction and spatial distribution of fibers.

Traditionally, researchers consider DPFC as a one phase for thermal conductivity studies and consider the whole structure as a one macro structure while microstructure of DPF is neglected in the analysis. On the contrary, RVE model consider the microstructure of DPF as one phase and improve the accuracy of predicted thermal conductivity of DPFC.

In general, thermal conductivity (k) is calculated experimentally as shown in Fig. 12 and using Eq. 3,

$$k = \frac{q.e}{s(T_1 - T_2)} \tag{3}$$

Whereas, 'q' is the rate of heat flow from the heater, e is the sample thickness, S the heated surface, T1 and T2 the temperatures of the lower and upper surfaces of the sample, respectively.

Haddadi and his Research team used several analytical models to determine the thermal conductivity of DPFC as a function of date palm fiber content (Haddadi et al. 2015). Most prominent models were Hashin and Shtrikman model and Hatta and Taya model (Bigg 1995; Wattanakul et al. 2011). Haddadi et al. developed RVE model based on commercial FEM software COMSOL to determine the thermal conductivity of DPFC.

In commercial FEM software, determinations of thermal conductivity of DPFC are considered as steady state analysis. To perform the steady state finite element analysis, simple cubic arrangement is considered to be a microstructure of matrix. As for DPF,

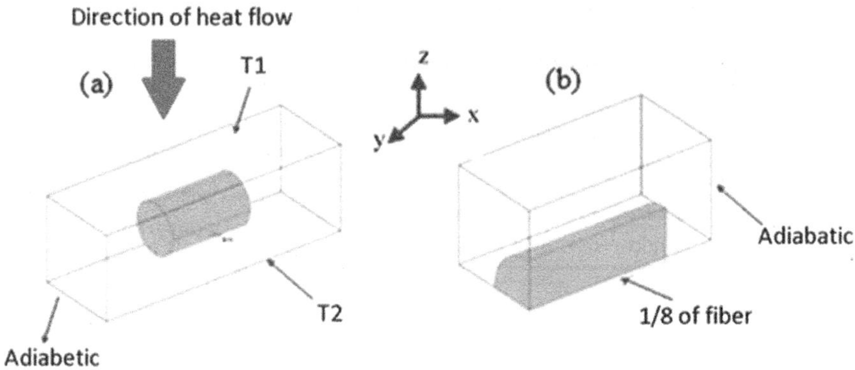

Fig. 13 Schematic representation of RVE elementary cell for thermal conductivity studies (a) Cube with centered full length fiber; (b) Cube with 1/8 part of the fiber

cylindrical arrangement is considered to be a microstructure. From morphological studies, the number of DPF placed inside the cube matrix can be varied based on the percentage of filler composition. For schematic representation, elementary cell is composed of centered fiber in a cube and length of the fiber is arranged parallel to the face of cube as shown in Fig. 13. For simplification, computation is performed on $1/8^{th}$ of the elementary cell and due to plane symmetry and steady state; the entire remaining plane has similar effects.

As for FEM, numerical calculation using Laplace's equation is performed on the elementary cell. There are two assumption sets for the boundary condition of elementary cell as shown in Fig. 13b.

1. Temperature of two faces perpendicular to Heat flow directions are isothermal as T1 and T2 respectively.
2. Temperature of the four faces parallel to heat flow directions are adiabatic.

In FEM, Thermal conductivity of the composite k is calculated from the input values of DPF dimension, filler volume ratio, temperature, boundary condition and material properties of DPF.

Thermal conductivity k of the composite is given by:

$$k = k_m \frac{Q}{\alpha(B+1)} \tag{4}$$

$$\alpha = L/D \tag{5}$$

where, k and k_m Thermal conductivity of matrix, L is the average fiber length and D is the fiber diameter.

The heat flux Q of each elementary cell is calculated by Commercial FEM software (COMSOL) using:

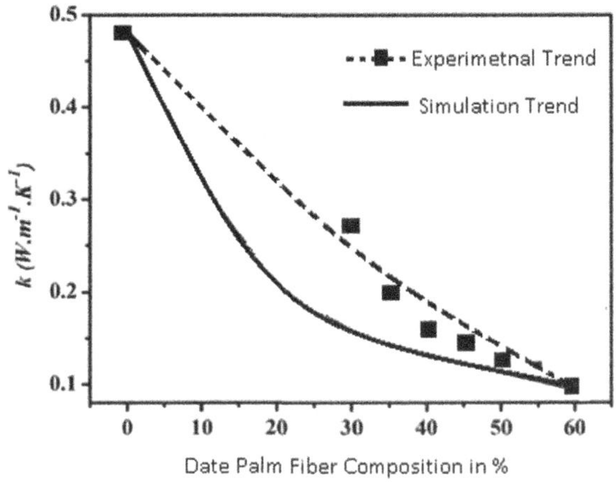

Fig. 14 Result trend of thermal conductivity of the DPFC verse fiber volume fraction

$$Q = \int_0^{(1+B\alpha)} \int_0^{(1+B)} \frac{dT}{dZ} dX dY \tag{6}$$

$$B = \left(\frac{\pi}{4\phi}\right)^{1/3} - 1 \tag{7}$$

where, 'ϕ' represents filler volume fraction.

Haddadi et al. have successfully fitted the experimental data with numerical FEM results as shown in Fig. 14. The result shows that thermal conductivity of the DPFC decreases with increase in date palm fiber content. As a fact, DPF is a lower thermal conductivity material than its polymer or ceramic matrix. SEM images of DPF as shown in Fig. 1a, b exhibits the porous structure of fiber and it contains multi-cellular fibers with central void. Addition of DPF content naturally increases the overall porosity of matrix, which eventually helps to decreases the density and overall thermal conductivity of the composite. Figure 14 shows that compared to analytical model, numerical FEM simulation results were more closed to experimental data.

7 Finite Element Analysis on the Water Diffusion Properties of DPFC

DPFC are made up of cellulose and hemicellulose, which has the tendency to absorb high level of water, technically termed as hydrophilic in nature. Such nature greatly affects the compatibility of DPFC with matrix (polymer and cement) and eventually reflects in the mechanical properties of the final composite structure.

There are three mechanisms which governs the water diffusion phenomenon in DPFC.

1. Water molecules diffuse into the micro gaps (amorphous region) among the polymer chains.
2. Diffusion of water molecules through capillary movement at the defected region of the interface between the fiber and matrix. This defect appears at the manufacturing stage due to poor wetting and impregnation.
3. Water molecules diffuse into the fiber through micro cracks developed on the matrix due to swelling of fibers.

There are many analytical models available in literature to determine the water diffusion kinetics of DPFC such as Fick's model, Langmuir model and Peleg Model. Most authors historically worked on Fick's model to determine the hydrothermal aging properties of natural fiber. According to Fick's model, amount of water molecules absorbed increases linearly with the square root of time and then gradually decreasesto attain the equilibrium. Fick's model is expressed in the Eq. 8

$$\frac{M(t)}{M_s} = 1 - \frac{8}{\pi^2}\sum_{n=0}^{\infty}\frac{1}{(2n+1)^2}exp\left(D\frac{(2n+1)^2}{h^2}\pi^2 t\right) \tag{8}$$

where M(t) is the moisture content at time t, M_s is the moisture content at the equilibrium, D is the diffusion coefficient, and h is the thickness of specimen.

To the knowledge of author, only Alioua and his team (Alioua et al. 2019) developed a FEM model to determine the water absorption properties of concrete filled with DPF using COMSOL Multiphysics. The FEM results are found to be in good agreement with the experimental values and maximum deviation reported between experimental and numerical values are 14%. Alioua (Alioua et al. 2019) considered DPF and concrete as a homogeneousdomain and RVE method is not practiced in their approach.

Although there are countable FEM simulation on other natural fiber composites (Boukettaya et al. 2018; Lei et al. 2019; Pan et al. 2019), there are no published studies on FEM simulation of water diffusion kinetics of DPFC using RVE methods. There is an open scope for researchers to develop FEM-RVE models in DPFC based on general natural fiber techniques in near future.

8 Finite Element Analysis on the Mechanical Properties of DPFC

To analyze the mechanical property of engineering structures are the primary task of any FEM study. As discussed earlier, classical FEM considered fiber reinforced composite model as one phase. Many researchers studied the mechanical behavior of date palm fiber in polymers and concretes as one phase (Kumar et al. 2014; Deli

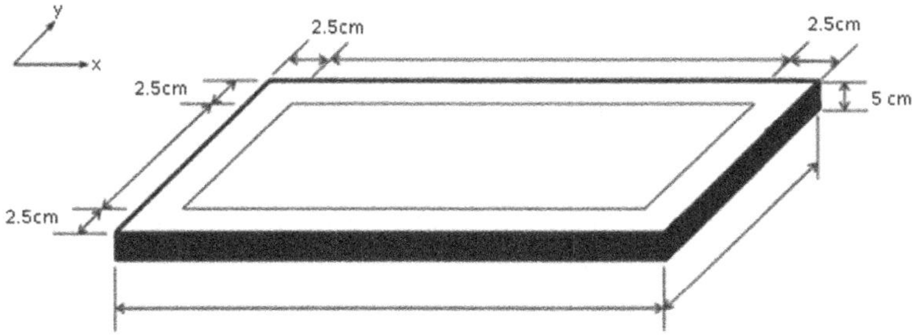

Fig. 15 Vibration test specimen model as per ASTM D3039

2016; Bouzouaid and Kriker 2017). Deli studied the natural frequency of DPF in polyester resin numerically and verified experimentally through vibration test (Deli 2016). Deli found that natural frequencies of the DPFC are increasing with increase in the volume fraction of fiber. The numerical results showed the deviation of 7.5% from the experimental results. Deli used ANSYS with 3D model of ASTM D 3039 samples as shown in Fig. 15.

In specific, experimental studies on concrete beam is a costly and time consuming process. Generally, sample conditioning period for concrete beam test is about 7 days to 90 days and predicting the behavior beam using FEM helps to save considerable amount of time. Consider that concrete beam is quasi brittle material and exhibits highly ductile stress strain relationship. At tension, it behaves linear elastic upto maximum tensile strength and then cracks. At compression, it behaves linear elastic only upto 30% of maximum compressive strength and then peaks to maximum compressive strength and crushing failure at ultimate strain.

Bouzouaid and Kriker studied the behavior of concrete beam with various DPF reinforcement range of 0.2%, 0.3%, 04% & 0.5% using ANSYS APDL (Bouzouaid and Kriker 2017) and validated the FEM results experimentally. Figure 16 shows the model and respective mesh of the concrete beam. Multilinear Isotropic Hardening Constants model in Ansys is used for the concrete beam (Dawari and Vesmawala 2014). Load deflection curve for the concrete beam is developed using FEM simulation and verified experimental at various filler ranger of DPF. Load–deflection curve denotes the deflection of structural beam with increasing flexural load. For all beams, numerical results showed the sudden increase in deflection value at first crack compared to experimental value. Failure load values obtained from numerical

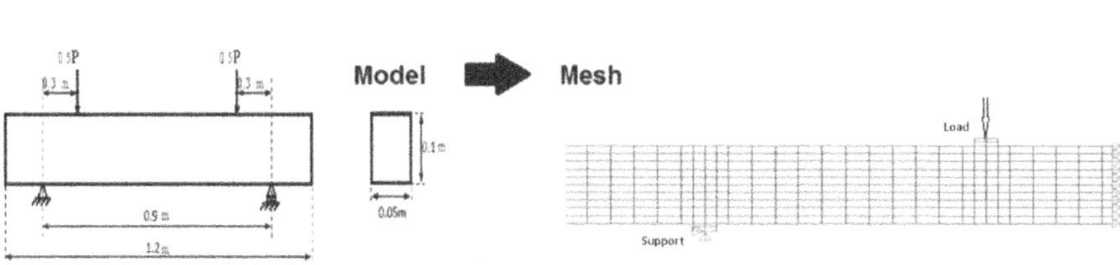

Fig. 16 FEM model for Beam with Hex mesh

simulation is always smaller than experimental results. The deviation of numerical result from experimental analysis is exhibited due to consider the whole concrete DPF composite structure as one phase. In practical, DPFC concrete has four phase, concrete matrix, steel reinforcement, DPF reinforcement and interface.

Though there is a deviation in results and the crack behavior and mode of failure on concrete beam showed similar visual pattern both at experiment and numerical simulation. The uncertainly in classical FEM simulation leads to address more focus on RVE based simulation which in turn can consider all the phases.

As for RVE based mechanical analysis on DPFC are very limited, notably Megahed and his team studied the micromechanical properties of DPF through RVE model using commercial software ABAQUS 6.12 (Megahed et al. 2019). Megahed predicted the elastic modulus of DPF with various volume fraction range from 20 to 50% in starch based hybrid composite. Length of the fiber is based on fiber aspect ratio. RVE model of DPF based hybrid composite is a rectangular prism with fiber enclosed at the center. Hexahedral mesh elements are used for the matrix and beam element with 2 nodes for fiber.

Megahed et.al has used the following assumptions,

- Cylindrical cross sections of the all fibers are considered to be of equal length to overcome the end point effect.
- Effect of voids in the composite is not considered
- The area of matrix is controlled by matrix volume fraction in which thickness and width are assumed to be equal.

In most cases, elastic modulus results of various DPFC using FEM showed close agreement with experimental results and within the tolerance range of 5%. For fewer cases, FEM results are quite higher than experimental values. Megahed et.al states that higher void percentage affects the mechanical performance of DPFC. Megahed et.al calculated the void percentages of each composite using Eq. 10 and found range from 1.2% to 4.6%. In general, void can initiate the crack nucleation mechanism in the composite and also allows the moisture to penetrate into the structure.

$$\rho_{\text{Theoretical}} = \rho_f V_f + \rho_m (1 - V_f) \tag{9}$$

$$\text{Voidfraction} = \frac{(\rho_{\text{Theoretical}} - \rho_{\text{Experimental}})}{\rho_{\text{Theoretical}}} \tag{10}$$

where, ρ_f and ρ_m are the density of fiber and matrix. V_f is the volume of fiber.

As void percentage isn't taken into consideration in the FEM analysis, actual volume fraction of fiber has no guarantee to be equal with that of RVE model. In addition, samples are prepared using manual compression molding process which leads to higher void percentage. Though there is a deviation in FEM results of DPFC on fewer samples and in all cases, it has direct connection to external factor like processing parameter etc. In conclusion, FEM using RVE model showed a close agreement with the experimental results.

9 Conclusion

Design and modeling of various numerical methods used in DPFC are reviewed and found that FEM simulation is the best choice to analyze the mechanical, thermal and diffusion coefficient of the DPFC. It should be emphasized that heart of the FEM analysis is to select a suitable mesh convergence technique for the simulation and in turn, it greatly affect the accuracy of the simulation. The study reveals that classical FEM method is outdated technique for most fiber based composites and in specific for DPFC. RVE based FEM method is the most suitable homogenization based multi scale computational technique for DPFC to analyze the effect of microstructure on thermal, diffusion and mechanical properties. As role of DPFC in construction industry, are gaining popularity, RVE based model could be a corner stone for new material ad application developments in this segments.

Surprisingly, until today, there are very limited studies published on DPFC based on RVE models and there is a scope for researchers to address this area for further improvements.

References

Agoudjil B, Benchabane A, Boudenne A, et al (2011) Renewable materials to reduce building heat loss: Characterization of date palm wood. Energy Build 43:491–497. https://doi.org/10.1016/j.enbuild.2010.10.014

Alawar A, Hamed AM, Al-Kaabi K (2009) Characterization of treated date palm tree fiber as composite reinforcement. Compos Part B Eng 40:601–606. https://doi.org/10.1016/j.compositesb.2009.04.018

Alioua T, Agoudjil B, Boudenne A (2019) Numerical modelling and experimental study of heat and moisture properties of a wall based on date palm fibers concrete. E3S Web Conf 85:1–6. https://doi.org/10.1051/e3sconf/20198502009

Alshammari BA, Saba N, Alotaibi MD et al (2019) Evaluation of mechanical, physical, and morphological properties of epoxy composites reinforced with different date palm fillers. Materials (Basel) 12. https://doi.org/10.3390/ma12132145

Bigg DM (1995) Thermal conductivity of heterophase polymer compositions BT—thermal and electrical conductivity of polymer materials. Springer, Berlin Heidelberg, Berlin, Heidelberg, pp 1–30

Boukettaya S, Alawar A, Almaskari F et al (2018) Modeling of water diffusion mechanism in polypropylene/date palm fiber composite materials. J Compos Mater 52:2651–2659. https://doi.org/10.1177/0021998317752228

Boukhattem L, Boumhaout M, Hamdi H et al (2017) Moisture content influence on the thermal conductivity of insulating building materials made from date palm fibers mesh. Constr Build Mater 148:811–823. https://doi.org/10.1016/j.conbuildmat.2017.05.020

Bouzouaid S, Kriker A (2017) Nonlinear analysis of concrete beams strengthened by date palm fibers. AIP Conf Proc 1814. https://doi.org/10.1063/1.4976239

Dawari VB, Vesmawala GR (2014) Application of nonlinear concrete model for finite element analysis of reinforced concrete beams. Int J Sci Eng Res 5:776–782

Deli AA (2016) Experimental and numerical investigation of date palm fiber effect on natural frequency of composite plate with different B . Cs. 1163–1174. https://doi.org/10.15680/IJIRSET.2016.0502002

Gheith MH, Aziz MA, Ghori W et al (2019) Flexural, thermal and dynamic mechanical properties of date palm fibres reinforced epoxy composites. J Mater Res Technol 8:853–860. https://doi.org/10.1016/j.jmrt.2018.06.013

Ghori W, Saba N, Jawaid M, Asim M (2018) A review on date palm (phoenix dactylifera) fibers and its polymer composites. IOP Conf Ser Mater Sci Eng 368. https://doi.org/10.1088/1757-899X/368/1/012009

Haddadi M, Agoudjil B, Benmansour N et al (2015) Experimental and modeling study of effective thermal conductivity of polymer filled with date palm fibers. Polym Compos 37:915–924. https://doi.org/10.1002/pc.23741

Hakeem KR, Jawaid M, Rashid U (2014) Biomass and bioenergy: processing and properties. Biomass Bioenergy Process Prop 9783319076:1–367. https://doi.org/10.1007/978-3-319-07641-6

Kumar A, Dwivedi A, Paliwal V, Patil PP (2014) Free vibration analysis of al 2024 wind turbine blade designed for Uttarakhand region based on FEA. Procedia Technol 14:336–347. https://doi.org/10.1016/j.protcy.2014.08.044

Lei Y, Zhang J, Zhang T, Li H (2019) Water diffusion in carbon fiber reinforced polyamide 6 composites: experimental, theoretical, and numerical approaches. J Reinf Plast Compos 38:578–587. https://doi.org/10.1177/0731684419835034

Megahed H, Emara M, Farag M, Wifi A (2019) Enhancement of the mechanical behavior of starch-palm fiber composites. By-Prod Palm Trees Their Appl 11:201–210. https://doi.org/10.21741/9781644900178-15

Mokhtari A, Kriker A, Guemmoula Y et al (2015) Formulation and characterization of date palm fibers mortar by addition of crushed dune sand. Energy Procedia 74:344–350. https://doi.org/10.1016/J.EGYPRO.2015.07.624

Noorunnisa Khanam P, AlMaadeed MA (2014) Improvement of ternary recycled polymer blend reinforced with date palm fibre. Mater Des 60:532–539. https://doi.org/10.1016/J.MATDES.2014.04.033

Pan Y, Xian G, Li H (2019) Numerical modeling of moisture diffusion in an unidirectional fiber-reinforced polymer composite. Polym Compos 40:401–413. https://doi.org/10.1002/pc.24664

Wattanakul K, Manuspiya H, Yanumet N (2011) Thermal conductivity and mechanical properties of BN-filled epoxy composite: effects of filler content, mixing conditions, and BN agglomerate size. J Compos Mater 45:1967–1980. https://doi.org/10.1177/0021998310393297

Date Palm Fiber Composites Properties

Characterization of Date Palm Fiber

Lobna A. Elseify and Mohamad Midani

Abstract Date palm (*Phoenix Dactylifera L.*) is a very rich source of cellulosic fibers. Cellulosic fibers are extracted from four different parts of the palm; midribs, spadix stems, leaflets, and mesh. The literature showed that the extracted date palm fibers have promising properties when compared to other natural fibers like flax, hemp and sisal. This chapter will review the physical, mechanical, chemical, morphological, and thermal properties of the date palm fibers. Moreover, a comparison between the properties of date palm fibers and other natural fibers was conducted.

1 Introduction

Date palm (Phoenix Dactylifera L.) (DP) is considered a very rich source of cellulosic fibers. Fibers could be extracted from four different parts; midribs, spadix stems, leaflets, and, mesh. Date palm is cultivated primarily for the fruits and not the fibers, and the four sources of date palm fibers are regarded as byproducts of annual pruning. Figure 1 shows the four date palm fiber sources. The midrib is sometimes referred to as rachis or petiole in the literature. Moreover, the mesh is sometimes referred to as the tree surface, sheath, or coir. Finally, the spadix stems are sometimes referred to as the fruit bearing branches. In this chapter the terms midribs, spadix stems, leaflets, and mesh will be used to refer to the four date palm fiber sources.

The properties of natural fibers are affected by two types of factors; controlled and uncontrolled factors. The uncontrolled factors include, natural defects in fibers, microfibril angle (MFA), and chemical composition of the plant. Whereas, the controlled factors include, cultivation factors, and extraction technique or parameters (Hakeem et al. 2014; Elseify et al. 2019). Date palm mesh fibers have been

L. A. Elseify (✉) · M. Midani
Department of Materials Engineering, German University in Cairo, New Cairo City, Egypt
e-mail: lobna.elseify@gmail.com

M. Midani
inTEXive Consulting, El Obour City, Egypt

Wilson College of Textiles, NC State University, Raleigh, NC, USA

M. Midani et al. (eds.), *Date Palm Fiber Composites*, Composites Science and Technology,
https://doi.org/10.1007/978-981-15-9339-0_8

Fig. 1 Date palm parts; spadix stems, leaflets, midribs and mesh

used for centuries in making ropes and twines, since they are already present in a fiber state requiring minimal extraction. Whereas, the other sources are very recently thought after as sources of cellulosic fibers, which was reflected on the inadequate extraction techniques and consequently high variability in the extracted fiber properties. However, a recent research succeeded in extracting good-quality long textile fibers from the date palm midrib which has the potential to replace other natural fibers in different applications (Elseify et al. 2018; Midani et al. 2018; Elseify et al. 2020). This chapter will review the physical, mechanical, chemical, morphological, and thermal properties of the date palm fibers, in addition to brief overview on the characterization techniques. Moreover, the chapter will provide a comparison between the properties of date palm fibers and other natural fibers.

2 Morphological Properties

This section will explain and show the different morphological features of DP fibers. The cross-sectional shape and surface morphology of DP fibers could be observed using a field emission scanning electron microscope (FESEM) with accelerating voltage 30 K.V. and magnification up to 1000000x. The fibers have to be gold sputtered using a sputter coater and then attached to a metal plate using a double-face adhesive tape. The section is divided into four subsections; each subsection being dedicated to one DP fiber source.

2.1 Midribs

The morphology of DP midrib fibers has been rarely studied in the literature; it still needs further investigation.

The DP midrib is composed of fibre vascular bundles embedded in a lignin matrix, with a circular cross section that can reach up to 1 mm in diameter (Bendahou et al. 2009; Elseify et al. 2020). These vascular bundles contain two large hollow lumens in addition to other considerable amount of hollow content. Furthermore, those vascular bundles are composed of elemental micro-fibrils of much smaller diameters in the range of 12 μm held together by hemicellulose. Figure 2a illustrates the building blocks of the DP midrib, showing the fibre vascular bundle and the elemental micro-fibrils (Elseify et al. 2020).

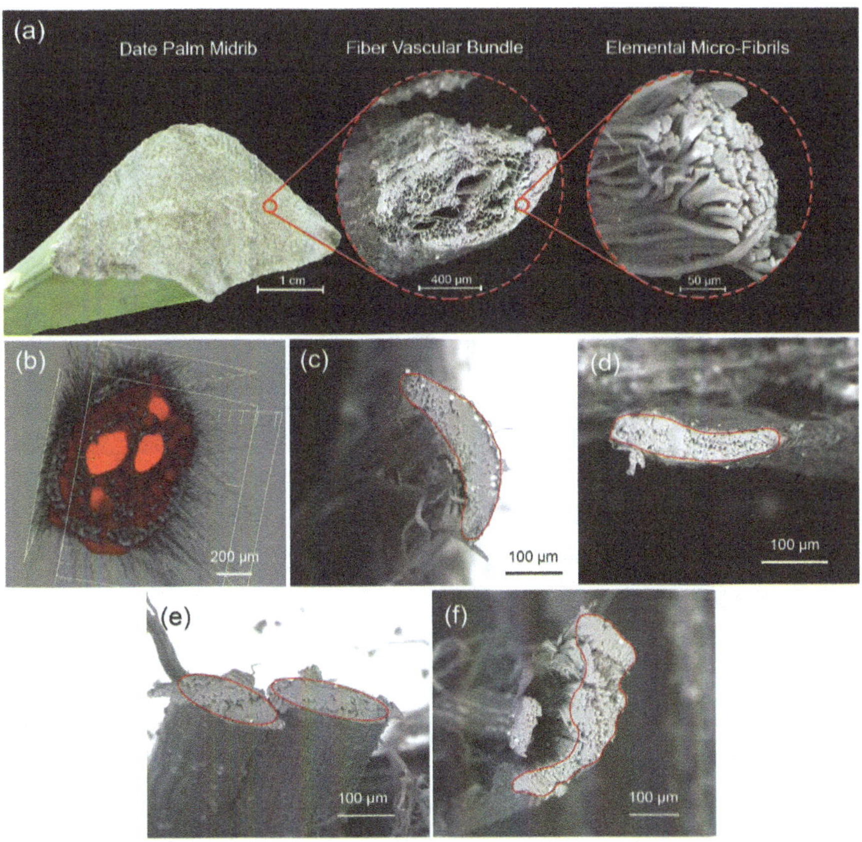

Fig. 2 DPM fibres cross section; **a** building blocks of the DPM, **b** CT scan of unfibrillated vascular bundle showing the amount of hollow content, **c** "C" shaped fibres, **d** flaky-shaped fibres, **e** elliptical fibres, and **f** irregular-shaped fibres (Elseify et al. 2020)

The cross sectional shape of the DP midrib fibers depends heavily on the severity of the treatment and the degree of fibrillation. Figure 2b shows date palm unfibrillated fiber vascular bundle; the fiber has two big voids (lumens) in addition to few smaller ones. Aside from the circular shape shown in Fig. 2b, the DP midrib fibers extracted had nearly four other cross-sectional shapes; C-shaped, flaky, circular, and elliptical. The fibres having the shape of a letter "C" are shown in Fig. 2c. Whereas, the fibers having the shape of flakes are shown in Fig. 2d and it seems that these fibres originally had C-shape but due to the chemical treatment the fibers were split into smaller flakes. The elliptical fibers cross-sectional shape is shown in Fig. 2e. Finally, Fig. 2f shows some irregularly shaped fibers. As for DP midrib longitudinal shape and the surface features, it was found that the untreated fiber surface has impurities and parenchyma cells on the surface as shown in Fig. 3a, b. Moreover, alkaline treated midrib fibers also have parenchyma cells on the surface and silica crystals as shown in Fig. 3c, d respectively (Elseify et al. 2020).

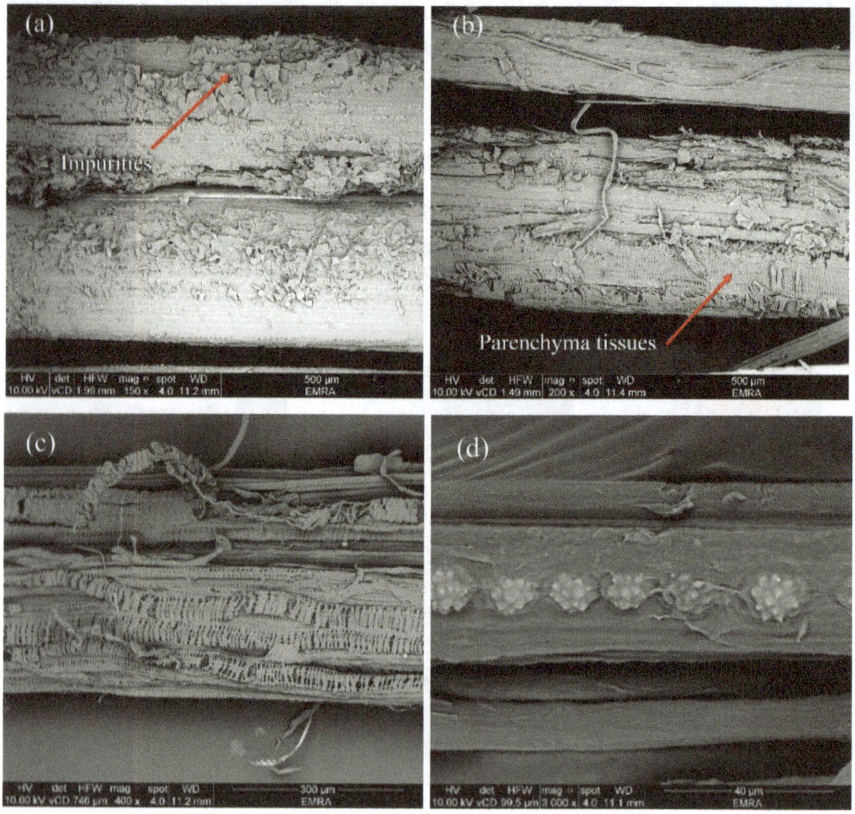

Fig. 3 SEM images in the longitudinal direction of **a, b** untreated DP midrib fibers and **c, d** NaOH treated DP midrib fibers showing parenchyma cells and silica crystals on the surface (Elseify et al. 2020)

2.2 Spadix Stems

The spadix stem fibers were investigated in the literature. The untreated fibers consist of fiber vascular bundles as shown in Fig. 4a, d in the longitudinal and cross sectional views respectively (Amroune et al. 2019). As for alkaline treated spadix stem fibers shown in Fig. 4b, it was noticed that the alkaline treatment had a positive effect in cleaning the fiber surface (Amroune et al. 2019). However, the cross-sectional view of untreated fibers, in Fig. 4c, showed an unfibrillated fiber vascular bundle (Amroune et al. 2019). By comparing the cross-sectional shapes of fibers extracted from the midribs and the spadix stems, it was noticed that they both contain large lumens; however, midrib fibers have larger lumens. The ratio of the largest lumen diameter to the vascular bundle diameter of the midrib in Fig. 2b was 0.28, compared to 0.1 for the spadix stem in Fig. 4c.

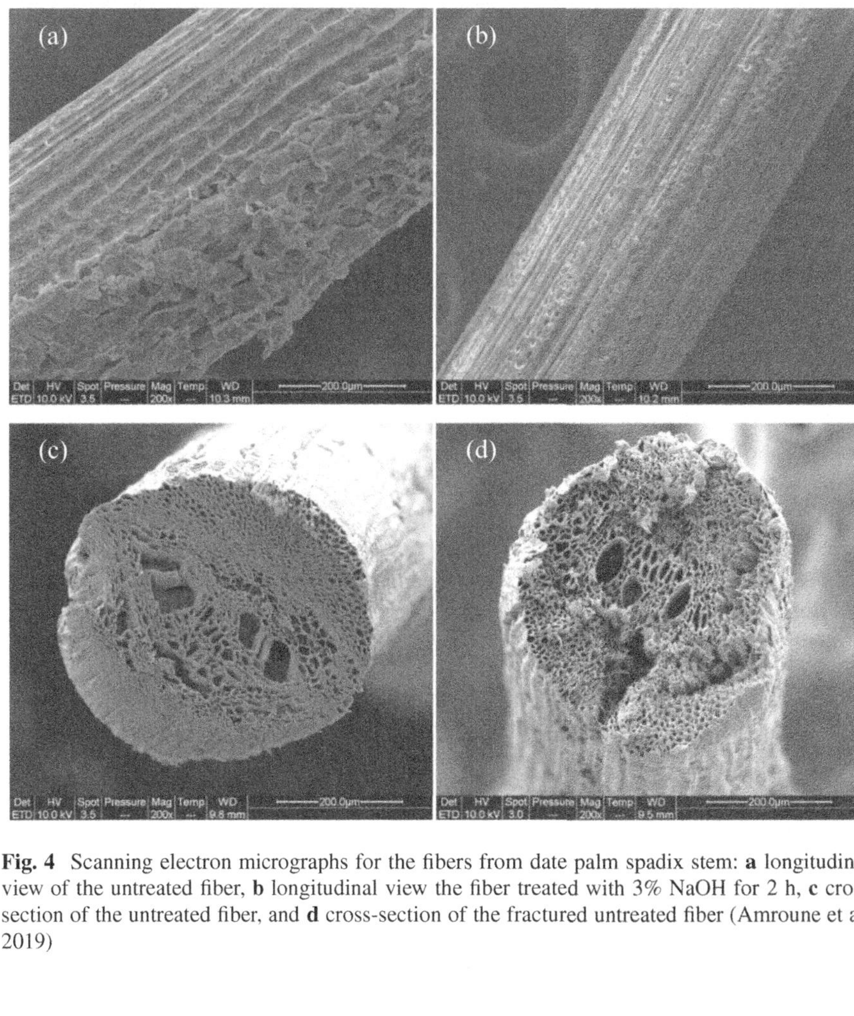

Fig. 4 Scanning electron micrographs for the fibers from date palm spadix stem: **a** longitudinal view of the untreated fiber, **b** longitudinal view the fiber treated with 3% NaOH for 2 h, **c** cross section of the untreated fiber, and **d** cross-section of the fractured untreated fiber (Amroune et al. 2019)

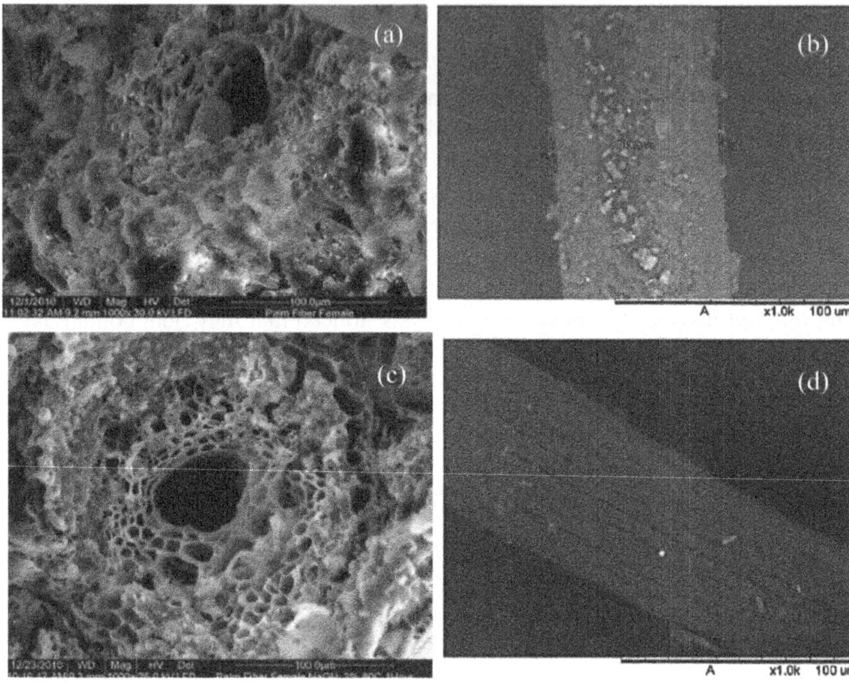

Fig. 5 SEM of DP leaflet fibers **a** untreated in cross-sectional view (AlMaadeed et al. 2014), **b** untreated in longitudinal view (Dehghani et al. 2013), **c** NaOH treated in cross-sectional view (AlMaadeed et al. 2014), and **d** NaOH treated in longitudinal view (Dehghani et al. 2013)

2.3 Leaflets

The morphology of the leaflet fibers is shown in Fig. 5. The cross-sectional shapes of untreated and NaOH treated fibers are shown in Fig. 5a, c respectively (Taha et al. 2007). The cross-sectional shape of the leaflet fibers is similar to those of the midribs and the spadix stems. They consist of large lumens surrounded by elemental fibers bounded together with lignin. Observing the fibers in the longitudinal view, it was noticed that the alkaline treatment has helped in cleaning the fibers surface and removing the impurities as shown in Fig. 5b, d respectively (Dehghani et al. 2013).

2.4 Mesh

DP mesh fibers were the most frequent type of date palm fibers investigated in the literature. This is probably because mesh fibers exist naturally in a sheath form which could be easily extracted. Figure 6 shows untreated DP mesh fibers. It was noticed that the cross-sectional shape of the fibers extracted from the mesh is different from

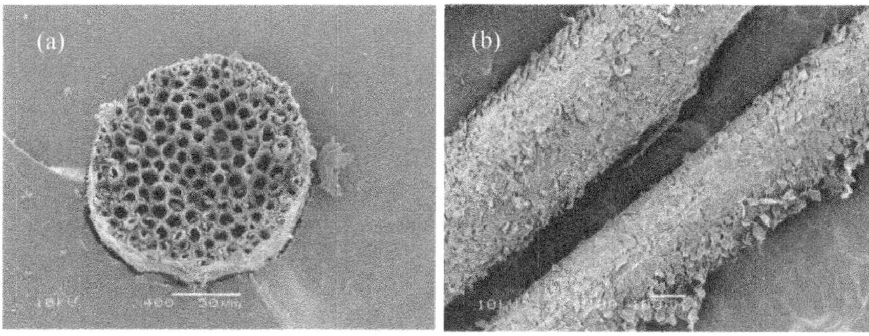

Fig. 6 SEM of untreated DP mesh showing **a** the cross-section and **b** the fiber's surface (Al-Khanbashi et al. 2005)

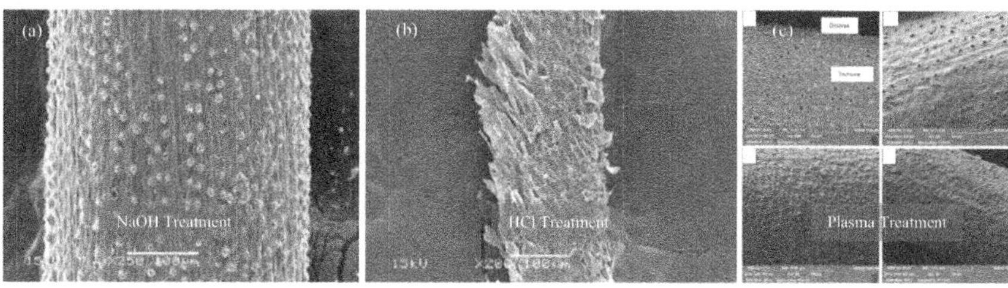

Fig. 7 SEM for DP mesh fiber treated with **a** 1% NaOH, **b** HCl (Alawar et al. 2009) and **c** plasma at different plasma power and exposure duration (Gholami et al. 2017)

those extracted from the midrib, the spadix stems, or the leaflets discussed in the previous subsections. The mesh fibers consist of equal-size small lumens unlike the other DP fibers.

The morphology of DP mesh fibers treated chemically with NaOH, HCl and plasma is shown in Fig. 7. It can be seen that the treatment has somehow succeeded in removing the surface impurities in case of NaOH and plasma shown in Fig. 7a, c respectively. As for the acidic treatment using hydrochloric acid shown in Fig. 7b, it can be seen that the surface was deteriorated due to the vigorous acidic attack (Alawar et al. 2009).

2.5 *DP Fibers and Other Natural Fibers*

It is very important to know where date palm fibers stand compared to other types of cellulosic fibers. Following is a comparison between date palm fibers and other types of natural fibers to show the similarities and differences between their morphological properties.

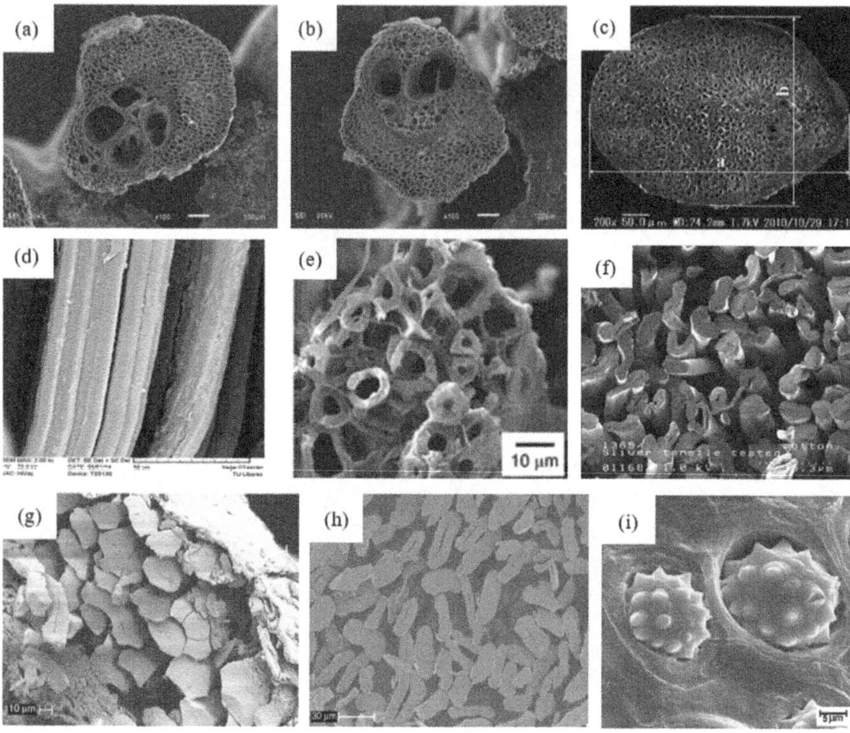

Fig. 8 SEM of **a**, **b** cross-sectional view of peach palm trunk (Diaz et al. 2016), **c** cross-section of coir fibers (Nam et al. 2012), **d** longitudinal view of jute fibers (Fangueiro and Rana 2018), **e** cross-sectional view of sisal fibers (Arenas and Asdrubali 2017), **f** cross-sectional of cotton fibers, **g** cross-section of flax fibers, **h** cross-section of ramie fibers (Kozłowski 2012) and **i** silica crystals on piassava fibers (d'Almeida et al. 2006)

It was found that DP fiber morphology is similar to fibers extracted from peach palm trunk as shown in Fig. 8a, b (Diaz et al. 2016). Peach palm fibers have large lumens surrounded by smaller fiber bundles similar to DP midrib fibers. The cross-sectional view of coir fibers, shown in Fig. 8c is similar to DP mesh fibers shown in Fig. 6; they both have small hollow fiber bundles (Nam et al. 2012). However, some natural fibers have completely different cross-sectional view when compared to DP fibers like cotton, flax and ramie shown in Fig. 8f–h respectively. As for sisal fibers, shown in Fig. 8d, e, they have cross-sectional and longitudinal shapes similar to DP fibers shown in Figs. 4, 5, and 6. The silica crystals observed on the surface of date palm midrib fibers, shown in Fig. 3d are quite similar to those observed on the surface of piassava fibers in Fig. 8i (d'Almeida et al. 2006).

The longitudinal views of palm fibers (*Archontophoenix alexandrae*) and kenaf fibers, shown in Fig. 9, are very similar to DP midrib fibers. They have surface scales and even palm fibers have some parenchyma cells on the surface, in Fig. 9d, just like date palm midrib fibers. (Fiore et al. 2015; Fangueiro and Rana 2018).

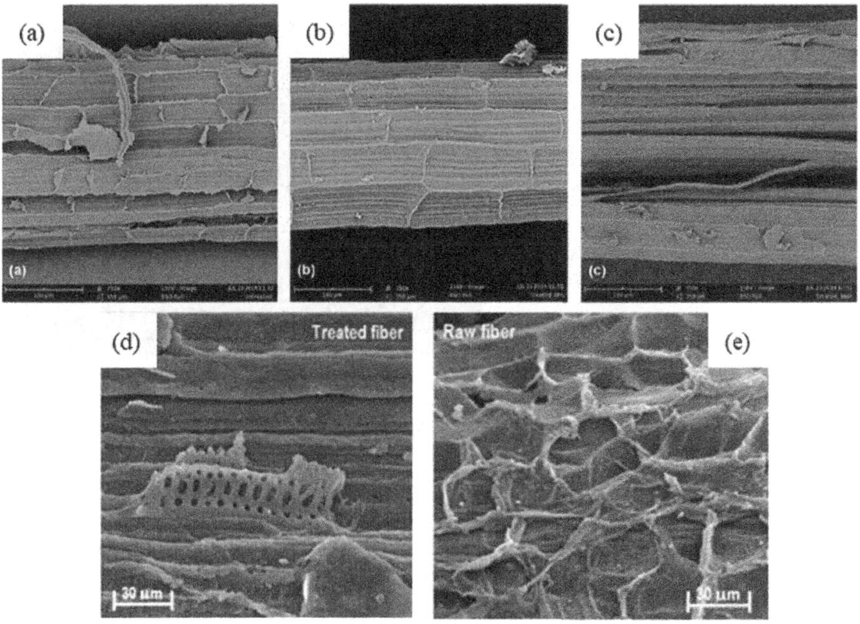

Fig. 9 SEM images of longitudinal views of **a** untreated, and **b**, **c** NaOH treated kenaf fibers (Fiore et al. 2015), and **d** treated and **e** untreated palm fibers (Fangueiro and Rana 2018)

3 Physical Properties

The physical properties of DP were investigated in the literature by calculating the fibers diameter, aspect ratio (length to diameter ratio) and density. The fibers diameter or cross-sectional area could be determined using SEM or optical microscope with an image analysis software. To observe the cross-sectional shape of fibers using optical microscope, cork method is used. Cork method is a technique used to view the cross-section of fibers by embedding the fibers in a cork using a thread and needle as shown in Fig. 10a. A typical shape of fibers cross-section observed using optical microscope is shown in Fig. 10b.

It was noticed that the extraction technique used had an impact on the properties. it was found that the mechanically extracted fibers (untreated) were the shortest; nearly less than 5 mm long for midribs, spadix stems, and leaflets (Khristova et al. 2005; Hassan et al. 2014; Mirmehdi et al. 2014). This is because the mechanical extraction technique used was based on chopping or milling the fibers (Saadaoui et al. 2013; Abdel-Aal et al. 2015). The short fibers result in lower fiber aspect ratio, and limits the processability of the fibers into yarns, woven preforms or nonwoven mats. Moreover, the reinforcement efficiency in natural fiber composites depends on the stress transfer between the matrix and the fibers, which is significantly influenced by the fiber aspect ratio. If the fiber aspect ratio is lower than a critical value, the

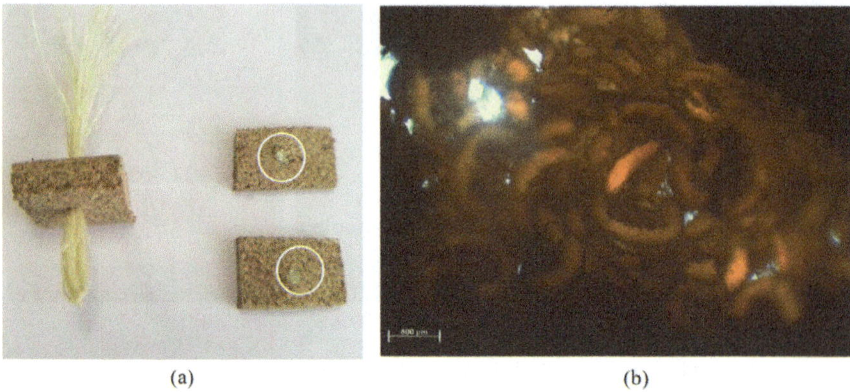

Fig. 10 **a** Embedded fibers using cork method and **b** typical cross-sectional view of embedded fibers bundle observed using optical microscope

fibers would be regarded as inclusions rather than reinforcements; hence, reducing the strength of the composite (Elseify et al. 2018).

On the other hand, the chemical treatment rendered longer fibers than the mechanically treated fibers. The length reached 300 mm for midrib fibers (Midani 2017; Midani et al. 2018; Elseify et al. 2020). As for spadix stems, the fibers length reached nearly 14 mm long which still could be regarded as short from a processability point of view (Ibrahim et al. 2014, 2017). Whereas, the diameter of the fibers was not affected by the extraction techniques as shown in Table 1.

The absolute density of date palm fibers was found to range between 1.178 and 1.45 gm/cm^3 for chemically treated fibers (Kriker et al. 2005; Elseify et al. 2020). The density calculation technique was not usually mentioned in the literature. However, the absolute density measurement could be performed in accordance with the ASTM D8171 – 18 Archimedes methods. The samples are in the form of compressed disks to eliminate all air voids. The disks are prepared by compressing the fiber bundles under high pressure and temperature nearly 35 bars and 180 °C respectively. The test is performed by measuring the specimen weight in air and another time in liquid, and from the difference in weights, the density is automatically calculated using a densometer (Elseify et al. 2020) as shown in Fig. 11. The physical properties of date palm fibers are shown in Table 1.

4 Mechanical Properties

4.1 *Single Fiber Tensile Properties*

The tensile strength, Young's modulus and strain at break of the single fiber could be determined using single fiber tensile test according to the standard test ASTM D

Table 1 Date palm fibers physical properties

Date palm part	Extraction technique	Diameter (μm)	Length (mm)	Aspect ratio	Density (gm/cm^3)	References
Midribs	Mechanical	13.7–510	1.1–1.69	66.3–95	0.13–0.3 (bulk)	Khristova et al. (2005), Saadaoui et al. (2013), Mirmehdi et al. (2014), Abdel-Aal et al. (2015)
	Chemical	10–100	1.01–300	45–3000	–	Agoudjil et al. (2011), Hegazy and Ahmed (2015), Jaber et al. (2016), Midani (2017)
		Cross-sectional area = 0.046–0.138 mm^2 Length = 300 mm			1.178–1.324 (abs.)	Elseify et al. (2018, 2020)
Spadix stems	Mechanical	23.4	0.61–0.81	26–34.6	–	Hassan et al. (2014)
	Biological (Enzymatic)	20.3–20.6	0.56–0.74	27.18–36.45	–	Hassan et al. (2014)
	Chemical	196 ± 120	13.8 ± 6.5	70–70.4	1.37 ± 0.009 (bulk)	Ibrahim et al. (2014, 2017)
Leaflets	Mechanical	460	3.42	7.43	–	Mirmehdi et al. (2014)
	Biological (Retting)	Cross-sectional area = 0.4105–0.5167 mm^2			0.99 (bulk)	Rao and Rao (2007)
	Chemical	8–10,000	1.5–50	150–190	–	Pandey and Ghosh (1995), Abu-Sharkh and Hamid (2004), Mohanty et al. (2014), Mohanty (2017)

(continued)

Table 1 (continued)

Date palm part	Extraction technique	Diameter (μm)	Length (mm)	Aspect ratio	Density (gm/cm^3)	References
Mesh	Mechanical	100–2,000	20–300	29–3000	0.6–0.7 (bulk)	Abdal-hay et al. (2012), Alajmi and Shalwan (2015), Al-Rifaie and Al-Niami (2016), Boukhattem et al. (2017), Boumhaout et al. (2017)
	Water separation	100–800	10–30	12.5–300	–	Tioua et al. (2017)
	Chemical	200–800	15–60	18.75–300	0.51–1.09 (bulk) 1.30–1.45 (abs.)	Kriker et al. (2005), Abdal-hay et al. (2012)

bulk bulk density, *abs.* absolute density

Fig. 11 Density samples preparation and measurement (Elseify et al. 2018)

3822 – 01. In which a single fiber is mounted between two paper frames then bonded together using cyanoacrylate adhesive. The frame's dimensions are 4 × 5 cm and the window centered inside is of dimensions 2.5 × 1 cm allowing 10 mm gauge length as shown in Fig. 12. The tensile test is conducted using a tensile tester at strain rate 0.5 mm/min with zero initial load. The load cell of the tensile tester should be less than 1 KN and at least 10 specimens should be tested per sample.

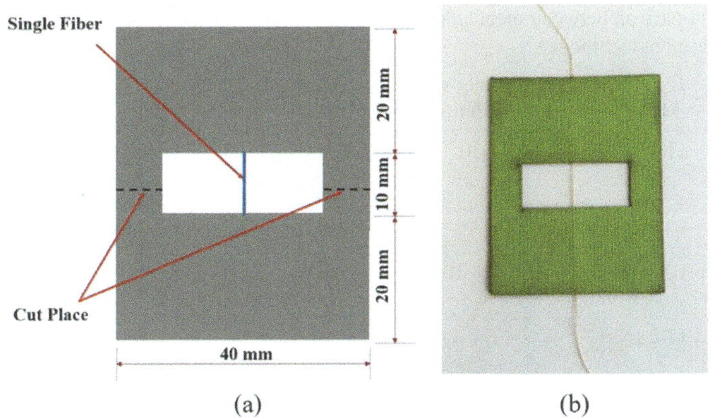

Fig. 12 **a** Schematic diagram of the cardboard frame used for SFTT and **b** real testing frame with a single fiber attached

The mechanical properties of date palm fibers have been investigated by several researchers, from all 4 sources of fibers. However, the standard test followed was not usually mentioned by researchers. Thus, we couldn't know if the test results are reliable or not. Tensile strength and modulus of elasticity are shown in Table 2. The mechanical properties of fibers were found to be significantly affected by their treatment.

From Table 2 it was noticed that the tensile strength of DP midrib fibers had a maximum of 452.8 MPa for NaOH treated fibers, while the untreated fibers had a maximum of 208 MPa. Spadix stems fibers had higher tensile strength than DP midrib fibers with maximum of 680 MPa. However, the modulus of elasticity wasn't as high as the tensile strength. As for DP leaflet fibers, the tensile strength results varied depending on the treatment; some researchers found that the tensile strength of leaflet fibers to be between 32 and 309 MPa (Rao and Rao 2007; AlMaadeed et al. 2013; Mohanty et al. 2014) and other researchers found it to be nearly equal to 1250 MPa (Tahri et al. 2016).

The tensile strain percentage was measured in the literature for the leaflets and mesh fibers only. The leaflets tensile strain was $4.33 \pm 0.41\%$ for untreated fibers (Mohanty et al. 2014) and 2.7% for fibers extracted with retting in water. However, the fibers treated with acrylic acid had tensile strain percentage of $5.84 \pm 0.89\%$ (Rao and Rao 2007; Mohanty et al. 2014). As for mesh fibers, the untreated fibers had tensile strain within the range of $10.3 \pm 1.57\%$ (Abdal-hay et al. 2012), and NaOH treated mesh fibers between 12.8 ± 2.4 and 14.34% (Taha et al. 2006; Abdal-hay et al. 2012). Moreover, the tensile strain was found to be 24%, 16.48%, 10.31%, and 14.66% for fibers treated by water retting, $NaC_{12}H_{25}SO_4$, CCl_4, and Toluene/methanol/acetone treatments respectively.

Table 2 Comparison between mechanical properties of date palm fibers from different parts with different treatment techniques

Date Palm Part	Tensile strength (MPa)	Young's modulus (GPa)	Treatment technique	References
Midribs	11.4–208	10–30	Untreated	Abdel-Rahman et al. (1988), Ghulman et al. (2017)
	105–452.8	1.59–10.5	NaOH	Elseify et al. (2020)
Spadix stems	117 ± 35	–	Untreated	Amroune et al. (2015)
	195–680	8.81–19	NaOH	Taha et al. (2007), Amroune et al. (2015), Ibrahim et al. (2017)
Leaflets	32.43 ± 2.23	6.48 ± 0.89	Untreated	Mohanty et al. (2014)
	33.33 ± 3.49 to 1246.7 ± 444.7	5.5 to 26.1 ± 11.2	NaOH	AlMaadeed et al. (2013), Mohanty et al. (2014), Tahri et al. (2016)
	309	11.32	Water Retting	Rao and Rao (2007)
	70.72 ± 5.64	16.34 ± 1.72	Acrylic acid	Mohanty et al. (2014)
Mesh	233 ± 27.1 to 5359	0.1782 to 7.15 ± 2.0	Untreated	Taha et al. (2006), Abdal-hay et al. (2012)
	215.8 ± 80 to 7073	2.75–11	NaOH	Taha et al. (2006), Abdal-hay et al. (2012), Alsaeed et al. (2013), Elbadry (2014), Shalwan and Yousif (2014), Tahri et al. (2016)
	459	1.91	Water Retting	Rao and Rao (2007)
	204.7 (55.1)	4.43 (1.43)	$NaC_{12}H_{25}SO_4$	Taha et al. (2006)
	180.6 (65.7)	5.244 (1.53)	CCl_4	Taha et al. (2006)
	226.8 (18.1)	4.46 (1.66)	Toluene/methanol/acetone	Taha et al. (2006)

4.2 XRD Analysis

The crystalline structure, of the extracted fibers could be examined using X-ray diffractometer at room temperature with a monochromatic CuKα radiation source ($\lambda = 0.154$ nm) in step-scan mode with a 2θ angle ranging from 5° to 60° with a step of 0.04 and a scanning time of 5.0 min. The crystallinity index could be calculated from the peak height method using Eq. (1) and the percent crystallinity using Eq. (2).

$$CI = \frac{(I_{002} - I_{am})}{I_{002}} \times 100 \tag{1}$$

$$\%Cr = \frac{I_{002}}{(I_{002} + I_{am})} \times 100 \tag{2}$$

where I_{002} is the peak intensity at $2\theta = 22°$ and I_{am} is the peak intensity at $2\theta = 16°$.

The XRD results of four typical curves of treated and untreated DP fibers are shown in Fig. 13; each curve corresponds to one of DP fiber sources. In Fig. 13a DP midrib fibers showed three main peaks at 2θ equal 16.5° (110 plane), 22.6° (200 plane) and 34.5° (004 plane) (Elseify et al. 2020). As for DP spadix stem fibers, the

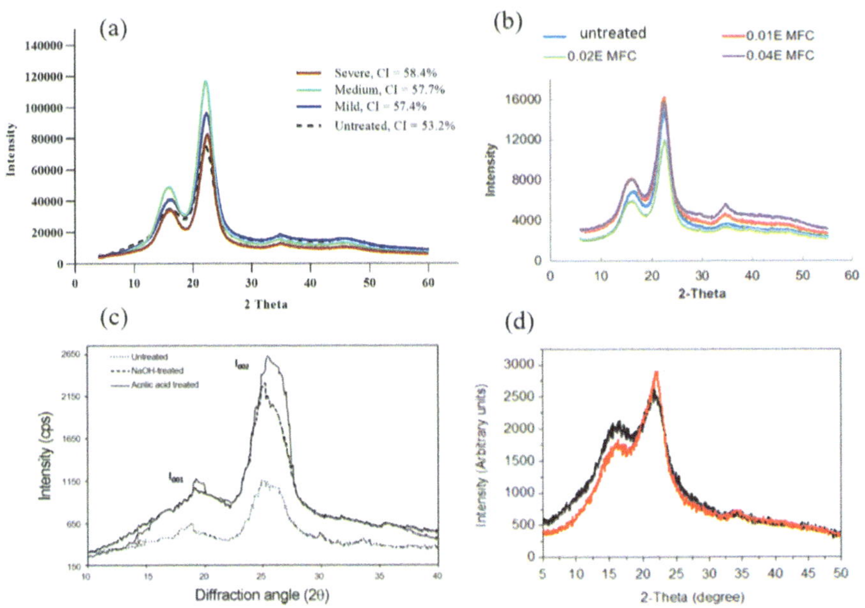

Fig. 13 XRD curves of **a** untreated and NaOH treated DP midrib fibers at different concentrations (Elseify et al. 2020), **b** untreated (0.0E MFC) and enzymatic treated DP spadix stems fibers at different concentrations (Hassan et al. 2014), **c** untreated and treated (NaOH and acrylic acid) DP leaflets fibers (Mohanty et al. 2014), and **d** NaOH treated (red) and untreated (black) DP mesh fibers (Abdal-hay et al. 2012)

Table 3 Date palm fibers crystallinity indices

Type of Fibers	Treatment	CrI (%)	References
Midrib	Raw	53.23	Elseify et al. (2020)
	NaOH	49.01–58.39	
Spadix stem	Raw	76	Hassan et al. (2014)
	Enzymatic	76–78	
Leaflet	Raw	45.53	Mohanty et al. (2014)
	NaOH	52.29	
	Acrylic acid	55.07	
Mesh	Raw	19.9	Abdal-hay et al. (2012)
	NaOH	38.5	

XRD patterns in Fig. 13b showed 2 peaks at 2θ equal 15° (101 and 10$\bar{1}$ planes) and 27° (002 plane) (Hassan et al. 2014). Considering the XRD curves of DP leaflets fibers, shown in Fig. 13c, it can be seen that 2 peaks exist at 2θ equal 19.2° (101 plane) and 25.4° (002 plane). Moreover, the leaflet fibers XRD curves showed that there isn't significant difference between the NaOH and acrylic treatments (Mohanty et al. 2014). Finally, XRD results of DP mesh fibers, Fig. 13d showed 2 peaks at 2θ equal 16° and 22° (Abdal-hay et al. 2012).

The crystallinity index (CrI) of DP fibers are presented in Table 3. It can be observed that the CrI of fibers extracted from the midribs and leaflets are within the same range. However, the CrI of spadix stem fibers was found to be the highest among the four fiber sources and the mesh fibers had the lowest CrI value. The values of the crystallinity indices were found to be affected with the severity of the treatments (Hassan et al. 2014; Elseify et al. 2020).

5 Chemical Properties

The chemical composition could be determined using TAPPI standard methods. The test is carried out to determine the weight percentages of cellulose, hemicellulose, and lignin following standards number T 203 cm-09, T 223 cm-10, and T 222 om-11 respectively. The chemical analysis results are presented in Table 4. The chemical constituents are compared with respect to their cultivar. It was noted that the weight percent of cellulose in date palm fibers ranges from 35 wt% till nearly 70 wt%. Whereas, the amount of lignin ranged from nearly 12 wt% to 36 wt%. Finally, hemicellulose weight percentage was sometimes as high as 30 wt% or low as 15 wt% (Amirou et al. 2013; Nasser 2014; Hegazy and Ahmed 2015; Nasser et al. 2016; Elseify et al. 2020). Therefore, it should be taken into consideration that the weight percentages of cellulose, hemicellulose, and lignin depend on the cultivar and the efficiency of the fiber extraction process.

Table 4 Chemical constituents of date palm fibers

Date palm part	Cultivar	Cellulose (wt%)	Lignin (wt%)	Hemicellulose (wt%)	References
Midribs	Barhi	44.14–69.31	15.3–29.6	15.38–29.93	Nasser (2014), Hegazy and Ahmed (2015), Elseify et al. (2020)
	Khodry	45.74 ± 1.7	29.37 ± 2.1	24.9 ± 2	Nasser (2014)
	Sukkari	41.5–47.17	26.68–30.19	22.3–28.16	Nasser (2014), Hegazy and Ahmed (2015), Nasser et al. (2016)
	Saqie	48.86	31.28	19.86	Hegazy and Ahmed (2015)
Spadix stems	Sukkari	43.05	29.47	27.48	Nasser et al. (2016)
Leaflets	Sukkari	47.14	36.73	16.13	Nasser et al. (2016)
	–	29–58	15.3	–	Pandey and Ghosh (1995), Saadaoui et al. (2013), Mirmehdi et al. (2014)
Mesh	Sukkari	47.5	39.86	12.64	Nasser et al. (2016)
	–	43–50	24–32	8–19	Taha et al. (2006), Saadaoui et al. (2013), Mekhermeche et al. (2016)
Trunk	Sukkari	39.37	30.32	30.31	Nasser et al. (2016)
	–	43.7	16.94	38.14	Amirou et al. (2013)

6 Thermal Properties

The thermal degradation analysis of DP fibers could be analyzed using a thermogravimetric analyzer. Where 15 mg of the fiber sample is heated from 30 to 600 °C with a heating rate of 10°C/min under a dried O_2 atmosphere and finally pyrolyzed from 600 to 700 °C at 50 °C/min. The apparatus records the mass loss as a function of temperature. The thermal degradation of date palm fibers was found to be similar to other natural fibers following four main stages (Ali and Alabdulkarem 2017; Elseify et al. 2019). The first stage corresponds to dehydration of fibers at a temperature between 35 and 130 °C. While the second stage is associated with the degradation of hemicellulose at temperature between 200 and 320 °C. As for the third stage, it is due to the degradation of cellulose at a temperature between 330 and 360 °C. Then finally, the fourth stage is due to the degradation of lignin which occurred at a temperature above 400 °C (Elseify et al. 2020). Figure 14a shows typical thermogravimetric analysis (TGA) curves showing the four degradation stages of date palm midrib fibers. While Fig. 14b shows the TGA curves of DP spadix stems treated with NaOH at different durations (Taha et al. 2007).

The possibility of using date palm fibers in the production of insulating panels was investigated by several researchers (Al-Sulaiman 2003; Benmansour et al. 2014). It was reported that the thermal conductivity of DP mesh fibers ranged between 0.475 and 0.0697 W/mK and that of the leaflets equals 0.087 W/mK (Agoudjil et al. 2011; Ali and Alabdulkarem 2017). Benmansour et al. reported that increasing the midrib fiber weight content increased the insulation capability of the reinforced panels (Benmansour et al. 2014). The thermal conductivity of DP fibers was found to be within an accepted comparable range to other natural fibers as shown in Fig. 15 (Al-Oqla and Salit 2017).

7 Date Palm Fibers and Natural Fibers

The fibers extracted from date palm still needs further investigation, and the fact that date palm has four distinct sources of fibers is still not considered among researchers. Moreover, the process of DP fiber extraction still needs further researching to optimize the process. However, there were few successful extraction attempts and the properties of these extracted fibers showed very promising results when compared to other natural fibers (Elseify et al. 2020). Table 5 shows the properties of natural fibers in comparison to date palm fibers.

As can be seen in the Table 5 the physical properties of date palm fibers are better than some of fibers presented. For instant the density of DP fibers, which reaches a maximum of 1.32 gm/cm^3, is lower than that of cotton, jute, flax, and hemp. Moreover, the fibers extracted from date palm midribs could reach 3-m long since fibers could be extracted with the full length of the midrib. This is considered an advantage when compared to other natural fibers that have original lengths less than 1 m.

Considering the chemical composition of DP fibers, it can be observed from the discussion in this chapter and from Table 5 that the amount of cellulose weight content can reach up to nearly 70% depending on the extraction process/treatment severity. Hence, larger amount of cellulose and lower amounts of hemicellulose and lignin could be achieved by optimizing the extraction process (Elseify et al. 2018, 2020).

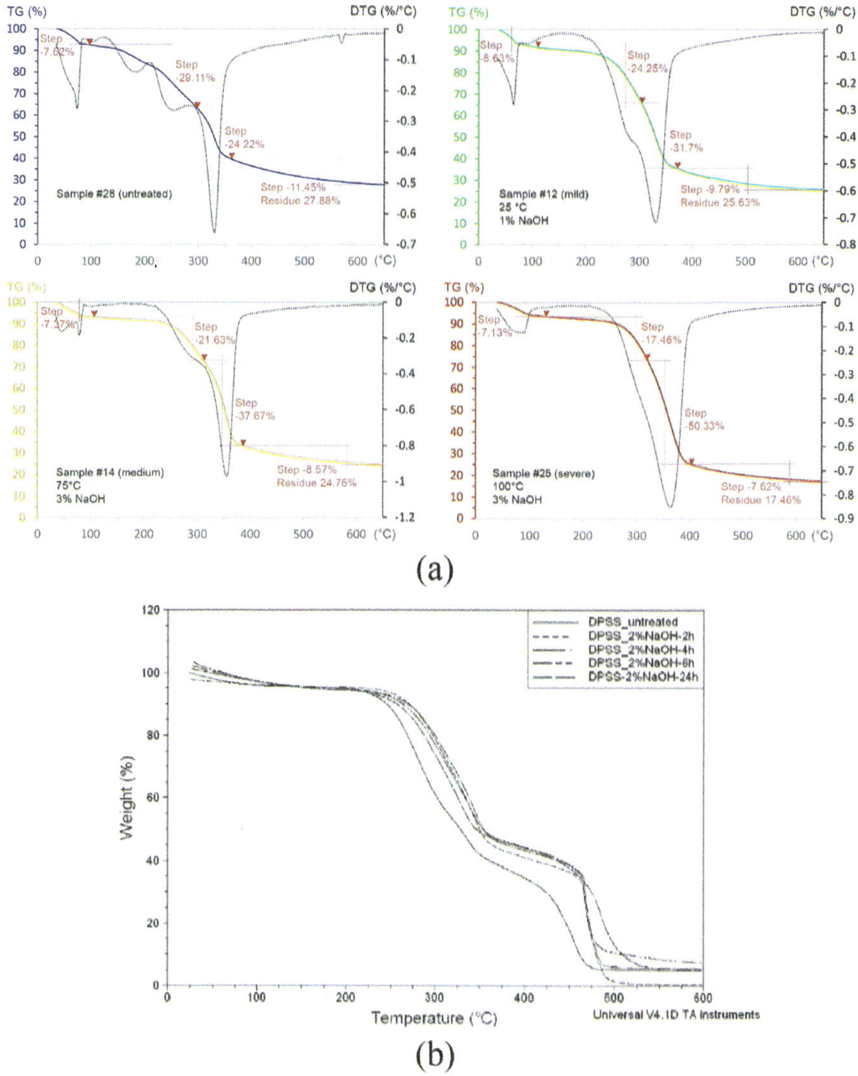

Fig. 14 TGA/DTG curves for **a** DP midrib fibers untreated and treated at different conditions (Elseify et al. 2020) and **b** DP spadix stem fibers treated with 2% NaOH solution at different durations (Taha et al. 2007)

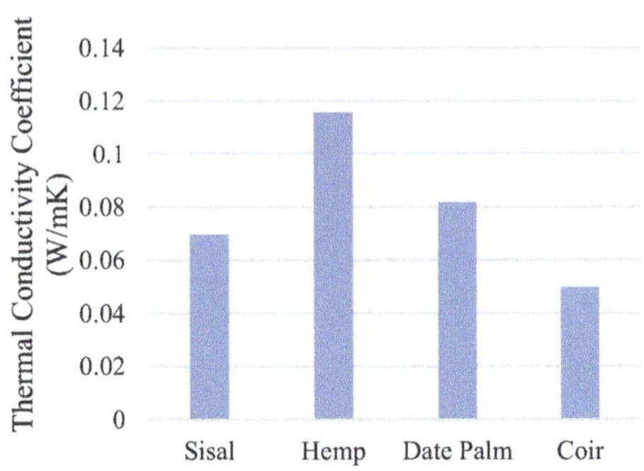

Fig. 15 Thermal conductivity coefficient values in W/mK of sisal, hemp, date palm and coir fibers (Al-Oqla and Salit 2017)

As for the mechanical properties, date palm fibers have comparable properties to other vegetable fibers. However, it should be noted that their fiber extraction is still not ideal. Hence, the reported properties do not represent the full potential of such fibers.

8 Conclusion

Date palm as a plant is widely available in the MENA region and it is originally cultivated for its fruit. Hence, the price of the fiber sources is very low compared to the other natural fiber sources. Fibers could be extracted from four different parts of the date palm; midribs, spadix stems, leaflets, and mesh. Date palm fibers have great potential to be utilized in various applications. The advantage of date palm fibers is that they have long lengths since the fiber sources are initially long. The literature showed that date palm fibers can have high cellulose purity up to 70%. The density of date palm is between 1.18 and 1.45 gm/cm^3 which is comparable to other widely used natural fibers. Date palm fibers can be used in automotive applications, in particleboard manufacturing and as a reinforcement to panels to be used as thermal or sound insulation. Date palm, as a biodegradable alternative to expensive natural fibers, have high potential of being used in various applications. Extracting and utilizing date palm fibers as composite reinforcement will help with the MENA region economic situations by creating an entire value chain that starts from the palm growers ending with the composite manufacturers.

Table 5 Comparison of date palm and other natural fibers properties

	Fiber	Physical properties			Chemical properties			Mechanical properties				References
		Density (gm/cm^3)	Diameter (μm)	Length (mm)	Cellulose (wt%)	Hemicellulose (wt%)	Lignin (wt%)	Tensile strength (MPa)	Young's modulus (GPa)	Specific Strength ($MPa/gm/cm^3$)	Elongation (%)	
Date palm	Midrib	1.18–1.32	10–510	1.01–300	41.5–69	15.4–30	15.3–31.3	105–452.8	30	79.55–343.03	–	Khristova et al. (2005), Abdel-Aal et al. (2015), Elseify et al. (2020)
	Spadix stem	–	20.3–200	0.56–19	43.05	27.48	29.47	195–680	8.81–19	–	–	Hassan et al. (2014), Nasser (2014), Ibrahim et al. (2017)
	Leaflets	–	8–10,000	3.42–50	29–58	16.13	15.3	33–309	5.5–26.1	–	–	Abu-Sharkh and Hamid (2004), Saadaoui et al. (2013), Mohanty et al. (2014), Nasser et al. (2016)
	Mesh	1.3–1.45	100–2000	10–300	39–43.7	17–30.32	31.3–38.2	215–459	2.75–11	165.38–353.08	–	Abdal-hay et al. (2012), Mirmehdi et al. (2014), Amroune et al. (2015)
Other fiber sources	Cotton	1.5–1.6	12–38	10–60	85–90	5.7	–	287–800	5.5–13	190–530	7–8	Baltazar-Y-Jimenez and Sain (2012), Ho et al. (2012), Khan et al. (2018), Mann et al. (2018)

(continued)

Table 5 (continued)

Fiber	Physical properties			Chemical properties			Mechanical properties				References
	Density (gm/cm^3)	Diameter (μm)	Length (mm)	Cellulose (wt%)	Hemicellulose (wt%)	Lignin (wt%)	Tensile strength (MPa)	Young's modulus (GPa)	Specific Strength ($MPa/gm/cm^3$)	Elongation (%)	
Jute	1.3–1.5	25–200	1.5–120	61–75.5	13.6–20.4	5–13	393–800	10–55	300–610	1.16–1.5	Baltazar-Y-Jimenez and Sain (2012), Ho et al. (2012), Khan et al. (2018), Mann et al. (2018)
Flax	1.4–1.5	40–600	5–900	70–75.2	8.6–20.6	2.2–5	345–1830	27–80	227–1220	2.7–3.2	Baltazar-Y-Jimenez and Sain (2012), Ho et al. (2012), Khan et al. (2018), Mann et al. (2018)
Hemp	1.4–1.48	10–500	5–55	70–75.1	2–22.4	3.5–8	550–1110	58–70	370–740	1.6	Pickering (2008), Baltazar-Y-Jimenez and Sain (2012), Al-Oqla and Sapuan (2014), Khan et al. (2018), Mann et al. (2018)
Ramie	1.4–1.5	18–80	40–1200	68–85	3–16.7	0.5	400–938	44–128	258–620	1.2–3.8	Pickering (2008), Baltazar-Y-Jimenez and Sain (2012), Khan et al. (2018), Mann et al. (2018)

(continued)

Table 5 (continued)

Fiber	Physical properties			Chemical properties			Mechanical properties				References
	Density (gm/cm^3)	Diameter (μm)	Length (mm)	Cellulose (wt%)	Hemicellulose (wt%)	Lignin (wt%)	Tensile strength (MPa)	Young's modulus (GPa)	Specific Strength ($MPa/gm/cm^3$)	Elongation (%)	
Sisal	1.2–1.5	8–200	900	47.6–78	10–17.8	8–14	468–855	9.4–28	349.3–610	3–7	Katiyar et al. (2019), Baltazar-Y-Jimenez and Sain (2012), Ho et al. (2012), Al-Oqla and Sapuan (2014), Khan et al. (2018), Mann et al. (2018)
Coir	1.1–1.46	10–460	20–150	32–43	<1	40–45	131–220	4–6	110–180	15–40	Baltazar-Y-Jimenez and Sain (2012), Ho et al. (2012), Al-Oqla and Sapuan (2014), Dicker et al. (2014), Khan et al. (2018), Mann et al. (2018)
Kenaf	1.2–1.45	12–37	0.4–11	45–57	21.5	8–13	295–930	53	641	1.6–6.9	Pickering (2008), Baltazar-Y-Jimenez and Sain (2012), Khan et al. (2018)

(continued)

Table 5 (continued)

Fiber	Physical properties			Chemical properties			Mechanical properties				References
	Density (gm/cm^3)	Diameter (μm)	Length (mm)	Cellulose (wt%)	Hemicellulose (wt%)	Lignin (wt%)	Tensile strength (MPa)	Young's modulus (GPa)	Specific Strength ($MPa/gm/cm^3$)	Elongation (%)	
Abaca	1.1–1.5	132–226	–	56–63.7	17.5	15.1	705–1041	9.8–14.8	267–937.8	3–12	Baltazar-Y-Jimenez and Sain (2012), Khan et al. (2018)
Oil Palm	0.7–1.55	–	–	60–65	–	11–19	248	3.2	25	14	Katiyar et al. (2019), Ho et al. (2012)
Bamboo	0.6–1.4	7–33	–	26–43	–	21–31	500–1000	35.9	454–1098.9	1.4	Katiyar et al. (2019), Pickering (2008), Baltazar-Y-Jimenez and Sain (2012), Dicker et al. (2014), Khan et al. (2018)

(continued)

Table 5 (continued)

Fiber	Physical properties			Chemical properties			Mechanical properties				References
	Density (gm/cm^3)	Diameter (μm)	Length (mm)	Cellulose (wt%)	Hemicellulose (wt%)	Lignin (wt%)	Tensile strength (MPa)	Young's modulus (GPa)	Specific Strength ($MPa/gm/cm^3$)	Elongation (%)	
Coconut	1.5	–	–	28–29	–	16–45	500	2.5	436	3.36	Saba et al. (2014), Khan et al. (2018)
Banana	1.35	–	–	60–65	6–8	5–10	529–914	8–32	444	3–10	Katiyar et al. (2019), Khan et al. (2018), Mann et al. (2018)
Pineapple	1.5	8–41	3–8	69.5	–	4.4	170–1627	60–82	–	2.4	Katiyar et al. (2019), Pickering (2008), Saba et al. (2014)
Bagasse	1.2	10–34	0.8–2.8	–	–	–	20–290	–	–	0.9	Katiyar et al. (2019), Pickering (2008)

References

Abdal-hay A, Suardana NPG, Jung DY et al (2012) Effect of diameters and alkali treatment on the tensile properties of date palm fiber reinforced epoxy composites. Int J Precis Eng Manuf 13:1199–1206. https://doi.org/10.1007/s12541-012-0159-3

Abdel-Aal MA, Nasser RA, Khan PR, Al-Mefarrej HA (2015) Anatomical and physical characterization of some lignocellulosic residues and their suitability for fibre production. J Environ Biol 36:537–542

Abdel-Rahman HH, Al-Juruf R, Ahmad F, Alam I (1988) Physical, mechanical and durability characteristics of date palm frond stalks as reinforcement in structural concrete. Int J Cem Compos Light Concr 10:175–181

Abu-Sharkh BF, Hamid H (2004) Degradation study of date palm fibre/polypropylene composites in natural and artificial weathering: mechanical and thermal analysis. Polym Degrad Stab 85:967–973. https://doi.org/10.1016/j.polymdegradstab.2003.10.022

Agoudjil B, Benchabane A, Boudenne A et al (2011) Renewable materials to reduce building heat loss: characterization of date palm wood. Energy Build 43:491–497. https://doi.org/10.1016/J.ENBUILD.2010.10.014

Al-Khanbashi A, Al-Kaabi K, Hammami A (2005) Date palm fibers as polymeric matrix reinforcement: fiber characterization. Polym Compos 26:486–497. https://doi.org/10.1002/pc.20118

Al-Oqla FM, Salit MS (2017) Materials selection for natural fiber composites. Woodhead Publishing, Elsevier, Cambridge, USA

Al-Oqla FM, Sapuan SM (2014) Natural fiber reinforced polymer composites in industrial applications: feasibility of date palm fibers for sustainable automotive industry. J Clean Prod 66:347–354. https://doi.org/10.1016/j.jclepro.2013.10.050

Al-Rifaie WN, Al-Niami M (2016) Mechanical performance of date palm fibre-reinforced gypsums. Innov Infrastruct Solut 1:18. https://doi.org/10.1007/s41062-016-0022-y

Al-Sulaiman FA (2003) Date palm fibre reinforced composite as a new insulating material. Int J Energy Res 27:1293–1297. https://doi.org/10.1002/er.957

Alajmi M, Shalwan A (2015) Correlation between mechanical properties with specific wear rate and the coefficient of friction of graphite/epoxy composites. Materials (Basel) 8:4162–4175. https://doi.org/10.3390/ma8074162

Alawar A, Hamed AM, Al-Kaabi K (2009) Characterization of treated date palm tree fiber as composite reinforcement. Compos Part B Eng 40:601–606. https://doi.org/10.1016/j.compositesb.2009.04.018

Ali ME, Alabdulkarem A (2017) On thermal characteristics and microstructure of a new insulation material extracted from date palm trees surface fibers. Constr Build Mater 138:276–284. https://doi.org/10.1016/j.conbuildmat.2017.02.012

AlMaadeed MA, Kahraman R, Noorunnisa Khanam P, Al-Maadeed S (2013) Characterization of untreated and treated male and female date palm leaves. Mater Des 43:526–531. https://doi.org/10.1016/j.matdes.2012.07.028

AlMaadeed MA, Nógellová Z, Janigová I, Krupa I (2014) Improved mechanical properties of recycled linear low-density polyethylene composites filled with date palm wood powder. Mater Des 58:209–216. https://doi.org/10.1016/j.matdes.2014.01.051

Alsaeed T, Yousif BF, Ku H (2013) The potential of using date palm fibres as reinforcement for polymeric composites. Mater Des 43:177–184. https://doi.org/10.1016/j.matdes.2012.06.061

Amirou S, Zerizer A, Pizzi A et al (2013) Particleboards production from date palm biomass. Eur J Wood Wood Prod 71:717–723. https://doi.org/10.1007/s00107-013-0730-3

Amroune S, Bezazi A, Belaadi A et al (2015) Tensile mechanical properties and surface chemical sensitivity of technical fibres from date palm fruit branches (Phoenix dactylifera L.). Compos Part A Appl Sci Manuf 71:98–106. https://doi.org/10.1016/j.compositesa.2014.12.011

Amroune S, Bezazi A, Dufresne A et al (2019) Investigation of the date palm fiber for green composites reinforcement: thermo-physical and mechanical properties of the fiber. J Nat Fibers 0478:1–18. https://doi.org/10.1080/15440478.2019.1645791
Arenas JP, Asdrubali F (2017) Handbook of ecomaterials. https://doi.org/10.1007/978-3-319-48281-1
Baltazar-Y-Jimenez A, Sain M (2012) Natural fibres for automotive applications. In: Handbook of natural fibres. Woodhead Publishing Limited, pp 219–253
Bendahou A, Habibi Y, Kaddami H, Dufresne A (2009) Physico-chemical characterization of palm from Phoenix Dactylifera-L, preparation of cellulose whiskers and natural rubber-based nanocomposites. J Biobased Mater Bioenergy 3:81–90. https://doi.org/10.1166/jbmb.2009.1011
Benmansour N, Agoudjil B, Gherabli A et al (2014) Thermal and mechanical performance of natural mortar reinforced with date palm fibers for use as insulating materials in building. Energy Build 81:98–104. https://doi.org/10.1016/j.enbuild.2014.05.032
Boukhattem L, Boumhaout M, Hamdi H et al (2017) Moisture content influence on the thermal conductivity of insulating building materials made from date palm fibers mesh. Constr Build Mater 148:811–823. https://doi.org/10.1016/j.conbuildmat.2017.05.020
Boumhaout M, Boukhattem L, Hamdi H et al (2017) Thermomechanical characterization of a bio-composite building material: mortar reinforced with date palm fibers mesh. Constr Build Mater 135:241–250. https://doi.org/10.1016/j.conbuildmat.2016.12.217
d'Almeida JRM, Aquino RCMP, Monteiro SN (2006) Tensile mechanical properties, morphological aspects and chemical characterization of piassava (Attalea funifera) fibers. Compos Part A Appl Sci Manuf 37:1473–1479. https://doi.org/10.1016/j.compositesa.2005.03.035
Dehghani A, Madadi Ardekani S, Al-Maadeed MA et al (2013) Mechanical and thermal properties of date palm leaf fiber reinforced recycled poly (ethylene terephthalate) composites. Mater Des 52:841–848. https://doi.org/10.1016/j.matdes.2013.06.022
Diaz JPV, Silva FDA, D'Almeida JRM (2016) Effect of peach palm fiber microstructure on its tensile behavior. BioResources 11:10140–10157. https://doi.org/10.15376/biores.11.4.10140-10157
Dicker MPM, Duckworth PF, Baker AB et al (2014) Green composites: a review of material attributes and complementary applications. Compos Part A Appl Sci Manuf 56:280–289. https://doi.org/10.1016/j.compositesa.2013.10.014
Elbadry EA (2014) Agro-residues: surface treatment and characterization of date palm tree fiber as composite reinforcement. J Compos
Elseify LA, Midani M, Hassanin A et al (2020) Long textile fibers extracted from date palm (Phoenix Dactylifera L.) midribs: effect of the alkaline treatment on the mechanical properties. Ind Crops Prod 115:112466
Elseify LA, Midani M, Shihata LA (2019) Review on cellulosic fibers extracted from date palms (Phoenix Dactylifera L.) and their applications. Cellulose 6. https://doi.org/10.1007/s10570-019-02259-6
Elseify LA, Midani M, Shihata LA (2018) Investigating the effect of the chemical treatment on the properties of a novel microfibrillated long date palm fibers. German University in Cairo (GUC)
Fangueiro R, Rana S (eds) (2018) Advances in natural fibre composites: raw materials, processing and analysis. Springer Nature
Fiore V, Di Bella G, Valenza A (2015) The effect of alkaline treatment on mechanical properties of kenaf fibers and their epoxy composites. Compos Part B Eng 68:14–21. https://doi.org/10.1016/j.compositesb.2014.08.025
Gholami M, Ahmadi MS, Tavanaie MA, Khajeh Mehrizi M (2017) Effect of oxygen plasma treatment on tensile strength of date palm fibers and their interfacial adhesion with epoxy matrix. Sci Eng Compos Mater. https://doi.org/10.1515/secm-2017-0102
Ghulman HA, Metwally MN, Alhazmi MW (2017) Study on the benefits of using the date palm trees residuals in Saudi Arabia for development of the non-traditional wooden industry. AIP Conf Proc 1814. https://doi.org/10.1063/1.4976231
Hakeem KR, Jawaid M, Rashid U (2014) Biomass and bioenergy processing and properties. Springer

Hassan ML, Bras J, Hassan EA et al (2014) Enzyme-assisted isolation of microfibrillated cellulose from date palm fruit stalks. Ind Crops Prod 55:102–108. https://doi.org/10.1016/j.indcrop.2014.01.055

Hegazy S, Ahmed K (2015) Effect of date palm cultivar, particle size, panel density and hot water extraction on particleboards manufactured from date palm fronds. Agriculture 5:267–285. https://doi.org/10.3390/agriculture5020267

Ho MP, Wang H, Lee JH et al (2012) Critical factors on manufacturing processes of natural fibre composites. Compos Part B Eng 43:3549–3562. https://doi.org/10.1016/j.compositesb.2011.10.001

Ibrahim H, Farag M, Megahed H, Mehanny S (2014) Characteristics of starch-based biodegradable composites reinforced with date palm and flax fibers. Carbohydr Polym 101:11–19. https://doi.org/10.1016/j.carbpol.2013.08.051

Ibrahim H, Mehanny S, Darwish L, Farag M (2017) A comparative study on the mechanical and biodegradation characteristics of starch-based composites reinforced with different lignocellulosic fibers. J Polym Environ 0:1–14. https://doi.org/10.1007/s10924-017-1143-x

Jaber MA, Hammadi KJ, Abdul Karem AA, Abd-Alrazak M (2016) Physical and mechanical properties of medium density fiber board. Asian J Appl Sci 04:972–978

Katiyar JK, Bhattacharya S, Patel VK, Kumar V (eds) (2019) Automotive tribology

Khan T, Hameed Sultan MT, Bin AAH (2018) The challenges of natural fiber in manufacturing, material selection, and technology application: a review. J Reinf Plast Compos 37:770–779. https://doi.org/10.1177/0731684418756762

Khristova P, Kordsachia O, Khider T (2005) Alkaline pulping with additives of date palm rachis and leaves from Sudan. Bioresour Technol 96:79–85. https://doi.org/10.1016/j.biortech.2003.05.005

Kozłowski RM (2012) Handbook of natural fibers—volume 1: types, properties and factors affecting breeding and cultivation

Kriker A, Debicki G, Bali A et al (2005) Mechanical properties of date palm fibres and concrete reinforced with date palm fibres in hot-dry climate. Cem Concr Compos 27:554–564. https://doi.org/10.1016/j.cemconcomp.2004.09.015

Mann GS, Singh LP, Kumar P, Singh S (2018) Green composites: a review of processing technologies and recent applications. J Thermoplast Compos Mater. https://doi.org/10.1177/0892705718816354

Mekhermeche A, Kriker A, Dahmani S (2016) Contribution to the study of thermal properties of clay bricks reinforced by date palm fiber. In: AIP conference proceedings. AIP Publishing, p 030004

Midani M (2017) Date palm fibre composites : a novel and sustainable material for the aerospace industry

Midani M, Elseify LA, Hassanin A, Hamouda T (2018) Optimizing the extraction process of a novel long fibrillated fibers from the midribs of date palm (Phoenix Dactylifera L.). In: 1st world conference on by-products of palm trees and their application

Mirmehdi SM, Zeinaly F, Dabbagh F (2014) Date palm wood flour as filler of linear low-density polyethylene. Compos Part B Eng 56:137–141. https://doi.org/10.1016/j.compositesb.2013.08.008

Mohanty JR (2017) Investigation on solid particle erosion behavior of date palm leaf fiber- reinforced polyvinyl pyrrolidone composites. J Thermoplast Compos Mater 30:1003–1016. https://doi.org/10.1177/0892705715614079

Mohanty JR, Das SN, Das HC, Swain SK (2014) Effect of chemically modified date palm leaf fiber on mechanical, thermal and rheological properties of polyvinylpyrrolidone. Fibers Polym 15:1062–1070. https://doi.org/10.1007/s12221-014-1062-6

Nam TH, Ogihara S, Kobayashi S (2012) Interfacial, mechanical and thermal properties of coir fiber-reinforced poly(lactic acid) biodegradable composites. Adv Compos Mater 21:103–122. https://doi.org/10.1163/156855112X629559

Nasser R, Salem M, Hiziroglu S et al (2016) Chemical analysis of different parts of date palm (Phoenix dactylifera L.) using ultimate, proximate and thermo-gravimetric techniques for energy production. Energies 9:374. https://doi.org/10.3390/en9050374

Nasser RA (2014) An evaluation of the used if midribs from common date palm cultivars grown in Saudi Arabia for energy production. Bioresource.com 9:4343–4357

Pandey SN, Ghosh SK (1995) The chemical nature of date-palm (phoenix dactylifera- l) leaf fibre. J Text Inst 86:487–489. https://doi.org/10.1080/00405009508658775

Pickering KL (ed) (2008) Properties and performance of natural-fibre composites. Woodhead Publishing Limited

Rao KMM, Rao KM (2007) Extraction and tensile properties of natural fibers: Vakka, date and bamboo. Compos Struct 77:288–295. https://doi.org/10.1016/j.compstruct.2005.07.023

Saadaoui N, Rouilly A, Fares K, Rigal L (2013) Characterization of date palm lignocellulosic by-products and self-bonded composite materials obtained thereof. Mater Des 50:302–308. https://doi.org/10.1016/j.matdes.2013.03.011

Saba N, Tahir PM, Jawaid M (2014) A review on potentiality of nano filler/natural fiber filled polymer hybrid composites. Polymers (Basel) 6:2247–2273. https://doi.org/10.3390/polym6082247

Shalwan A, Yousif BF (2014) Influence of date palm fibre and graphite filler on mechanical and wear characteristics of epoxy composites. Mater Des 59:264–273. https://doi.org/10.1016/j.matdes.2014.02.066

Taha I, Steuernagel L, Ziegmann G (2006) Chemical modification of date palm mesh fibres for reinforcement of polymeric materials. Part I: examination of different cleaning methods. Polym Polym Compos 14:767

Taha I, Steuernagel L, Ziegmann G (2007) Optimization of the alkali treatment process of date palm. Compos Interfaces 14:669–684

Tahri I, Devin IZ, Ruelle J, Brosse N (2016) Extraction and characterization of fibers from palm tree 11:7016–7025. https://doi.org/10.15376/biores.11.3.7016-7025

Tioua T, Kriker A, Barluenga G, Palomar I (2017) Influence of date palm fiber and shrinkage reducing admixture on self-compacting concrete performance at early age in hot-dry environment. Constr Build Mater 154:721–733. https://doi.org/10.1016/j.conbuildmat.2017.07.229

Date Palm Fiber Composites: Mechanical Testing and Properties

Tamer Hamouda and Ahmed H. Hassanin

Abstract Mechanical Testing and evaluation of the date palm reinforced composites will be discussed in this chapter in order to understand their behavior. This chapter will focus on tensile, flexural and impact properties of several date palm composite materials. Hybrid date palm fiber composite with other natural fibers and other materials such as polyester, fiber glass, and carbon fibers will be discussed in this chapter. On the other hand, comparison between date palm fiber reinforced composite using thermoplastic and thermoset matrix will be presented in order to evaluated the interfacial adherence between the date palm fibers and different resin types. From environmental point, date palm reinforced composite with recycled polymer matrix will be discussed in order to draw some attention to the recyclability of the date palm composite materials.

1 Introduction

Natural fibers such as date palm fibers are considered an alternative reinforcement for the Fiber Reinforced Polymer (FRP). This is due to their attractive properties such as high specific strength, low cost, sustainability, accepted mechanical properties and biodegradable (Mahdavi et al. 2010), which make the date palm composite widely used in different applications (Satyanarayana et al. 2009; Khanam and AlMaadeed 2014). In this section, date palm fiber composite testing including different types of mechanical testing such as tensile strength, stiffness, impact resistance, and flexural rigidity will be discussed.

T. Hamouda (✉)
Textile Research Division, National Research Centre, 33 El Bohouth st., Dokki 12622, Egypt
e-mail: tehamoud@ncsu.edu

A. H. Hassanin
Materials Science and Engineering, Egypt Japan University for Science and Technology (EJUST), New Borg El-Arab City, Alexandria, Egypt

Department of Textile Engineering, Alexandria University, Alexandria, Egypt

M. Midani et al. (eds.), *Date Palm Fiber Composites*, Composites Science and Technology,
https://doi.org/10.1007/978-981-15-9339-0_9

2 Hybrid Date Palm Fiber-Reinforced Composite

Sandwich composites made of polyester resin reinforced with hybrid materials made of date palm fibers and glass fiber were evaluated (Mahdavi et al. 2010). Samples were prepared with a hybrid fiber weight fraction of 30 and 70% matrix. Date palm leaf stalk fiber mat, palm leaf sheath fiber mat, and its mixture were used to fabricate the hybrid reinforcements. All-natural fibers were treated with 8% $KMnO_4$ to enhance the date palm fiber properties. Samples were fabricated using a compression molding technique. Tensile test, flexural test, and impact test were performed according to ASTM standards D638, D 790, and D256 respectively. The test result shows that the average tensile strength of the hybrid date palm/glass fibers was between 126 and 159 MPa, the average flexural strength was between 62.6 and 208 Mpa, and the average impact strength was between 0.4 and 0.94 J/mm. Results also indicate that the alkali treatments enhanced the matrix wetting by separating the fiber bundles into smaller fibers.

Another study evaluated the mechanical properties of Hybrid composite using date palm fibers and flax fibers as a reinforcement with a biodegradable matrix made of starch (Ibrahim et al. 2013). The hybrid reinforcement ratio of date palm and flax fibers ranged between 20 and 80 wt%. The main objective of this study is to present a fully biodegradable fiber-reinforced composite using biodegradable matrix and natural fibers as a reinforcement. Thermoplastic starch prepared using 30 wt% glycerin and 20 wt% distilled water under temperature ranged between 60 and 80 °C according to Hulleman et al. (1998). Date palm fibers were initially retted for 2 days in water at room temperature. Fibers were cut into 20–30 mm and then mechanically treated followed by chemical treatments with 5% NaOH for 3 h at a temperature of 90 °C. Treated fibers were washed several times and neutralized with acetic acid to remove any traces of NaOH. Composite samples were prepared by hot pressing the emulsified thermoplastic starch with hybrid fibers with fiber contents ranged between 20 and 80 wt%. Results indicate that increasing fiber content to 50% caused increases in the tensile strength and modulus to reach 31 MPa and 2.8 GPa respectively. However, when fiber contents increased to 60 and 80 wt% the tensile strength and modulus decreased, this behavior can be attributed to the increased porosity (Fig. 1).

Hybrid laminates of date palm fiber, polyester and carbon fiber mats with epoxy resin were evaluated for their mechanical properties (Raghavendra et al. 2018). Date palm fibers extracted from fronds were cut into 5 mm length using the mechanical crushing machine. extracted fibers were chemically treated with 10% NaOH for 8 h followed by washing. Laminates of DPF + Polyester mat (NP), DPF + polyester + carbon mats (NPC), and DPF + carbon mats (NC) were prepared by vacuum bagging techniques. Polyester and carbon mates were bi-directional weave of 70gsm. Laminates prepared with a thickness of 4 mm. The tensile test result's value indicated that NP samples exhibit the highest tensile strength value of 43.41 MPa and NPC sample exhibit the lowest value of 32.92 MPa, while the tensile modulus results showed that NPC samples exhibit the highest value of 26.37 MPa. the study reported

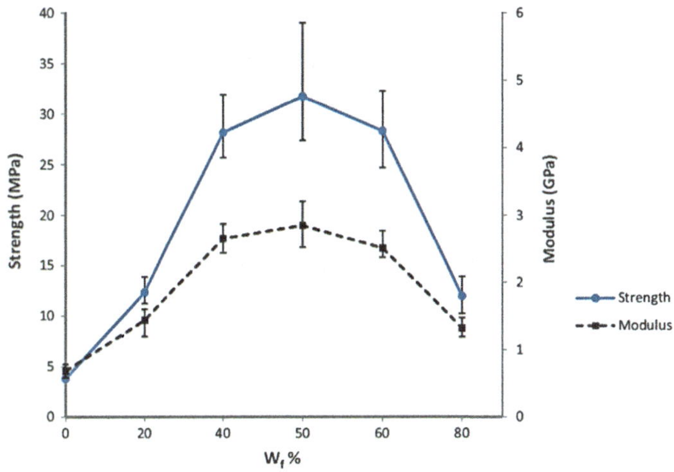

Fig. 1 Hybrid composite tensile and modulus properties (Ibrahim et al. 2013)

that the highest value of the tensile strength of NP sample is attributed to the good adhesion between DPF and polyester mat in the epoxy matrix. However, NP sample exhibited a low tensile modulus value of 15.60 MPa due to the ductile behavior of polyester mat. On the other hand, the highest flexural strength value of 84.83 MPa is recorded for the NP sample. This value is higher than the flexural strength of pure resin by 67.14% higher. The study attributed this improvement to the better adhesion between the polyester mat and DPF. The lowest value is recorded for the NPC samples with a value of 22.54 MPa.

The effect of fiber length of the reinforced natural fibers (coir and date palm) on the composite mechanical properties was investigated (Sharath et al. 2016). Coir fibers and date palm fibers were extracted and treated with 5% NaOH and washed and then dried in hot air. Coir fibers and date palm fibers were cut in different length namely 3, 6, 9, 12, and 15 mm and mixed in 50/50 ratio. Composite samples prepared with hand lay-up technique using epoxy resin. Tensile test results depict that as the fiber length increased the tensile strength increased until the fiber length reached 12 mm. Recorded values show increasing by 3.94% in the tensile strength of the fibers with 3 mm length compared to unreinforced samples. This increase reached 24.87% when the fiber length of the reinforcements reached 12 mm. The same trend was observed in tensile modulus, whereas the modulus value of the unreinforced samples was 3.5 GPa and increased to 3.9 GPa for the reinforced samples with fibers length of 12 mm. This may be attributed to the enhancement in the effective contact area of fibers with the matrix. It also observed that the tensile strength and tensile modulus values decreased when the fiber length of the reinforcement reached 15 mm. Flexural strength results also show the same trends, as the flexural strength value of the unreinforced samples was 27.7 MPa and increased to reach 81.75 MPa with samples reinforced with a fiber length of 12 mm. A significant decrease in the flexural strength was observed when the fiber length reached 15 mm as it decreased to 48.2 MPa. This may be attributed to the crimped fiber due to its length. Unlike the tensile and flexural test results, impact strength results increased as the fiber length increased. Impact strength values increased from 2.09 kJ/m^2 for the unreinforced samples and reached 6.41 kJ/m^2 for the sample reinforced with a fiber length of 15 mm.

3 Date Palm Fiber Reinforced Thermoplastic Composite

Wood-plastic composite made of recycled plastic and date palm leaves waste was investigated as an environmentally friendly artificial wood (Binhussain and El-Tonsy 2013). In this investigation, three different types of plastic were used; Polycarbonate (PC), Polystyrene (PS), and Polyvinyl chloride (PVC). All recycled plastics were pretreated with a coupling agent to enhance the adherence between the palm fibers and the plastic matrices. Wood Plastic Composite (WPC) samples were fabricated by cutting the dried date palm leaves into length ranged between 0.63 and 4.0 mm using a shredder machine, waste plastic was also shredded into 1–4 mm length. A mixture of palm fibers and waste plastic with a ratio of 1:1 was prepared and pretreated with a coupling agent and mixed in a mixer to ensure the homogeneity of the mixed particles. Mixed particles then extruded in different temperature profiles and compressed with a hydraulic press. Tensile test results and tensile modulus results showed that PC-mix exhibits the lowest values of 2.88 MPa and 1.13 GPa respectively, while the PS-mix showed the highest tensile strength value of 4.82 MPa and PVC-mix showed the highest tensile modulus value of 1.31 GPa. This is due to the inherently high stiffness, strength, and lower elongation of the PVC and PS plastics. Impact results showed different trends as the PC-mix exhibit the highest impact strength value of 1.95 kJ/m^2. This due to the highest ductility of the PC-mix compared to PVC and PS-mix. The high elongation at break of the PC-mix (5.8%) indicates the high ductility, which results in high impact resistance compared to PVC and PS-mix. Flexural modulus results showed that PS-mix has the highest values of 1.91 GPa, while the PC-mix exhibits the lowest flexural modulus value of 1.68 GPa.

Another study investigated the mechanical properties of date palm leaf fibers (DPLF) with recycled polyethylene terephthalate (PET) as a matrix (Dehghani et al. 2013). DPLFs were surface treated in order to enhance the mechanical properties of prepared composite materials. twin-screw extruder and Injection molding machine were used to prepare the composite samples. This study also investigated the effect of the DPLF ratio on the final mechanical and thermal properties of the prepared composite. The particle size of the DPLF in this study was 75 μm and the surface was treated with 8% sodium hydroxide. DPLFs were grafted with 1.8% maleic anhydride as a coupling agent to enhance the mechanical properties of the prepared composite. DPLF ratios were 5, 10, and 15% with 10 phr maleic anhydride. The Prepared mixture of DPLF, PET, and coupling agents was mold injected to prepare the composite test specimens. Test results show that as the DPLF ratio increased the tensile strength values increased, while the highest value of the tensile modulus was recorded at the

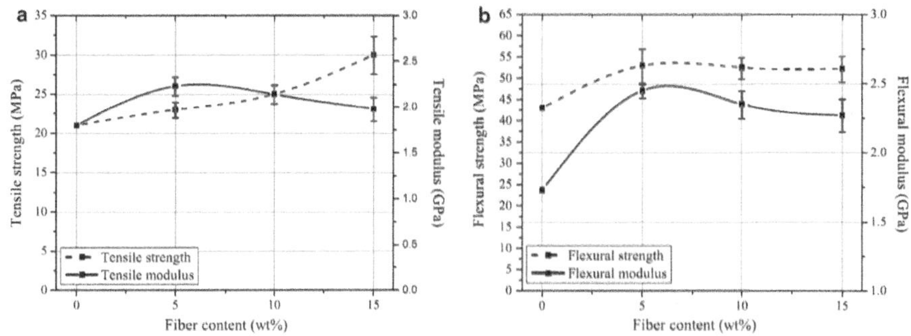

Fig. 2 Mechanical properties of recycled PET reinforced with DPFL (Dehghani et al. 2013)

DPFL ratio of 5% and the value decreased as the DPFL ratio increased (Fig. 2). Flexural strength results showed that reinforcing the recycled PET with DPFL has a significant effect on enhancing the flexural strength regardless of the ratio of the DPFL. Although the increasing ratio of the DPFL from 5 to 15% has caused a reduction in flexural modulus of the final composite. This reduction may be due to the lower mechanical properties of the coupling agents. Impact results show that as the DPFL ratios increased from 5 to 15% the impact strength increased from 11.3 to 13.8 kJ/m^2, although all these results are lower than the prepared PET with coupling agents without reinforcements which were 18.4 kJ/m^2. This may attribute to the increase in the number of stress concentration points after adding DPFL.

4 Date Palm Fiber Reinforced Thermoset Composite

The effect of date palm fiber length on the flexural properties of polyester composite was studied in Al-Kaabi and Al-Khanbashi (2005). Date palm fiber length 0.5, 1, 2, and 3 cm were used after treatments with detergent, 5% NaOH, and bleaching with dioxin solution. Composite samples of the polyester resin reinforced with raw date palm fiber with different lengths and with date palm fibers treated with different treatments were tested for flexural and impact properties. Results showed that raw DPF/polyester resin with fiber length 2 cm exhibits the highest flexural strength among all tested fiber length. Results also depict that the composite reinforced with date palm fibers treated with 5% NaOH showed the highest value of flexural strength of 70 MPa. The study also reported that composite made of polyester resin with 2 cm date palm fiber length treated with 5% NaOH exhibit the highest flexural strength value with fiber weight fraction of 9%. Regarding the impact strength, the study indicated that as the fiber weight fraction increased the impact strength increased, and the highest impact strength recorded was 12 kJ/m^2 when the fiber weight fraction was 9% and the fiber length was 2 cm and the fibers were alkali treated.

Table 1 Mechanical properties of date palm fiber/epoxy composite (Saba et al. 2019)

Samples	Tensile strength (MPa)	Tensile modulus (GPa)	Elongation at break (%)
Pure epoxy resin	20.5	0.51	0.91
40% DPF	21.4	0.61	1.16
50% DPF	25.7	1.54	1.41
60% DPF	24.3	1.32	1.34

Composite of epoxy reinforced with date palm stem fibers was investigated for its tensile and impact properties. Composite samples were prepared with different fiber loading namely 40, 50, and 60 wt% (Saba et al. 2019). Samples were molded by pouring the premix into a stainless-steel mold and left to cure at room temperature. Table 1 shows the recorded results of tensile strength and modulus. Results indicate that adding the date palm fiber as a reinforcement, increased the tensile strength and the tensile modulus of the pure epoxy. However, when the date palm fiber loading reaches 60%, a decrease in the mechanical properties was observed. this was attributed to the poor wetting, improper mixing, and agglomeration, which may result in a notable decrease in the fiber-matrix adhesion. Elongation at break results also follows the same trend as tensile strength and tensile modulus. Moreover, results indicated that date palm fiber loading of 50% caused higher impact strength of 98.71 J/m, while increasing the fiber loading to 60% caused a reduction in the impact strength and the recorded value was 70.62 J/m. This reduction may be attributed to the poor adhesion between date palm fibers and the matrix.

Treated date palm fibers mixed with polyester resin to produce a date palm fiber-reinforced composite samples (Sadik et al. 2017). Date palm fibers extracted from date palm fronds were used in this study. Fibers were mixed with polyester resin in two different ratios namely 30 and 50%. The date palm fibers were cut to obtain a 10 mm length. Prior to composite fabrication, date palm fibers were chemically treated with 10% NaOH for 6 h followed by washing and drying. Prepared composite samples were tested for its tensile, flexural and impact properties. Results indicated that all the mechanical properties of reinforced composite samples are improved compared to unreinforced samples. Also, results revealed that samples prepared with 30% date palm fiber contents exhibit better tensile strength compared to samples prepared with 70% fiber contents.

Using different date palm fillers as reinforcement was studied (Alshammari et al. 2019). Date palm fibers were extracted from different tree parts namely; leaf stalk (A), fruit bunch stalk (AA), leaf sheath (G), and tree trunk (L). Date palm fibers were ground to a size ranged between 0.8 and 1.0 mm and used as a filler. Composite samples were fabricated using epoxy resin by hand lay-up technique. Mixture of 50 wt% date palm fiber and 50 wt% epoxy resin was poured in a steel mold and then hot pressed under 110 °C for 10 min followed by the cold press for 5 min. tensile, flexural, and impact tests were conducted on prepared samples to evaluate

Table 2 Mechanical properties of DPF reinforced composite (Alshammari et al. 2019)

Samples	Tensile strength (MPa)	Tensile modulus (GPa)	Flexural strength (MPa)	Flexural modulus (GPa)	Energy absorption (%)	Impact strength (J/m)
A	26.45	1.42	77.65	4.55	48	62.65
AA	40.12	2.88	110.16	6.34	30	99.45
G	36.17	2.77	98.46	5.54	39	87.34
L	28.44	1.66	89.43	5.12	45	71.07
Epoxy	20.60	0.57	32.11	2.27	50	45.71

the mechanical properties of produced composite reinforced with DPF filters. Results revealed that the incorporation of DPF with epoxy resin has improved the mechanical properties of the prepared composite compared to pure epoxy samples as shown in Table 2. Flexural strength values depict a significant improvement in AA sample results, as the improvement reached 247% compared to pure epoxy. These results revealed that AA has a significant attribution in enhancing the toughness and stiffness of the reinforced composite samples. Tensile and impact strength results also show the same trend, where AA exhibit the higher values among all DPF fillers reinforced composite followed by G, L, and A. These results may attribute to the stiffer nature of the DPF filler compared to pure epoxy resin which agrees with other studies (Arpitha and Yogesha 2017). However, sample AA shows the lowest energy absorption value of 30% and the highest value was recorded for the pure epoxy resin of 50%. Energy absorption represents the area under the stress–strain curve and it represents material toughness (Mallick 2012). These observations of the lower energy absorption of the AA samples may be attributed to the internal damage during impact load due to a higher amount of cellulose content in DPF filler.

Different loading ratio of date palm fiber in the production of date palm fiber-reinforced composites was investigated to enhance the flexural and thermal properties of the final composite (Gheith et al. 2019). Composite samples were prepared with different fiber loading namely 40, 50, and 60 wt%. Date palm fibers were ground to a size ranged between 0.8 and 0.01 mm and moisture content kept between 6 and 8%. Samples were prepared using the hand lay-up technique. Table 3 shows the recorded results of flexural strength and flexural modulus. Results revealed that reinforcing the composite with different date palm fiber loading enhances the flexural

Table 3 Mechanical properties of date palm fiber/epoxy composite (Gheith et al. 2019)

Samples	Flexural strength (MPa)	Flexural modulus (GPa)
Pure epoxy resin	26.15	2.26
40% DPF	28.60	2.30
50% DPF	32.64	3.28
60% DPF	27.83	2.94

properties of the prepared composite samples. Pure epoxy samples showed the lowest flexural strength and modulus value of 26.15 MPa and 2.26 GPa. The highest values of flexural strength and modulus were recorded for 50% DPF loading. This value shows enhancement by 122.7 and 141.6% in the flexural strength and flexural modulus. Results also indicated that, as the DPF loading ratio increased the flexural strength and modulus increased up to 50%. When DPF loading reached 60% a reduction in the flexural properties was observed. Several studies attribute this behavior to the critical loading point (Saba et al. 2015; Hammood 2015). This may due to the insufficient matrix to impregnate the high ratio of reinforcement fibers. This causes a poor interfacial adhesion between the matrix and the reinforcement.

Effect of several parameters on the date palm fiber composite mechanical properties such as production technique, resin types, fiber to resin ratio and fiber orientation were studied (Al-Sulaiman 2002). Used fabrication techniques are namely; vacuum bagging under approx. 0.95 bar, autoclave at 8 bar, and Vulcan press at 23–32 bar. Two resin types were used namely Phenolic resin and Bisphenol resin. The two selected fiber to resin ratio is 72:25 for the Bisphenol resin and 65:35 for the Phenol resin. Three different fiber orientations were used, these orientations are unidirectional leaves stacked at 12 mm length with 300 mm long, unidirectional fibers' shredded leaves at 1.5 mm and 300 long, and short fibers at 1 and 3 mm with 5 mm width. Tensile strength results showed that unidirectional fiber laminate with Bisphenol resin exhibit the highest tensile strength values of 152.3 MPa at 32 bar pressure and the lowest tensile strength value of 8.4 MPa is recorded for the short fiber with 1.5 mm length and Phenolic resin. Results also depict that the high tensile strength values are recorded for the Bisphenol resin under the fabrication pressure of 32 bar.

Mechanical properties of eight different types of date palm residues from the same environment were evaluated (Almi et al. 2015). This evaluation aims to asses the usage of each type individually or as a mixture in the fabrication of fiber-reinforced composite. The used eight types of date palm fibers are; Petiole, Rachis, Leaflets, Thorns, Spathe, Bunch, Pedicels, and Fibrillium. The results show that Rachis variety exhibits high values of tensile strength and tensile modulus of 213 MPa and 8.5 GPa respectively. A study reports that Rachis variety exhibit high cellulose content and less rate of porosity. The lowest tensile strength and tensile modulus values of 86 MPa and 3 GPa were observed for the Pedicels variety.

Several studies were carried out to evaluate the effect of chemical treatments on mechanical properties on date palm fiber-reinforced composite. These studies evaluate the effect of different chemical treatments, chemical concentration, and treatment duration. Chemical treatment using HCL, NaOH, and CH_3COOH were evaluated for their effect on the mechanical properties of date palm fiber-reinforced composites (Abdellah et al. 2019). Chemically treated date palm fibers were chopped into small particles and mixed with epoxy resin. A mixture of date palm fibers and epoxy resin was poured into 10 mm mold and then pressed under 5 tons until curing occurred. The flexural test was conducted on fabricated composite samples to acquire flexural strength and elastic modulus. Table 4 depicts the flexural strength and elastic modulus of treated date palm fibers using different chemical treatments. The flexural strength value of 18.2 MPa was observed for the 50% NaOH treatments for 2 h.

Table 4 Effect of chemical treatments on date palm fiber composite (Abdellah et al. 2019)

Condition	Tensile strength (MPa)			Tensile modulus (GPa)		
	HCL	NaOH	CH_3COOH	HCL	NaOH	CH_3COOH
10% (1 h)	11.5	12.52	14.2	1.6	16	3.14
10% (2 h)	12.7	12.6	15.35	2.4	1.5	3.9
20% (1 h)	11.4	12.6	11.7	8.35	10.7	2.5
20% (2 h)	12.6	13.4	14.1	3.07	5.17	9.01
50% (1 h)	15.2	16.7	12.2	0.8	15.6	5.7
50% (2 h)	17.9	18.2	15.3	12.1	6.34	1.7

Results also depict that as the treatment duration increased from 1 to 2 h the flexural strength increased. Results also revealed that HCL and CH_3COOH has a significant effect on the flexural strength at 10 and 20% concentration, while NaOH exhibit a higher effect on enhancing the flexural strength at 50% concentration.

5 Conclusion

Date palm fibers reinforced composite shows a high potential to replace and/or substitute regular fiber-reinforced composites due to their good mechanical properties and acceptable properties such as high specific strength, low cost, sustainability, accepted mechanical properties and biodegradable. Several researches showed that treating date palm fibers with $KMnO_4$ has a significant effect on enhancing the mechanical properties of the final composite. The biodegradable composite was fabricated using date palm fiber as a reinforcement and thermoplastic starch as a matrix material. The thermoplastic matrix is used to fabricate a replacement to the wood-plastic composite using data palm fibers as a reinforcement and polyethylene as a matrix. Investigated mechanical properties of the reinforced composite with date palm fibers showed a high potential of using such composite in different applications.

References

Abdellah MY, Seleem Abo-El Hagag A, Marzok WW et al (2019) Tensile and fracture properties of chemically treatment date palm tree fibre reinforced epoxy. Int J Mech Mechatron Eng 19:90–99

Al-Kaabi K, Al-Khanbashi A (2005) Date palm fibers as polymeric matrix reinforcement: DPF/polyester composite properties. Polym Compos 26:604–613

Almi K, Benchabane A, Lakel S, Kriker A (2015) Potential utilization of date palm wood as composite reinforcement. J Reinf Plast Compos 34:1231–1240

Alshammari BA, Saba N, Alotaibi MD et al (2019) Evaluation of mechanical, physical, and morphological properties of epoxy composites reinforced with different date palm fillers. Materials 12:1–17

Al-Sulaiman FA (2002) Mechanical properties of date palm fiber reinforced composites. Appl Compos Mater 9:369–377
Arpitha G, Yogesha B (2017) An overview on mechanical property evaluation of natural fiber reinforced polymers. Today Proc 4:2755–2760
Binhussain MA, El-Tonsy MM (2013) Palm leave and plastic waste wood composite for out-door structures. Constr Build Mater 47:1431–1435
Dehghani A, Ardekani SM, Al-Maadeed MA et al (2013) Mechanical and thermal properties of date palm leaf fiber reinforced recycled poly (ethylene terephthalate) composites. Mater Des 52:841–848
Gheith MH, Abdel Aziz M, Ghori W et al (2019) Flexural, thermal and dynamic mechanical properties of date palm fibres reinforced epoxy composites. J Mater Res Technol 8:853–860
Hammood SA (2015) Effect of erosion on water absorption and morphology for treated date palm fiber-reinforced polyester composites. Int J Mech Mechatron Eng 15:108–114
Hulleman DSH, Janssen PFH, Feil H (1998) The role of water during plasticization of native starches. Polymer 39:2043–2048
Ibrahim H, Farag M, Megahed H et al (2013) Characteristics of starch-based biodegradable composites reinforced with date palm and flax fibers. Carbohyd Polym 101:11–19
Khanam PN, AlMaadeed MA (2014) Improvement of ternary recycled polymer blend reinforced with date palm fibre. Mater Des 60:532–539
Mahdavi S, Kermanian H, Varshoei A (2010) Comparison of mechanical properties of date palm fiber-polyethylene composite. BioResources 5:2391–2403
Mallick KP (2012) Failure of polymer matrix composites (PMCs) in automotive and transportation applications. In: Failure mechanisms in polymer matrix composites. Woodhead Publishing, Cambridge
Raghavendra S, Sivaram NM, Sadik T et al (2018) Mechanical properties of hybrid composites with date palm fibre reinforcement. Adv Mater 7:78–81
Saba N, Paridah M, Jawaid M (2015) Mechanical properties of kenaf fibre reinforced polymer composite: a review. Constr Build Mater 76:87–96
Saba N, Alothman OY, Almutairi Z et al (2019) Date palm reinforced epoxy composites: tensile, impact and morphological properties. J Mater Res Technol 8:3959–3969
Sadik T, Sivaram NM, Senthil P (2017) Evaluation of mechanical properties of date palm fronds polymer composites. Int J ChemTech Res 10:558–564
Satyanarayana KG, Arizaga CGG, Wypych F (2009) Biodegradable composites based on lignocellulosic fibers—an overview. Prog Polym Sci 34:982–1021
Sharath PG, Kishor SB, Vinod B, Naveen A (2016) Effect of fiber length on the mechanical properties of coir and wild date palm reinforced epoxy composites. IJSTE 2:555–559

Evaluation and Comparison of Date Palm Fibers with Other Common Natural Fibers

Faris M. AL-Oqla

Abstract The developed awareness toward sustainability raised the importance of developing and using green products of natural fibers and natural fiber composites instead of the conventional products. However, this is a long path, and an enormous effort should be paid in order to enhance the performance properties of these green composites and to spread their use for new applications. The performance properties including the mechanical properties (stiffness, strength, etc.), the physical properties (density, appearance, etc.), thermal stability, moisture resistance, degradability and others, making the selection of the most fit fiber types for industrial application is a very complex task. Improving these properties of natural fiber composites can be achieved by the appropriate selection of the composite materials constituents through reliable decision-making methods. Such methods should also evaluate the green composites in combined evaluation criteria. Date palm fibers were compared and evaluated against other common natural fibers used for composites. It was shown that date palm fibers have a potential for natural fiber composites with various environmental conditions. Such reliable evaluations of natural fibers would help finding the potential of the new composites for various industrial applications.

Keywords Sustainable design · Composite reliability · Moisture content · Natural fibers · Date palm fibers · Green composites · Fiber selection

1 Introduction

Appropriate material selection became a key factor to achieve a sustainable design with higher customer satisfaction attributes. Applying new materials in a particular industry is limited by numerous constrains and factors (AL-Oqla 2017; AL-Oqla and Salit 2017b; AL-Oqla and Sapuan 2014b; Alaaeddin et al. 2018). Thus, picking the most fit material type for a certain application is a complex problem where proper decisions have to be taken using instructive pairwise judgments, which are the core

F. M. AL-Oqla (✉)
Department of Mechanical Engineering, Faculty of Engineering, The Hashemite University, Zarqa 13133, Jordan
e-mail: fmaloqla@hu.edu.jo

M. Midani et al. (eds.), *Date Palm Fiber Composites*, Composites Science and Technology,
https://doi.org/10.1007/978-981-15-9339-0_10

of the decision making process in diverse engineering applications (AL-Oqla and Hayajneh 2007; Al-Oqla and Omar 2012, 2015; AL-Oqla et al. 2018; AL-Oqla and Sapuan 2015). The present tremendous needs for achieving both sustainability and better environmental performance reveal the eligibility of the available natural fibers and their composites for different industrial applications (Al-Oqla et al. 2019a, 2016b).

1.1 Date Palm Fibers

Date palm fibers are considered of the most available natural fibers. With more than 120 million trees spread all over the world, about 80 million in the Arabic region only (Al-Kaabi et al. 2005; AL-Oqla and Salit 2017a). Plenty of date palm fibers are available with good compatibility with polymers to produce sustainable bio-composite products for different industrial applications, especially in automobile industry. The annual production of date palm fibers exceeds the production of coir by over than 40 times, it also exceeds hemp and sisal production by 20 and 10 times, respectively (AL-Oqla and Sapuan 2018a). The date palms have a cultural dimension as they were involved in the daily lives of millions of people over thousands of years. Thus, they have social acceptance, positive view, and most importantly, governmental support. Furthermore, the date palm trees produce dates (very valuable food fruit), as well as they provide raw materials for various home accessories and furniture. Overall, the date palm brings good revenue for economies of many countries (Al-Shahib and Marshall 2003) as the production amount of dates fruits in some of Mediterranean countries indicates the importance of the date palm trees to the local economies as well as their availability (AL-Oqla et al. 2014a; Chandrasekaran and Bahkali 2013).

1.2 Properties of Fibers

The different properties of the natural fibers dominate their use in composites. Mechanical, physical, chemical, and other properties can either encourage or discourage the use of the natural fibers with different polymer materials to form compatible composites. For same type of fibers, the properties can widely vary due to many factors as the location of the fiber on the plant, the soil quality, the harvest timing, the weather conditions, the fertilization, as well as the extraction and drying processes.

1.3 Chemical Composition of Date Palm Fibers

The chemical composition of the natural fibers is of a great importance in controlling different fiber characteristics as the degradability and recyclability, the moisture absorption, and the fungi attack. Cellulose and lignin are the most important structural components in the natural fibers. The cellulose in the natural fiber is normally aligned along its length as slender crystalline micro fibrils. The cellulose nature and its crystallinity can determine the natural fibers reinforcing efficiency (AL-Oqla et al. 2017, 2019b; Al-Oqla and El-Shekeil 2019; Almagableh et al. 2017; Fares et al. 2019; Ilyas et al. 2018; Sadrmanesh et al. 2019). On the other side, lignin is the hydrocarbon polymer, which gives the fiber its rigidity as well as it helps in water transportation along the plant. Many researchers have studied the date palm fibers regarding their different characteristics such as the chemical compositions, the degradability, the effect of the extraction and drying processes, etc. Examples of these studies includes the structural characterization of cellulose, cellulose oxidizations, effect of lignin content, NaOH treatments, effects of H_3PO_4 and KOH in carbonization of lignocellulose, membrane-bound peroxidase characterization from date palm leaves, biochemical analysis at the developing stages of embryogenesis, and biosorption characteristics of phosphates from aqueous solution onto date palm fibers (Agoudjil et al. 2011; Ahmed et al. 1995; Al-Senaidy and Ismael 2011; Al Eid 2006; Alawar et al. 2009; Chao and Krueger 2007).

1.4 Physical Properties of Date Palm Fibers

Beside the chemical characteristics of the natural fibers, the physical properties include the density of fiber, its length (L), its diameter (D), its aspect ratio (L/D), and its thermal conductivity are of great importance in determining the capability of using specific natural fibers with certain polymer materials in order to form appropriate composites for particular applications. For instance, the fiber density is considered as one of the dominating parameters in evaluating and selecting the fibers in the natural fiber composites (NFCs) (Al-Oqla and El-Shekeil 2019; Alaaeddin et al. 2019a, b). The lower density of fiber may lead to considerable reduction in the composite weight makes it very appropriate in many industries as the automotive and the aerospace industries where the fuel consumption is directly related (Al-Oqla et al. 2019a). The date palm density is ranged between (0.9–1.2 g/cm^3) which is below the densities of many other types of natural fibers. It is believed that various other applications can get benefit from the light weight of composites like that of sports, furniture, military applications as well as others. Beside the density, the aspect ratio (L/D) plays a crucial role in determining some of the important characteristics of the NFCs (AL-Oqla et al. 2015b, c; AL-Oqla and Sapuan 2018a; Aridi et al. 2016; Miwa and Horiba 1994). For high aspect ratio, the NFCs will generally be stronger and stiffer than the matrix material alone (Aridi et al. 2016; Etaati et al. 2014). Whereas fibers of short aspect ratio will be preferred for large-scale production where the cost will be reduced and the NFCs will have isotropic properties. The aspect ratio of the date palm fibers is moderate in comparison to other types of natural fibers (AL-Oqla et al. 2014a). Fortunately, this will increase the possibilities of producing both continuous and discontinues fibers, which in turns, expands the use of date palm fibers over wide range of applications. Several researches investigated the influence of the density of the date palm fibers, its length, its diameter, and its aspect ratio on the different NFCs characteristics (Abdal-hay et al. 2012; AL-Oqla et al. 2014b; AL-Oqla and Sapuan 2014a, 2015; Al-Khanbashi et al. 2005; Alsaeed et al. 2012; Bendahou et al. 2008).

1.5 Mechanical Properties of Date Palm Fibers

The mechanical properties of the date palm fibers are influenced by many characteristics including the fiber structure, the cell dimensions, the chemical composition, the microfibrillar angle and defects. For example, fibers having high cellulose content, long cell length, high cellulose polymerization degree, and low microfibrillar angle are normally possessing excellent mechanical properties as high Young's modulus and high tensile strength (AL-Oqla et al. 2015d; Aridi et al. 2017; Das et al. 2014; Rashid et al. 2014; Sapuan et al. 2013).

2 Moisture Content Criterion

The moisture amount in green fibers is a critical factor for the suitability of the fibers to make Biocomposites. The moisture absorption characteristic of the natural fibers has a strong reflect on their mechanical properties. However, inconsistencies in the results of these properties were reported (AL-Oqla and Salit 2017b; AL-Oqla et al. 2014c). Hence, predicting the fiber performance under wet conditions will be a complex task. Nevertheless, a recent promising research has developed a systematic evaluation tool based on the moisture content criterion (MCC) in the natural fibers to evaluate their capabilities (AL-Oqla et al. 2014c). This novel tool is designed to predict some of the desired distinctive characteristics in the natural fibers under the influence of the moisture absorption phenomenon. The behaviors of different types of natural fibers commonly used in industry were investigated regarding the MCC. Date palm fibers were investigated as well. The investigations revealed that MCC is able to predict the relative drop in the performance properties of fibers individually due to the moisture absorption effect. The prediction of the MCC is strongly close to the results obtained by many researchers in the literature (AL-Oqla et al. 2014c). This proves the MCC as an efficient and very useful evaluation tool that can be used to predict the relative fiber behaviors with less magnitude of errors compared to the experimental measurements. Hence, MCC is considered as a great tool that verifies the natural fiber behaviors regarding specific characteristics systematically, leads to more reasonable decisions, and enhances the natural fiber selection process that leads to better sustainable products in industry.

2.1 Moisture Absorption and Natural Fibers Capabilities

The performance characteristics of the NFCs depend entirely on the chemical and the physical compositions of the inherent constituents (Al-Oqla and Sapuan 2018b; Alaaeddin et al. 2019c) The NFCs are often exposed during their lifetime to various circumstances including the hygroscopic environment (Wang et al. 2006). In wet conditions, the natural fibers absorb plenty of moisture due to their hydrophilic nature leading to significant changing in the fibers mechanical performance. The natural fibers hydrophilic behavior is serious for its great role in reducing the mechanical performance as it weakens the bonding between the fillers and the matrix in NFCs (Azwa et al. 2013; Sapuan et al. 2013; Wang et al. 2006). Therefore, the moisture absorption behavior is considered as one of the most drawbacks in the natural fibers characteristics for its direct influence on NFCs causing several negative effects as the mechanical performance reduction, accelerating the degradation in fibers, swelling, fungi attacking, and cracking (Alaaeddin et al. 2019d; Azwa et al. 2013; Sapuan et al. 2013). Both hemicellulose and moisture content in the natural fibers are main factors controlling the moisture absorption capabilities, the rate of the biological and the thermal degradation, as well as the flammability (Azwa et al. 2013). Although the impact of the moisture absorption behavior in the natural fibers has been studied for the NFCs, inconsistent results were reported in the literature (Célino et al. 2014). Some research works concluded that the moisture absorption may lead to an improvement or a reduction in the properties of the NFCs depending on the type of their constituents (Azwa et al. 2013). Other works reported that moisture absorption may improve a particular mechanical property for a certain type of fibers and reduce it for another type (Azwa et al. 2013; Placet et al. 2012; Symington et al. 2009). Hence, no clear rule is available regarding the moisture absorption effect on the NFCs. Therefore, predicting the relative performance of the NFCs in wet conditions is a complicated process, accordingly, selecting the proper fiber type for a certain NFC in wet applications is an extremely sophisticated process but very important. This in order can help designers to properly evaluate and select the most appropriate natural fiber types to expand their sustainable design possibilities in one hand, and to contribute to the selection process of natural fibers and their composites on the other (AL-Oqla et al. 2014b).

2.2 Moisture Content Criterion Methodology

In order to improve the natural fiber selection method, the relative performance of many types of natural fibers including the date palm fibers have to be shown based on the moisture content criterion. An effective evaluation criterion developed by (AL-Oqla et al. 2014c) is demonstrated in the following (See Table 2 for detailed steps). First, comparisons between different pairs of fiber types with or without MCC are accomplished. Second, for MCC, the performance of each fiber type is normalized with respect to the rest types, then, each performance is initiated relative to fiber moisture content. Third, other normalized values are obtained according to the previous normalized values obtained relative to the fiber moisture content. Forth, the fiber relative performances before and after MCC are controlled with respect to the superlative values. The steps of the methodology are presented in Fig. 1. At last, the variations in the fiber performances that are already calculated in normalized modes before and after MCC are obtained. These calculated variations provide a strong tool to show the behavior of each fiber concerning a certain property. For example, fibers with positive variations in their performances after applying the MCC will be more considered in design for sustainable products. In contrasts, those fibers with negative variations in their performances will not be recommended for industry. To elaborate more, fibers with negative variations in their normalized performance with respect to a certain property indicate a decrement in the performance of that property due to water absorption compared to other types of fibers. Such decrement in performance is normally due to the inherent moisture content in fibers that encourages mechanical instabilities, aging, swelling, and fungi decay. This comparison methodology regarding different evaluation criteria is now considered the main guide to improve the selection process for producing sustainable NFCs products for different industrial applications.

Start better evaluation of natural fibers

Step 1: Select fibers to be evaluated

Step 2: Select a property to be evaluated

Step 3: Find the average value of each fiber regarding that property

Step 4: Calculate the normalized values found in step 3 with respect to the best beneficial one

Step 5: Find the fibers average moisture content values

Step 6: Calculate the ratios of the values found in step 3 with respect to the values found in step 5

Step 7: Calculate the normalized values of that found in step 6 with respect to the best one

Step 8: Find the differences between each value found in step 7 and their corresponding values in step 4

Step 9: The best choice of fibers regarding the considered property is the one with the highest positive value (for beneficial properties) or highest negative (for non-beneficial) in step 8

Step 10: Repeat steps 2-9 to find the best fiber type which is the best regarding the most desirable properties simultaneously

Fig. 1 Steps of the methodology

2.3 Investigations Without MCC

Evaluating the natural fiber capabilities at the design level based on the MCC can help in finding the comparative worsening in the fiber performance as a result of water absorption. It can also help in improving the fibers selection process to select the most appropriate type for a certain design. The date palm fibers are good competitors with other types of natural fibers with respect to the moisture content. That is; 5–12.5% of its weight from dry to wet conditions. Table 1 lists different properties of the date palm fibers beside many other types commonly used in industry as coir, flax, hemp, jute, and sisal (AL-Oqla et al. 2014b; AL-Oqla and Sapuan 2014b; Dittenber and GangaRao 2011; Pilla 2011).

The judgments between these fibers were accomplished with respect to some important characteristics in order to show the effectiveness of the MCC evaluation tool already discussed (AL-Oqla et al. 2014c). Also, to show the competitiveness of the date palm fibers compared to other types listed in Table 1 for various applications. As such fibers have wide variations in their possessions reflected from the dissimilarities in the fiber potentials, the average values were considered for comparison purposes. Figure 2 shows the comparison between the diverse fibers according to their moisture contents (AL-Oqla et al. 2014c). Differences in the fibers moisture contents are obvious. This reflects the fibers' long-time performances in wet conditions. Only one fiber type (coir) has less moisture content than date palm. This is due to the date palm resistance to: swelling, fungi, and moisture absorption compared to other fibers (flax, hemp, jute, and sisal).

The date palm moisture content is about two thirds of the moisture content in jute, and about one half of that in sisal. Hence, it is expected that the date palm performance will be higher than these types of fibers under the wet conditions. Another criterion, namely elongation-to-break property, is illustrated in Fig. 3, where the different types of fibers are compared with respect to the elongation-to-break property. This property is very important in the impact loading applications. Regarding this property, date

Table 1 Properties of the date palm fibers beside many other commonly used types in industry

Fiber type	Coir	Date palm	Flax	Hemp	Jute	Sisal
Density (g/cm^3)	1.15–1.46	0.90–1.20	1.40–1.50	1.40–1.50	1.30–1.49	1.33–1.50
Tensile strength (MPa)	95–230	97–275	343–2000	270–900	320–800	363–700
Tensile modulus (GPa)	2.8–6.0	2.5–12.0	27.6–103.0	23.5–90.0	8.0–78.0	9.0–38.0
Elongation to break (%)	15.0–51.4	2.0–19.0	1.2–3.3	1.0–3.5	1.0–1.8	2.0–7.0
Moisture content (wt%)	8.0	5.0–12.1	8.0–12.0	6.2–12.0	12.5–13.7	10.0–22.0

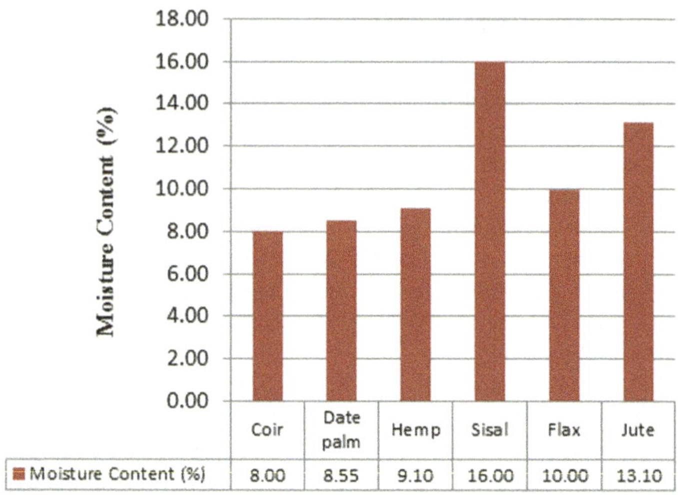

Fig. 2 Moisture contents for various natural fibers (average values) (AL-Oqla et al. 2014c)

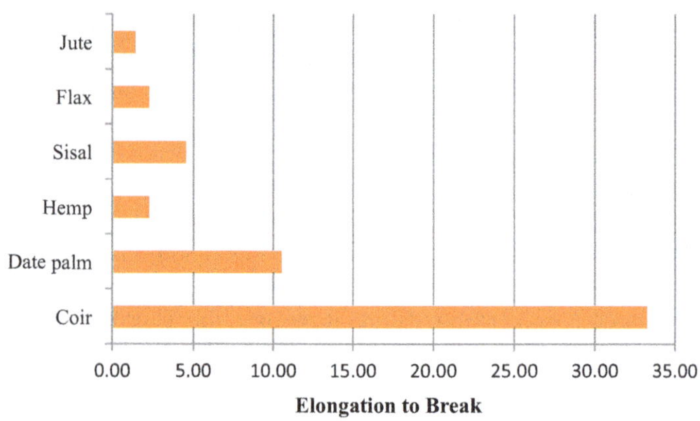

Fig. 3 Elongation-to-break properties of various fibers (AL-Oqla and Sapuan 2018a)

palm surpasses flax, jute, hemp, and sisal fibers. It is obvious that the elongation-to-break of the date palm is more than twice the elongation-to-break in sisal, and much more than that in flax, jute, and hemp.

2.4 Investigations Utilizing MCC

Although studying the natural fibers performances regarding the mechanical properties in individual can be useful for the evaluation and selection process, combined

Table 2 Fibers relative performance calculations regarding the elongation to break property with the MCC tool

	Coir	Date palm	Hemp	Sisal	Flax	Jute
Moisture content (MC) (%)	8.000	8.550	9.100	16.000	10.000	13.100
Elongation to break (EB) (%)	33.200	10.500	2.250	4.500	2.250	1.400
Normalized EB (NOR EB)	1.000	0.316	0.068	0.136	0.068	0.042
EB/MC	4.150	1.228	0.247	0.281	0.225	0.107
Normalized EB/MC (NOR EB/MC)	1.000	0.296	0.060	0.068	0.054	0.026
NOR(EB/MC)-NOR(EB)	0.000	−0.020	−0.008	−0.068	−0.014	−0.016

evaluations are more informative and provide more realistic design decisions. Though, before going more in evaluating the best type of fibers in a certain design, especially, in wet conditions, the behavior of these fibers in the presence of the water absorption phenomenon should be investigated first. Hence, the NFCs produced from these fibers will be more reliable and their future performance under wet conditions will be accurately predicted and accounted for in the design stage. To elaborate more, it is very necessary to recognize if the fibers will perform the same under the wet conditions (where the water absorption phenomenon occurs) after their use in the NFCs products. In order to predict the relative performance of the different types of fibers regarding a certain property (for example, the elongation-to-break), the following procedure is followed. First, the fiber values of this property should be normalized with respect to the best value (i.e. 33.2 in coir fibers) as shown in Table 2.

Second, the ratio of this property (elongation-to-break) to the moisture content in each fiber type is calculated individually (EL/MC). These ratios are then normalized with respect to the best ratio (4.15 in Coir). Finally, the normalized values calculated in the first step are subtracted from those calculated in the previous step as seen in Table 2 (i.e. nor EB/MC − nor EB). This detailed procedure can efficiently predict the relative performance of the fibers regarding a certain property in the MCC conditions. For this example, where the elongation-to-break property is investigated for the different types of natural fibers, it was found that all types have deterioration in performance relative to coir fibers, however, with different magnitudes. Moreover, the elongation-to-break performance of the coir fibers is the best, followed by hemp, whereas sisal fibers are the worst type affected by the moisture absorption. The results of the applied methodology for this criterion is demonstrated in Fig. 4.

It is worth to say that this conclusion was not clear before applying the MCC (Fig. 3). It can be concluded from the MCC that even some types of fibers are seemed to have better performance with respect to a certain property as the case with sisal fibers (which are having better performance regarding the elongation-to-break property than flax, jute, and hemp), they are considered inappropriate as they yield high deterioration in performance due to the moisture absorption. Furthermore, the performance of some types of fibers as jute and flax are expected not to

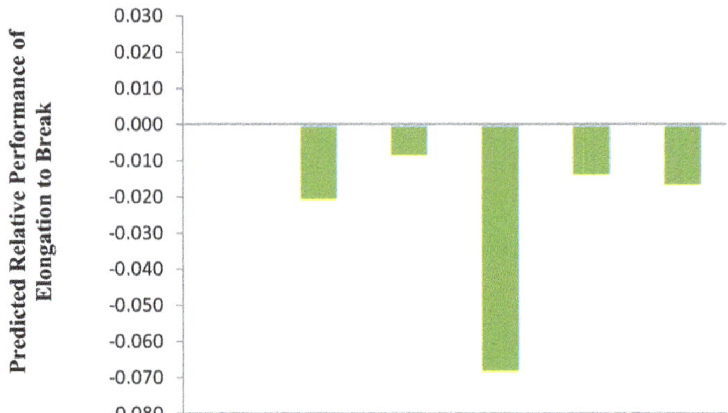

	Coir	Date palm	Hemp	Sisal	Flax	Jute
Dviation between normalized values	0.000	-0.020	-0.008	-0.068	-0.014	-0.016

Fig. 4 The results of the applied MCC methodology for elongation-to-break

highly respond to the moisture absorption compared to other types as the date palm fibers. The reason behind that is the original low performance of these types (with respect to the elongation-to-break property) in comparison to the date palm fibers as depicted in Fig. 3. Further comparisons can be conducted between the natural fibers regarding other performance properties by following similar procedures as discussed in the preceding. The relative performance of the tensile strength property for the natural fibers considered here under the moisture absorption is accomplished and demonstrated in Fig. 5.

The results of the MCC methodology revealed that the performance of some types of fibers regarding the tensile strength property will be deteriorated due to the influence of the moisture absorption as sisal and jute with reduction of 15% and 10% of their original values, respectively. On contrary, MCC predicted that other fibers as coir, date palm, and hemp will undergo an improvement in their performance (highest in hemp of 5%). This conclusion from the MCC is completely agreed with the results reported in the literature (Abral et al. 2014; Alamri and Low 2012; Azwa et al. 2013; Dhakal et al. 2007; Kuciel et al. 2014; Sapuan et al. 2013). In similar manner, applying the MCC to predict the fibers performance regarding the tensile modulus revealed that sisal and jute will undergo a reduction in their performance due to the moisture absorption by 13.5% and 15.6%, respectively. Whereas, coir, date palm, and hemp fibers are predicted to improve their performance due to moisture absorption as shown in Fig. 6 (highest in hemp of 8.6%).

Fig. 5 The relative performance of the tensile strength property considering MCC

3 CEMCEST Method

The behavior of the natural bio-composites depends on their inherent constituents. In current industry, limited criteria are available concerning the agro waste natural fibers. A simple and an efficient technique developed by (AL-Oqla et al. 2014b) called combined multi-criteria evaluation stage technique (CMCEST) can be adopted to improve the evaluation and selection processes of the agro waste natural fibers for NFCs. The standards influencing the evaluation process of the agro waste fibers are integrated and separated into order stages: single-evaluation-criterion (SEC), combined-double-evaluation-criterion (CDEC), combined-triple-evaluation-criterion (CTEC), etc. Such stages are combined according to the mechanical, physical, and economic characteristics of the fibers. They can be extended further to include other characteristics. The schematic illustration of the methodology is illustrated in Fig. 6.

The efficiency of this method (CMCEST) is demonstrated by investigating fibers of coir, date palm, oil palm, kenaf, jute, and hemp, simultaneously. The evaluation and selection of the proper type of fibers for the polymeric composites will be more precise and reasonable as more helpful information will be revealed using this combined method. The CMCEST method is able to scope new types of natural fibers with excellent potential for NFCs. It can also provide more information on the capabilities of the available fibers. Besides, the CMCEST method can guide to better practice regarding the management of the agro waste. This combined technique revealed that date palm fibers are very promising due to the very good characteristics obtained by CTEC. These fibers provide cheap, reasonable, and green alternatives to the conventional materials for different applications.

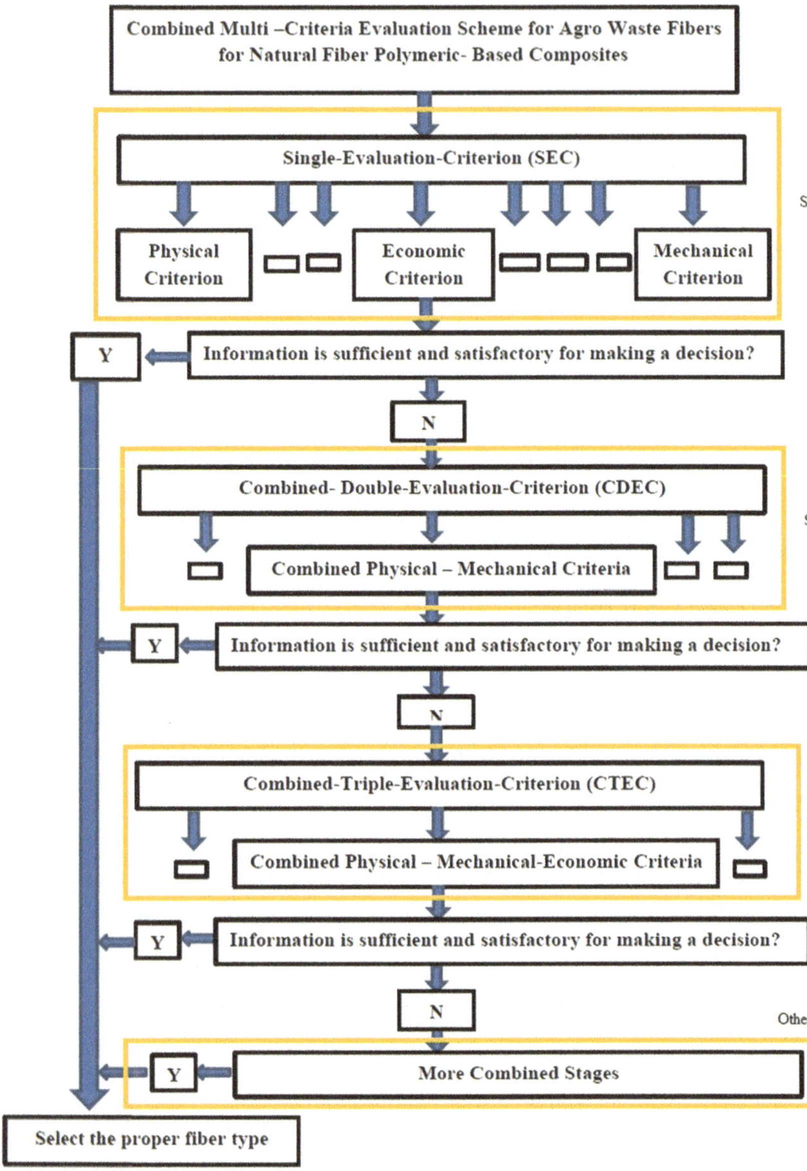

Fig. 6 The schematic illustration of the CEMCEST methodology (AL-Oqla et al. 2014b)

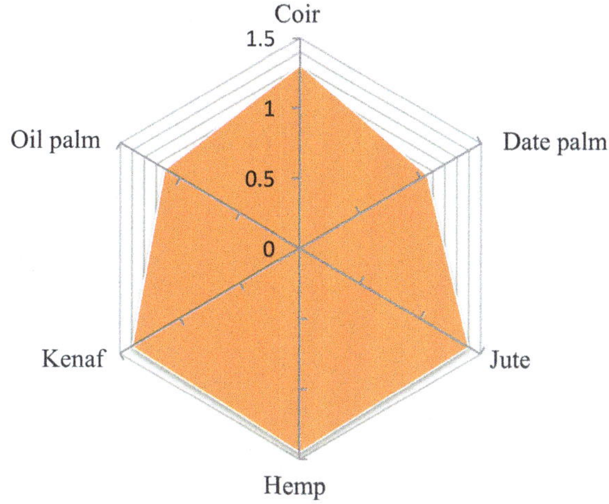

Fig. 7 Densities of natural fibers

3.1 Comparisons Based *on Single-Evaluation Criterion (SEC)*

3.1.1 Comparison Based on a Single Physical Evaluation Criterion

The densities of many types of natural fibers are listed for comparison in Fig. 7. It is obvious that the density of the date palm fibers is the smallest among other types.

This encourages the manufacturers to consider the date palm fibers for light applications as in automobile and aerospace industries. Fig. 7 also shows the similarity in densities of jute, kenaf, and hemp. Meaning, these three types are holding the same priorities regarding this characteristic. Although the date palm is found superior over other types of fibers regarding this SEC, considering other characteristics may lead to different conclusions.

3.1.2 Comparison Based on a Single Economic Evaluation Criterion

A comparison regarding another characteristic is introduced in Fig. 8 where the cost ratio of different types of fibers is illustrated (AL-Oqla et al. 2014b).

The cost of the date palm fibers is the cheapest among all other considered fibers. Hence, this SEC adds more competencies to the date palm fibers. In addition, a variation in cost is obvious between other fibers. Therefore, this SEC delivers more information that will be helpful in evaluating the fibers. However, this criterion alone cannot lead to an appropriate decision in the fibers' evaluation and selection for the NFCs. In other words, other characteristics need to be considered before delivering the realistic results (AL-Oqla et al. 2015a, c, e, 2016a, b). These characteristics can be studied in separate (using many SECs) or together (using combined evaluation criteria).

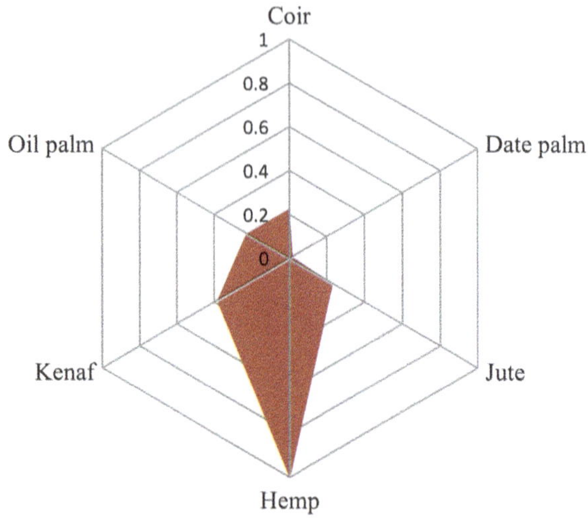

Fig. 8 Cost ratio of different types of fibers

3.1.3 Comparison Based on a Single Mechanical Evaluation Criterion

Another SEC is conducted for the fibers regarding their tensile strength as depicted in Fig. 9 (AL-Oqla et al. 2014b).

It is very obvious the wide variations among the different types of fibers regarding this mechanical property. The tensile strength for jute, kenaf, and hemp fibers exceeds 500 MPa, whereas for coir, date palm, and oil palm fibers it is less than 200 MPa. Regarding this criterion, hemp is the most favorite type, and fibers as date palm are not recommended. It is clear that this SEC leads to different decision in fiber evaluation

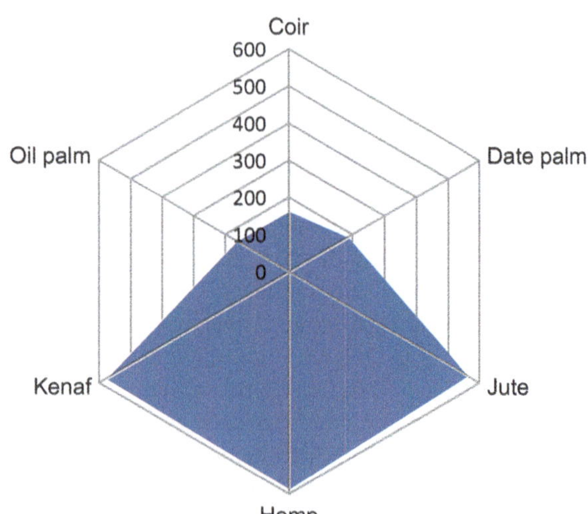

Fig. 9 Tensile strength of natural fibers

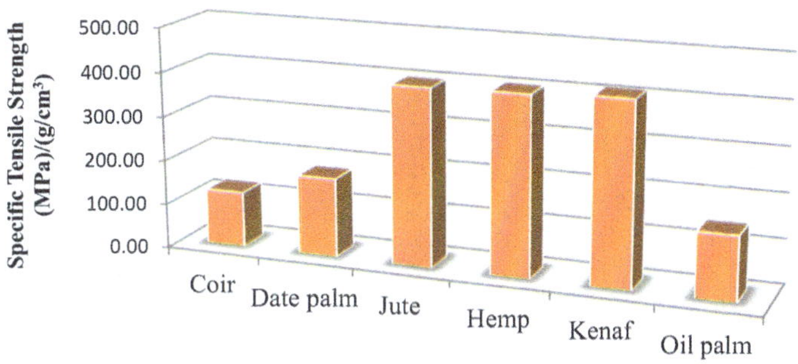

Fig. 10 Tensile strength relative to the density for each type of fibers

comparing to the previous two SECs adding more complexity to the evaluation of the fibers. Allowing for more SECs would resulted in more contradictions rather than guiding to the most suitable fiber evaluation for a certain NFCs application.

3.1.4 Comparisons Based on Combined Double-Evaluation Criterion (CDEC)

Joint mechanical-physical condition

As indicated from the name of this criterion, the properties of the fibers are investigated together in pairs in a combined manner. For example, evaluating the fibers regarding the tensile strength and the density together can be performed using the CDEC. That is; the tensile strength relative to the density for each type of fibers (specific strength) is calculated and illustrated in Fig. 10.

Based on the CDEC, kenaf is found as the best fiber type among all others, whereas date palm fibers lead coir and oil palm fibers. It is worth to remind that a slight variation in the fibers order has occurred when applying the CDEC comparing to the SEC regarding the tensile strength property where hemp was in the front. As other properties are still required to be evaluated using other SECs or CDECs, which may lead to more contradictions, higher combined evaluation stages will be necessary.

3.1.5 *Comparisons* Based on Combined Triple-Evaluation Criterion (CTEC)

Joint mechanical-physical-economic condition

The CTEC evaluate the fibers with regarding to three distinct properties simultaneously. Combining mechanical, physical, and economical properties in a single criterion to evaluate the natural fibers leads to more reasonable and realistic decisions. That is, evaluating these properties simultaneously can change the precedency

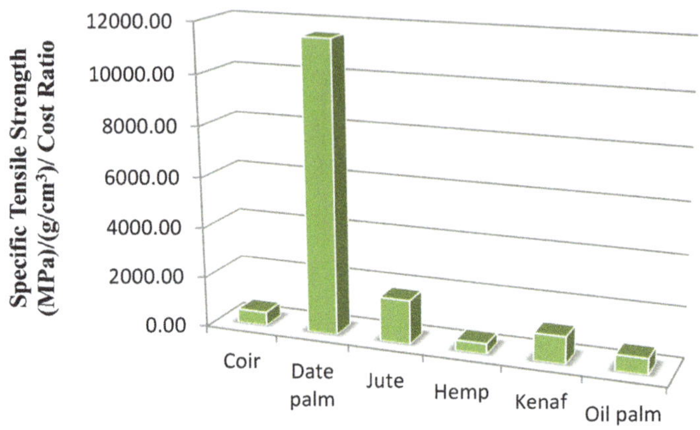

Fig. 11 Comparison based on the tensile strength, density, and the cost ratio

and appropriateness of the natural fibers for a certain NFC application. Integrating the economic property with other properties in CTEC is important in ensuring sustainable low cost productions. For example, judgments the fibers based on the tensile strength (mechanical property), the density (physical property), and the cost ratio (economical property) are accomplished and demonstrated in Fig. 11. The specific tensile strength (tensile strength/density) is calculated and divided by the cost ratio for each fiber type.

It is very obvious that the specific tensile strength to cost ratio for the date palm fibers is 5 times higher than that of jute. Hence, it can be concluded from this criterion that the date palm fibers are the most appropriate for the NFCs applications. In similar manner, replacing the tensile strength by other mechanical properties as the tensile modulus, or the elongation-to-break in a CTEC would yield the results (AL-Oqla et al. 2014b). Such results declare the priority to the date palm as the most appropriate fiber type among all other types to be involved in wide sustainable applications due to its reasonable low cost and eco-friendly characteristics. It is good to notice that further evaluation criteria are not expected to change this priority of the date palm due to the wide gap between the date palm value calculated and the next closest value. Besides, using date palm fibers in NFCs products can lead to superior types of composites with respect to the mechanical and the economic characteristics. In addition, using date palm fibers can reduce the environmental agro wastes, and contribute to enhance the environmental performance.

4 Conclusions

Developing the evaluation and selection processes of the natural fibers for sustainable products of NFCs is increasingly demanding and more efforts should be spent to attain this objective. As the behaviors of the natural fibers can vary under many conditions, accurate prediction of those behaviors will help in the fibers' selection process and will bring more realistic and confident decisions regarding implementing the fibers in a particular NFC for a certain industrial application. Date palm fiber was compared and evaluated for natural fiber composites considering wide other commonly used fibers. It was shown that date palm was potential for natural fiber composites with various environmental conditions. An example of such conditions is the wet environment where the water absorption occurs in fibers. The moisture content criterion can help in predicting the behavior of fibers in such circumstances. The moisture content criterion was able to predict the improvements and the deteriorations in the fibers' performance due to the moisture absorption. Hence, it can be adopted as a reliable evaluation tool for fibers in wet applications. Comparing to the measurement procedures of the fibers' performance in experimental work, the MCC predict the performance without cumulative errors. Besides, new types of fibers with potential characteristics can be discovered and involved in service in wide applications using the MCC criterion. Moreover, the integrated evaluation criteria scheme was also in favor of date palm fiber as a potential type for natural fiber composites. Date palm fibers showed very competitive behavior in wet conditions comparing to other types. Their use for wet applications is recommended.

References

Abdal-hay A, Suardana NPG, Jung DY, Choi K-S, Lim JK (2012) Effect of diameters and alkali treatment on the tensile properties of date palm fiber reinforced epoxy composites. Int J Precis Eng Manuf 13:1199–1206

Abral H, Kadriadi D, Rodianus A, Mastariyanto P, Arief S, Sapuan SM, Ishak MR (2014) Mechanical properties of water hyacinth fibers–polyester composites before and after immersion in water. Mater Des 58:125–129

Agoudjil B, Benchabane A, Boudenne A, Ibos L, Fois M (2011) Renewable materials to reduce building heat loss: characterization of date palm wood. Energy Build 43:491–497

Ahmed IA, Ahmed AWK, Robinson RK (1995) Chemical composition of date varieties as influenced by the stage of ripening. Food Chem 54:305–309

Al-Kaabi K, Al-Khanbashi A, Hammami A (2005) Date palm fibers as polymeric matrix reinforcement: DPF/polyester composite properties. Polym Compos 26:604–613

AL-Oqla FM (2017) Investigating the mechanical performance deterioration of Mediterranean cellulosic cypress and pine/polyethylene composites. Cellulose 24:2523–2530

AL-Oqla FM, Almagableh A, Omari MA (2017) Design and fabrication of green biocomposites. In: Green biocomposites. Springer, Cham, Switzerland, pp 45–67

AL-Oqla FM, Alothman OY, Jawaid M, Sapuan S, Es-Saheb M (2014a) Processing and properties of date palm fibers and its composites. In: Biomass bioenergy. Springer, Cham, Switzerland, pp 1–25

AL-Oqla FM, Sapuan MS, Ishak MR, Aziz NA (2014b). Combined multi-criteria evaluation stage technique as an agro waste evaluation indicator for polymeric composites: date palm fibers as a case study. BioResources 9:4608–4621

AL-Oqla FM, Sapuan S, Ishak MR, Nuraini AA (2014c) A novel evaluation tool for enhancing the selection of natural fibers for polymeric composites based on fiber moisture content criterion. BioResources 10:299–312

Al-Oqla FM, El-Shekeil Y (2019) Investigating and predicting the performance deteriorations and trends of polyurethane bio-composites for more realistic sustainable design possibilities. J Clean Prod 222:865–870

AL-Oqla FM, Hayajneh MT (2007) A design decision-making support model for selecting suitable product color to increase probability. In: Design challenge conference: managing creativity, innovation, and entrepreneurship. Yarmouk University, Amman, Jordan

Al-Oqla FM, Hayajneh MT, Fares O (2019a) Investigating the mechanical thermal and polymer interfacial characteristics of Jordanian lignocellulosic fibers to demonstrate their capabilities for sustainable green materials. J Clean Prod 241:118256

AL-Oqla FM, Hayajneh MT, Fares O (2019b) Investigating the mechanical thermal and polymer interfacial characteristics of Jordanian lignocellulosic fibers to demonstrate their capabilities for sustainable green materials. J Clean Prod 118256

Al-Oqla FM, Omar AA (2012) A decision-making model for selecting the GSM mobile phone antenna in the design phase to increase over all performance. Prog Electromagn Res C 25:249–269

Al-Oqla FM, Omar AA (2015) An expert-based model for selecting the most suitable substrate material type for antenna circuits. Int J Electron 102:1044–1055

AL-Oqla FM, Omar AA, Fares O (2018) Evaluating sustainable energy harvesting systems for human implantable sensors. Int J Electron 105:504–517

AL-Oqla FM, Sapuan SM, Ishak MR, Nuraini AA (2015a) Selecting natural fibers for industrial applications. In: Postgraduate symposium on biocomposite technology Serdang, Malaysia

AL-Oqla FM, Sapuan MS, Ishak MR, Nuraini AA (2015b) Decision making model for optimal reinforcement condition of natural fiber composites. Fibers Polym 16:153–163

AL-Oqla FM, Sapuan MS, Ishak MR, Nuraini AA (2015c) Selecting natural fibers for bio-based materials with conflicting criteria. Am J Appl Sci 12:64–71

AL-Oqla FM, Sapuan S, Anwer T, Jawaid M, Hoque M (2015d) Natural fiber reinforced conductive polymer composites as functional materials: a review. Synth Met 206:42–54

AL-Oqla FM, Sapuan S, Ishak M, Nuraini A (2015e) Predicting the potential of agro waste fibers for sustainable automotive industry using a decision making model. Comput Electron Agric 113:116–127

AL-Oqla FM, Salit MS (2017a) Material selection of natural fiber composites using the analytical hierarchy process. In: Materials selection for natural fiber composites. Woodhead Publishing, Elsevier, Cambridge, USA, pp 169–234

AL-Oqla FM, Salit MS (2017b) Materials selection for natural fiber composites. Woodhead Publishing, Elsevier, Cambridge, USA

AL-Oqla FM, Sapuan S (2015) Polymer selection approach for commonly and uncommonly used natural fibers under uncertainty environments. JOM 67:2450–2463

AL-Oqla FM, Sapuan S (2018a) Investigating the inherent characteristic/performance deterioration interactions of natural fibers in bio-composites for better utilization of resources. J Polym Environ 26:1290–1296

Al-Oqla FM, Sapuan S (2018b) Natural fiber composites. In: Sapuan SM, Ishak MR, Sahari J, Sanyang ML (eds) Kenaf fibers and composites. CRC Press

AL-Oqla FM, Sapuan S, Ishak MR Nuraini A (2016a) A decision-making model for selecting the most appropriate natural fiber—polypropylene-based composites for automotive applications. J Compos Mater 50:543–556

AL-Oqla FM, Sapuan S, Jawaid M (2016b) Integrated mechanical-economic–environmental quality of performance for natural fibers for polymeric-based composite materials. J Nat Fibers 13:651–659

AL-Oqla FM, Sapuan SM (2014a) Date palm fibers and natural composites. In: Postgraduate symposium on composites science and technology 2014 & 4th postgraduate seminar on natural fibre composites 2014, Putrajaya, Selangor, Malaysia, 28 Jan 2014

AL-Oqla FM, Sapuan SM (2014b) Natural fiber reinforced polymer composites in industrial applications: feasibility of date palm fibers for sustainable automotive industry. J Clean Prod 66:347–354

Al-Senaidy AM, Ismael MA (2011) Purification and characterization of membrane-bound peroxidase from date palm leaves (*Phoenix dactylifera* L.). Saudi J Biol Sci 18:293–298
Al-Shahib W, Marshall RJ (2003) The fruit of the date palm: its possible use as the best food for the future? Int J Food Sci Nutr 54:247–259
Al-Khanbashi A, Al-Kaabi K, Hammami A (2005) Date palm fibers as polymeric matrix reinforcement: fiber characterization. Polym Compos 26:486–497
Al Eid SM (2006) Chromatographic separation of fructose from date syrup. Int J Food Sci Nutr 57:83–96
Alaaeddin M, Sapuan S, Zuhri M, Zainudin E, AL-Oqla FM (2018) Properties and common industrial applications of Polyvinyl fluoride (PVF) and Polyvinylidene fluoride (PVDF). In: IOP conference series: materials science and engineering. IOP Publishing, p 012021
Alaaeddin M, Sapuan S, Zuhri M, Zainudin E, Al-Oqla FM (2019a) Photovoltaic applications: status and manufacturing prospects. Renew Sustain Energy Rev 102:318–332
Alaaeddin M, Sapuan S, Zuhri M, Zainudin E, AL-Oqla FM (2019b) Physical and mechanical properties of polyvinylidene fluoride—short sugar palm fiber nanocomposites. J Clean Prod 235:473–482
Alaaeddin M, Sapuan S, Zuhri M, Zainudin E, AL-Oqla FM (2019c) Polymer matrix materials selection for short sugar palm composites using integrated multi criteria evaluation method. Compos Part B: Eng 107342
Alaaeddin M, Sapuan S, Zuhri M, Zainudin E, AL-Oqla FM (2019d) Lightweight and durable PVDF–SSPF composites for photovoltaics backsheet applications: thermal, optical and technical properties. Materials 12:2104
Alamri H, Low IM (2012) Effect of water absorption on the mechanical properties of nano-filler reinforced epoxy nanocomposites. Mater Des 42:214–222
Alawar A, Hamed AM, Al-Kaabi K (2009) Characterization of treated date palm tree fiber as composite reinforcement. Compos Part B: Eng 40:601–606
Almagableh A, Al-Oqla FM, Omari MA (2017) Predicting the effect of nano-structural parameters on the elastic properties of carbon nanotube-polymeric based composites. Int J Perform Eng 13:73
Alsaeed T, Yousif B, Ku H (2012) The potential of using date palm fibres as reinforcement for polymeric composites. Mater Des 43:177–184
Aridi N, Sapuan S, Zainudin E, AL-Oqla FM (2016) Investigating morphological and performance deterioration of injection-molded rice husk–polypropylene composites due to various liquid uptakes. Int J Polym Anal Charact 21:675–685
Aridi N, Sapuan S, Zainudin E, AL-Oqla FM (2017) A review of rice husk bio-based composites. Curr Org Synth 14:263–271
Azwa Z, Yousif B, Manalo A, Karunasena W (2013) A review on the degradability of polymeric composites based on natural fibres. Mater Des 47:424–442
Bendahou A, Kaddami H, Sautereau H, Raihane M, Erchiqui F, Dufresne A (2008) Short palm tree fibers polyolefin composites: effect of filler content and coupling agent on physical properties. Macromol Mater Eng 293:140–148
Célino A, Fréour S, Jacquemin F, Casari P (2014) The hygroscopic behavior of plant fibers: a review. Front Chem 1:1–12
Chandrasekaran M, Bahkali AH (2013) Valorization of date palm (Phoenix dactylifera) fruit processing by-products and wastes using bioprocess technology—review. Saudi J Biol Sci 20:105–120
Chao CT, Krueger RR (2007) The date palm (Phoenix dactylifera L.): overview of biology, uses, and cultivation. HortScience 42:1077–1082
Das B, Chakrabarti K, Tripathi S, Chakraborty A (2014) Review of some factors influencing jute fiber quality. J Nat Fibers 11:268–281
Dhakal H, Zhang Z, Richardson M (2007) Effect of water absorption on the mechanical properties of hemp fibre reinforced unsaturated polyester composites. Compos Sci Technol 67:1674–1683
Dittenber DB, GangaRao HV (2011) Critical review of recent publications on use of natural composites in infrastructure. Compos Part A: Appl Sci Manuf 43:1419–1429

Etaati A, Pather S, Fang Z, Wang H (2014) The study of fibre/matrix bond strength in short hemp polypropylene composites from dynamic mechanical analysis. Compos Part B: Eng 62:19–28
Fares O, AL-Oqla FM, Hayajneh MT (2019) Dielectric relaxation of mediterranean lignocellulosic fibers for sustainable functional biomaterials. Mater Chem Phys
Ilyas R, Sapuan S, Ishak M, Zainudin E (2018) Development and characterization of sugar palm nanocrystalline cellulose reinforced sugar palm starch bionanocomposites. Carbohydr Polym 202:186–202
Kuciel S, Jakubowska P, Kuźniar P (2014) A study on the mechanical properties and the influence of water uptake and temperature on biocomposites based on polyethylene from renewable sources. Compos Part B: Eng 64:72–77
Miwa M, Horiba N (1994) Effects of fibre length on tensile strength of carbon/glass fibre hybrid composites. J Mater Sci 29:973–977
Pilla S (2011) Handbook of bioplastics and biocomposites engineering applications. Scrivener Publishing, Salem
Placet V, Cisse O, Boubakar ML (2012) Influence of environmental relative humidity on the tensile and rotational behaviour of hemp fibres. J Mater Sci 47:3435–3446
Rashid B, Leman Z, Jawaid M, Ishak MR, Al-Oqla FM (2014) Eco-friendly composites for brake pads from agro waste: a review
Sadrmanesh V, Chen Y, Rahman M, Al-Oqla FM (2019) Developing a decision making model to identify the most influential parameters affecting mechanical extraction of bast fibers. J Clean Prod 238:117891
Sapuan S, Pua F, El-Shekeil YA, AL-Oqla FM (2013) Mechanical properties of soil buried kenaf fibre reinforced thermoplastic polyurethane composites. Mater Des 50:467–470
Symington MC, Banks WM, West D, Pethrick R (2009) Tensile testing of cellulose based natural fibers for structural composite applications. J Compos Mater 43:1083–1108
Wang W, Sain M, Cooper P (2006) Study of moisture absorption in natural fiber plastic composites. Compos Sci Technol 66:379–386

Effect of Environmental Conditions on Date Palm Fiber Composites

Sherif Mehanny, Hamdy Ibrahim, Lamis Darwish, Mahmoud Farag, Abdel-Halim M. El-Habbak, and Emad El-Kashif

Abstract During the last few decades, raising environmental consciousness incited many governments, institutions, and firms to reduce fossil fuel consumption. Carbon footprint and resource nonrenewability are among the main motives to mitigate dependence on oil resources. On the other hand, palm trees number exceeded 120 million globally, which poses enormous amount of biowastes to get exploited per annum. Palm reinforced composites are consistently grabbing more attention, due to their availability (especially in the Middle East and southern Asia), and competitive mechanical and physical properties. Moreover, palm fibers can be successfully incorporated into polymeric and ceramic matrices; undergoing multitude of fabrication process including hot pressing, injection molding, extrusion, and mortar casting. Nevertheless, property versus age is still the critical factor to decide whether the candidate material will accommodate the application requirements. Weathering, moisture uptake, acidic (or alkaline) affinity, thermal influence, microbial, fungal, and termites resistance are examples of environmental influences on palm composite when it ages with time. In this chapter, we will attempt to throw light on the response of palm composites to aforementioned physical, chemical, and biological environmental influences. Morphological, gravimetric, mechanical (tension, flexure, compression, and impact tests), and dimensional responses were mainly measured versus time to emulate realistic operation of palm composite products. It was established in literature covered by this chapter that adding palm fibers improved composite antiaging behavior against UV rays. Palm composites tensile strength

S. Mehanny (✉) · A.-H. M. El-Habbak · E. El-Kashif
Mechanical Design & Production Department, Faculty of Engineering, Cairo University, Building 14, Giza, Egypt
e-mail: smehanny@gmail.com; smehanny@cu.edu.eg

H. Ibrahim
Mechanical Engineering Department, College of Engineering and Computer Science, University of Tennessee at Chattanooga, Chattanooga, USA

L. Darwish
Biomedical Engineering Department, Faculty of Engineering, Helwan University, Cairo, Egypt

M. Farag
Mechanical Engineering Department, School of Science and Engineering, American University in Cairo, Cairo, Egypt

M. Midani et al. (eds.), *Date Palm Fiber Composites*, Composites Science and Technology,
https://doi.org/10.1007/978-981-15-9339-0_11

deteriorated due to aging in moist environment, contrary to impact strength. Additionally, palm significantly decreases aging shrinkage and increases chemical resistance of mortar. Fungal and termite attacks to palm composites can be dealt with through some chemical treatments that doubled the composite durability. Summing up, palm fibers were found to be beneficial to decelerate aging effect of both polymeric palm composites and concrete palm composites. This antiaging behavior is promoting more versatility for palm based composites and more likelihood to substitute synthetic fibers.

Keywords Composite · Eco-friendly · Aging · Durability · Residues · Valorisation · Weathering

1 Introduction

In the last few decades, raising environmental consciousness incited many governments, institutions, and firms to reduce fossil fuel consumption. Carbon footprint and resource nonrenewability are amongst the main motives to mitigate dependence on oil resources. In the same context, lignocellulosic biomass emerged as a reliable alternative for petroleum based synthetic plastics (Mehanny et al. 2012). Lignocellulosic biomass is estimated to be the most widely spread biopolymer on earth. Global annual production of this biopolymer is roughly calculated to be 1.3×10^{10} metric tons (Kumar et al. 2008). Lignocellulosic biomass encompasses: (1) agriculture wastes (palm residues, empty fruit bunch, straw, bagasse, corncob, and stover), (2) forest wastes (branches, unwanted stems, and withered leaves), (3) food wastes, and 4) industrial wasted (waste paper, and demolished wood) (Mehanny et al. 2016). Plethora of the low cost lignocellulosic residues and their capacity to act as a "carbon sink", encouraged their use in more than 200 applications, including construction materials, moderate strength composites, adhesives, packaging, coatings, dental fillings, implants, and drug delivery (Kumar et al. 2008; Kumar et al. 2009; Mehanny et al. 2019). Palm trees number exceeded 120 million in the entire globe, 70% amongst them are located in the Middle East region. Egypt comprises more than 12 million palm trees, distributed in all governorates (El-Juhany et al. 2010). Each year, the problem of getting rid of palm residues (fronds, leaflets, coir, and fruit bunch), during the trimming season becomes obtrusive. Each palm tree can yield 30–50 kg. In other words, Egypt will face accumulation of palm residues tonnage of 240,000–600,000 per year. So far, palm residues are majorly piled and combusted in open air, leaving large carbon footprint for no benefit, which is severe environmental problem. Ultraviolet rays, water uptake, acidic/alkaline effect, thermal degradation, thermites and fungal attacks represent environmental factors, Table 1 summarizes the response of the palm-based composites to those factors. These environmental effects play major role in determining the durability of the palm-based composites. In this book chapter, the concept of each environmental phenomenon will be thoroughly explained. The effect of chemical, physical, and biological factors

Table 1 Effects of different environmental aging factors on palm reinforced composites

Environmental factor	Material	Measured response	Findings 1	Findings 2	References
Accelerated/natural weathering	Palm/Polypropylene	• Melting point • Max. stress • Max. strain	• Natural weathering: Almost zero change after 3 months	• PP Strength (34 MPa): Decreased by 50% in 9 months • Palm/PP composite strength (28 MPa): Decreased by 11%	Abu-Sharkh and Hamid (2004)
Accelerated Weathering	Oil Palm/Jute/Phenolic resin	• Max. stress • Max. strain	• Palm/epoxy composite strength (22.6 MPa): Decreased 6% in 100 h.		Jawaid et al. (2016)
Accelerated Weathering	Palm Fruit Bunch/PVC	• Flexure Strength • Impact • SEM, IR	Palm improved stiffness for weathered/non-weathered samples.		Bakar et al. (2005)
Thermal aging Photochemical aging	Palm/Polypropylene	• Mech prop • IR, UV	• Thermo-oxidation: The incorporation of palm fibers did not cause any improvements	• Photodegradation The incorporation of palm fibers did not cause any improvements	Chollakup et al. (2017)
Thermal aging High energy aging Moisture aging Biological aging	Treated/Untreated palm/Phenol formaldehyde	• Mech. prop • SEM	Water aging decreased the flexural strength of untreated palm composite by 50%	Water aging increased the impact strength of palm composites	Sreekala et al. (2004)
Moisture aging Seawater exposure	Palm/concrete	• Dimension stability • Moisture resistance	1% of fiber decreased concrete expansion by 30%		Machaka and ElKordi (2017)
Moisture aging	Palm/concrete	• Mech prop • Dimension stability	• Tried 0, 0.5, 1% fiber: 0.5% fibers showed max. comp strength for mortar	0.5% fibers showed max. flexural strength for mortar	Ozerkan et al. (2013)

(continued)

Table 1 (continued)

Environmental factor	Material	Measured response	Findings 1	Findings 2	References
Acidic attack Alkaline attack Moisture aging	Palm/concrete	• Acidic Resistance • Alkaline resistance	• Tried 0, 5, 10, 15% fiber: 15% of fibers yielded the highest swelling (1%)	15% of fibers yielded the lowest shrinkage and density 15% manifested higher resistance to alkaline and acidic attack	Kareche et al. (2019)
Thermal aging Wear	Palm/epoxy	• Thermal resistance • Wear resistance	• Tried 30, 50, 70% fiber: 30% showed the highest wear resistance		Shuhimi et al. (2016)
Hydrothermal aging	palm	• Fiber hydrophilicity	• Fiber treated with 0, 2, and 4% stearic acid: 4% stearic acid showed the lowest hydrophobicity to palm fibers		Jain et al. (2019)
Alkaline treatment (Fibers)	Palm/concrete	• Alkaline resistance	• Different alkali treatments: $Ca(OH)_2$ had a more destructive effect on fibers	NaOH had less destructive and more localized.	Kriker et al. (2008)
Moisture aging	Palm/epoxy	• Water uptake	• Tried 0, 10, and 20% fibers: 20% of fibers showed the highest moisture uptake		Leman et al. (2008)
Soil burial	Palm/polyester	• Mech. deterioration • Mass deterioration	• After 12 months of soil exposure: flexural strength was reduced from 41.6 MPa to 27.4 MPa	Modulus of elasticity was reduced from 3.85 GPa to 2.69 GPa	Hill and Abdul (2000)

(continued)

Table 1 (continued)

Environmental factor	Material	Measured response	Findings 1	Findings 2	References
Termites attack Fungi attack	Palm/phenolic resin treatment	• Weight loss	• After 4 weeks of exposure to the white-rot fungi: Weight loss of the treated specimens was between 1.71 and 8.99% compared to 16.9% for the untreated ones.	• After 4 weeks of exposure to termites: Treated samples showed significantly higher termite mortality (52–100%) and a lower weight loss (3.22–9.58%) compared to those for the untreated samples; 18% termite mortality and 27.94% weight loss	Bakar et al. (2013)
Termites attack Fungi attack	Palm/phenolic resin	• Mass deterioration	• After exposure to termites' attack for 12 weeks: The weight loss of the treated oil palm stem plywood (10.7% for outer veneer and 15.8% for inner veneer) was significantly lower than that for the untreated ones (19.2% for outer veneer and 23.9% for inner veneer)	• After exposure to Pycnoporous sanguineus fungi for 12 weeks: The phenolic-treated oil palm stem plywood specimens showed low weight loss (13%) compared to untreated ones (34%)	Loh et al. (2011)
Fungal attack	Palm/polypropylene	• Mech. deterioration • Mass deterioration	The weight loss of the treated date palm particles (1.08%) was less than that for the untreated ones (5.43%)		Nasser et al. (2017)

on palm fibers will be reviewed. Furthermore, different testing techniques that were employed in determining (quantifying) environmental aging effects, will be presented in addition to highlighting results in accordance to measurement methods. In other words, framework that we will follow in this chapter is to explain each environmental phenomenon, then how to measure its effect, and finally results obtained from those measuring techniques. The objective of this work is to throw light on different environmental effects, influencing performance of palm based composites; and provide experienced/inexperienced audience with extensive knowledge background that can be useful in any composite application subjected to aging load.

2 Weathering and Ultraviolet Effect

2.1 Concept and Scope

Weathering effect is a significant parameter to estimate operational life time for composite panels. Additionaly, measuring weathering effect aids us to assess material performance throughout operational life, especially for those materials undergoing outdoors exposure. Weathering effect comprises ultraviolet (UV) rays, atmospheric humidity, and raised temperature. UV rays and their influence on composite materials will be the main core of this section, as other effects releavant to weathering (humidity and temp.) will be highlighted down the line. UV is major form of energy conveyed by sun rays. As explained elsewhere (Jawaid et al. 2016), when UV rays hit natural fiber reinforced polymer composite, groups like chromophoric, quinones, and hydroperoxy radicals are resulted from degrading lignin. Mentioned groups catalyze photodegradation of binding polymeric matrix. Degradation which take the form of surface oxidation, polymer scission, and breaking of tieing molecules between crystalline segmants. Those types of material deterioration necessates mechanical and gravametric degradation (Fig. 1).

2.2 Measuring Methods

Weathering effect can be measured by two routes, natural weathering test emulates real operation of composite material outdoors. As shown in Fig. 2a, composite panels are placed on inclined boards facing sunlight, panels are supposed to be exposed to natural sun light, humidity, and temperature. To cope with prolonged test period, and to allow tuning humidity and temperature, accelerated weathering method was developed (Fig. 2b). Composite panels are situated in bench top or larger scale chambers (accelerated weathering equipment), hence, it can be loaded with UV rays, emitted from lamps (which must be tuned to UV wave lengths), humidity and tunable temperature, coupled with a fan to ensure uniform temp. and humidity distribution.

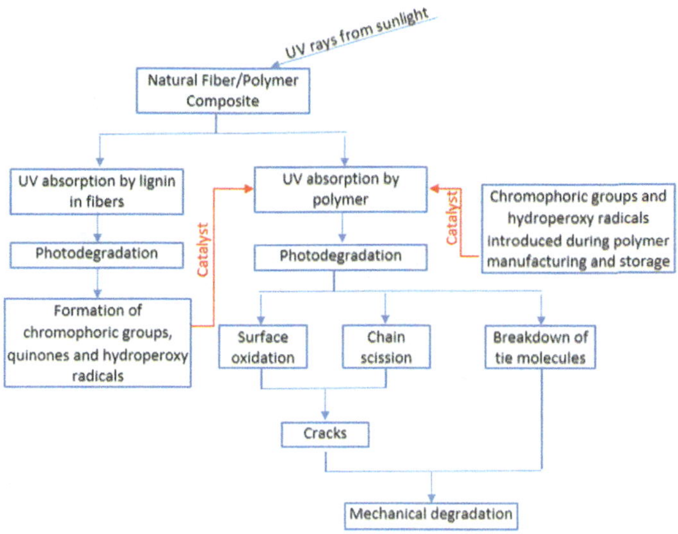

Fig. 1 Photodegradation of natural fiber reinforced polymer, cited from Jawaid et al. (2016)

2.3 *Results and Discussions*

In this section, we will showcase most relevant work to weathering of palm based composites, and attempt to interpret results and trends. Sharkh et al. (Abu-Sharkh and Hamid 2004) found that palm fibers significantly reduced negative weathering effect in polypropylene composite. For instance, strength of PP decreased from 34 MPa by 50% in 9 months, whereas Palm/PP composite exhibited minimal strength decrease from 28 MPa by 11% (Table 2). Elongation and melting temperature (T_m) were in line with previously explained behavior, palm fibers composite showed almost zero decrease in elongation and T_m, unlike pure PP.

Accelerated (artificial) weathering to the same type of materials, yielded same expected trends; palm constituted composites manifested minimal weather degradation in comparison to pure PP (Table 3). Deterioration in properties in attributed to reduction of molecular weight of PP during weathering process.

Jawaid et al. (2016) results prove that amid the three investigated ingredients; namely: palm, jute, and epoxy, palm was the most recalcitrant to deteriorate as a result of weathering. Tensile strength of palm/epoxy composite (22.6 MPa) decreased by 6% during 100 h accelerated weathering. Strength of jute epoxy composite (45.5 MPa) decreased by 44%, which is substantially worse than palm composites. Stiffness was consistent with aforementioned strength results (Table 4).

Bakar et al. (2005) investigated effect of palm fiber content on mechanical properties of PVC composites. Flexural and impact strength decreased with increasing fiber content, due to higher properties of PVC than palm fibers. Generally, there wasn't

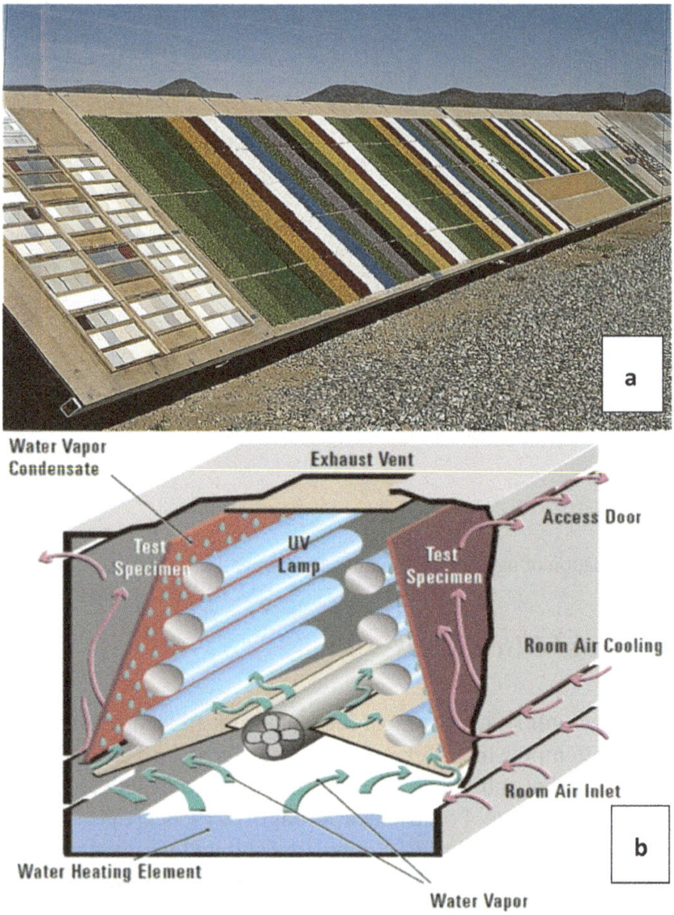

Fig. 2 Outdoors (natural) weathering testing setup (**a**), cited from (ATLAS-MTS 2020a), and indoors artificial weathering equipment (**b**), cited from ATLAS-MTS (2020b)

Table 2 Natural weathering effect on mechanical and physical properties of polypropylene (PP) and 40% wt. palm reinforced PP composite, results adopted from Abu-Sharkh and (Hamid 2004)

	0 month	3 months	9 months
Tensile strength (MPa)			
PP	34	33	17
Palm/PP	28	25.5	25
Elongation (%)			
PP	11	11.1	3.2
Palm/PP	4.5	5	4.5
T_m (°C)			
PP	163.5	161.8	155.8
Palm/PP	162.1	162	161.8

Table 3 Accelerated (artificial) weathering effect on mechanical and physical properties of polypropylene (PP) and 40% wt. palm reinforced PP composite, results adopted from Abu-Sharkh and Hamid (2004)

	0 h	750 h	2000 h
Tensile strength (MPa)			
PP	34	22	6
Palm/PP	28	26	24
Elongation (%)			
PP	11	4.1	2.2
Palm/PP	4.5	4.4	3.9
T_m (°C)			
PP	163.5	156	136
Palm/PP	163	161	160

Table 4 Accelerated (artificial) weathering effect on mechanical properties of epoxy and 40% wt. palm or jute reinforced epoxy composite, results adopted from Jawaid et al. (2016)

	0 h	100 h
Tensile strength (MPa)		
Epoxy	20.6	19.2
Palm/Epoxy	22.6	21.2
Jute/Epoxy	45.5	25.5
Elongation (%)		
Epoxy	16.2	1.5
Palm/Epoxy	12	2.1
Jute/Epoxy	11.3	1.4
Stiffness (GPa)		
Epoxy	2	1
Palm/Epoxy	2.2	2.2
Jute/Epoxy	3.9	2.8

any significant weathering effect for PVC and palm/PVC composites. Photodegradation of PVC takes place through two simultaneous competent actions, which are polymeric chain scission and cross linking. Chain scission reduces molecular weight while cross linking has counter effect. It seems that cross linking boosts up flexural strength, however it leads to embrittlement which necessitates impact strength and stiffness reduction (Table 5).

In sum, palm fibers drastically improved mechanical and physical properties of weathered PP and epoxy composites. For PVC composites, palm increases modulus of elasticity of weathered samples. Moreover, despite fact that palm didn't improve Flexural or impact strength, its positive impact on decreasing fabrication temp. and density cannot be ignored.

Table 5 Accelerated (artificial) weathering effect on mechanical properties of PVC and palm reinforced PVC composite, results adopted from Bakar et al. (2005)

	0 h	504 h
Flexural strength (MPa)		
PVC	81	91
Palm/PVC (20 phr)	79	79
Palm/PVC (40 phr)	66	66
Impact strength (KJ/m^2)		
PVC	8	7.8
Palm/PVC (20 phr)	6.8	6.5
Palm/PVC (40 phr)	5.7	5.4
Flexural stiffness (GPa)		
PVC	3.5	3
Palm/PVC (20 phr)	4	3.6
Palm/PVC (40 phr)	4.5	3.8

3 Water/Moisture Effect on Palm-Based Composites

3.1 Concept and Scope

Natural fiber composites are sensitive to water and humidity. The high content of hydroxyl group (OH) in the natural fibers is the main cause of its hydrophilic behavior. Although natural plant fibers are a viable replacement candidate to man-made fibers, the deterioration of their mechanical properties due to exposure to water and moisture from the surrounding environment represents a major impediment to their use. Deterioration in the natural fiber/ matrix adhesion arises due to the hydrophilicity nature of these natural fibers that leads to fiber swelling. Different fiber treatment techniques have been studied in literature to find the optimum technique that provides the most adequate protection against the high water/moisture absorption nature of these fibers. These techniques include silane treatment, alkali treatment, cyclic hornification, treatment with maleic and acetic anhydride, stearic acid treatment, and graft copolymerization. These treatment techniques proved to reduce the moisture susceptibility of the natural fibers. Moreover, besides improving the physical and mechanical properties of the fibers, these treatments enhance the fiber/matrix bonding. Therefore, the water and humidity absorption characteristics are considered one of the most important physical properties to determine the amount of water absorbed by composites reinforced with treated/untreated palm fibers at different conditions and times. The water absorption can be defined as the mobility of liquids in porous solids due to surface tension acting in the capillaries. The capacity of water and moisture absorption is directly related to the presence of voids and the fiber/matrix bonding. The increase in the composite weight after exposure to water or moisture is attributed to the trapping of water inside the voids within the composite. Moreover, moisture diffusion as a part of the physical properties rely on factors such as void/pore content, humidity, volume fraction of fiber, temperature, and the viscosity of matrix (Hill and Abdul 2000; Jain et al. 2019; Obi 2015; Ozerkan et al. 2013; Radzi et al. 2019; Teo et al. 2007).

4 Measurement Methods

- Water Immersion

specimens of palm fiber reinforced composites are first dried to a constant weight in an electric oven. The initial weight of the samples is recorded as W_0. The samples are then either totally immersed in tab water (Teo et al. 2007; Radzi et al. 2019), distilled water (Sreekala et al. 2004; Leman et al. 2008), or deionized water that has been sterilized by boiling (Hill and Abdul 2000), or one surface of the samples is submerged in 3–5 mm of water according to the ASTM C1403 (ASTM 2012b) (Ozerkan et al. 2013). Figure 3 shows a schematic diagram for a typical cylindrical desiccator. All these tests are performed at room temperature. The samples kept in water for either 360 min (Ozerkan et al. 2013), 72 h (Teo et al. 2007), 7 days (Radzi et al. 2019), or 12 months (Hill and Abdul 2000). The weight of the samples after immersion is then measured and recorded as W_1 along with the corresponding elapsed time. Before weighting, the samples are wiped dry by an absorbent lint free towels to get rid of the under-surface moisture. The percent of water absorption is calculated and recorded for each time interval by using Eq. (1). The percent weight change is recorded at constant time intervals until the samples reach the moisture equilibrium stage. The last value of percent moisture absorption is denoted as the moisture equilibrium content.

$$Water\ Absorption\,\% = \frac{W_1 - W_0}{W_0} \times 100 \tag{1}$$

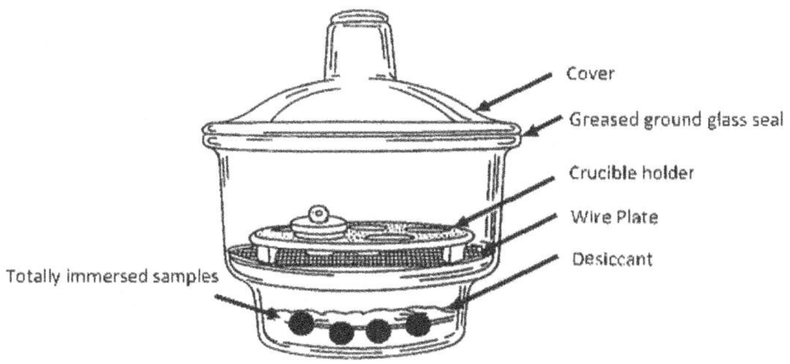

Fig. 3 Schematic diagram for a typical cylindrical desiccator, adopted from (http://www.ecs.umass.edu/cee/reckhow/courses/572/572bk11/572BK11.html)

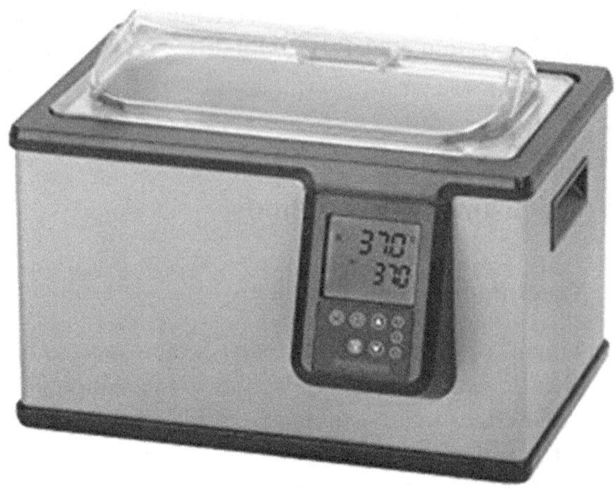

Fig. 4 A digital water bath, adopted from, cited from (https://www.polyscience.com/products/general-purpose-water-baths/digital-water-baths)

In (https://www.polyscience.com/products/general-purpose-water-baths/digital-water-baths) the samples are immersed in a container filled with cold tab water. The tests are done to measure the time required for the samples to disperse in water.

- Water immersion at different temperatures

Different temperatures are employed to simulate the different environmental conditions and their effect on the water absorption of palm fiber reinforced composites. In Jain et al. (2019), four different temperatures 0, 25, 50, 75 °C are chosen. For maintaining the 0 °C temperature inside water, a cryostat water bath was employed. The other temperature conditions are maintained by using a digital water bath (Fig. 4). The gain in weight is measured periodically. The water absorption in this case is also calculated by using Eq. 1.

In Sreekala et al. (2004), the samples are kept at boiling water for 2 h and then taken out and air dried. The effect of this test on the physical and mechanical properties of the samples is further studied.

- Exposure to humidity

The first step in the moisture content determination experiment is weighting the samples. Following drying the samples in oven, these dried samples are placed inside desiccators and allowed to reach the room temperature. Then the samples are taken out from the desiccators, wiped free of surface moisture, and weighted. The moisture content on wet basis is calculated according to Eq. 2 (Obi 2015).

$$Moisture\,content(wet\,basis) = \frac{Final\,weight - Initial\,weight}{Initial\,weight} \times 100\% \quad (2)$$

- Exposure to humidity at different temperatures

This test is performed to explore the combined effect of humidity at different temperatures. The date palm reinforced samples were placed inside a humidity chamber with a forced air circulation with the temperature control constancy of ±1.0 °C. The different temperatures are 25, 50 and 75 °C with the relative humidity of 75%. The experiments are conducted for an average of 20 days for each temperature and the gain in weight is measured periodically. The gain in weight due to moisture absorption is calculated by Eq. 1 (Jain et al. 2019).

4.1 Results and Discussion

- Water immersion

Water absorption tests were performed (Teo et al. 2007), in order to investigate the durability of lightweight concrete incorporating solid waste oil palm shell (OPS) as coarse aggregate. Their results suggested that the water absorption of OPS concrete was comparable to other light weight concretes.

The effect of water absorption on the roselle (RF)/sugar palm fiber (SPF) reinforced thermoplastic polyurethane hybrid composites had been investigated in Radzi et al. (2019). Their results showed that increasing the SPF content led to a decrease in the water absorption of the hybrid RF/SPF composites. The lowest recorded water absorption level in this study was for the composites reinforced with single SPF as shown in Fig. 5. RST-1, RST-2, RST-3 are composites reinforced with 75%/25%, 50%/50%, 25%/75% RF/SPF, respectively. SPFT is the composite reinforced with only SPFs.

The effect of soaking Crystic 471 PALV polyester composites reinforced with treated and untreated oil palm empty fruit bunch (efb) fiber mats was explored in Hill and Abdul (2000). The effect of water absorption on the flexural properties of the composites is shown in Table 6. Their results showed that the acetylation of oil palm empty fruit bunch fibers could be considered as the most effective fiber treatment technique in terms of reducing the rate and extent of water uptake. Thus, this technique proved to enhance the fiber/matrix interfacial bonding due to the increase in the hydrophobicity of the treated fibers.

The water absorption tests conducted in Ozerkan et al. (2013) showed that increasing the content of the alkali treated date palm fibers (DPFs) above 0.5% in the cement mortar composites decreased the water absorption capacity as shown in Fig. 6. The authors attributed this increase in the hydrophobicity of the DPFs to the removal of the hemicellulose and lignin by the alkali treatment.

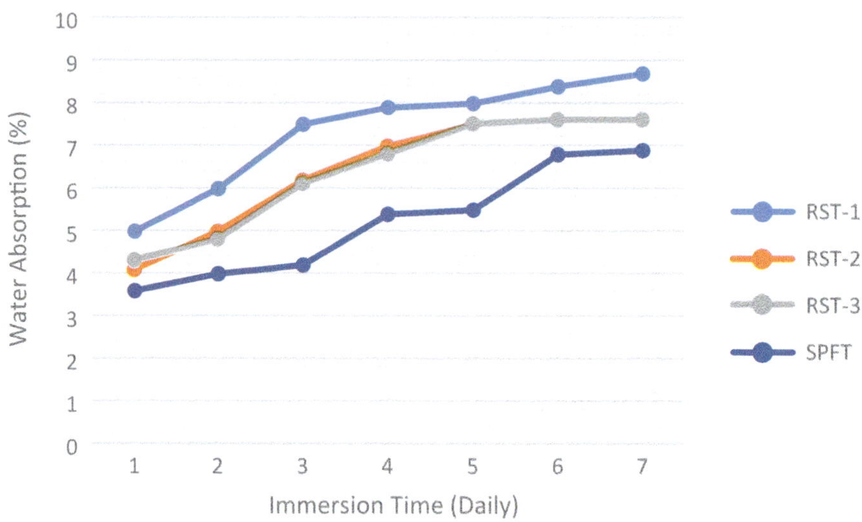

Fig. 5 Water absorption of RF/SPF hybrid composites, cited from Radzi et al. (2019)

Table 6 Variation in flexural properties during water soaking of at 20 °C, cited from Hill and Abdul (2000)

Fiber	Duration (Months)	Unmodified	Acetylated	Silane	Titanate	Resin
σ_f (MPa)						
efb	0	41.6	41.8	38.8	38.3	50.9
	3	38.7	44.2	39	37.7	49.8
	6	37.2	40.6	37.9	36.9	49.7
	12	35.3	39.5	36.7	35.5	49.5
E_f (GPa)						
efb	0	3.85	4.60	4.46	4.04	3.23
	3	3.49	4.67	4.52	3.73	3.22
	6	3.19	4.41	4.29	3.65	3.08
	12	2.92	4.31	4.04	3.29	3.02

- Water immersion at different temperatures

Treating the palmyra natural fibers with 4% stearic acid proved to be the most effective treatment to improve the hydrophobicity of these fibers. The reduction in water uptake of the epoxy composites reinforced with 4% stearic acid treated fibers can be attributed to the development of the hydrophobic fatty acid chain on the surface of the surface of the composites during the esterification reaction (Jain et al. 2019).

Elevating the temperatures affected the water diffusion behavior. The rate of water uptake increased with increasing the temperature. The authors in Jain et al. (2019)

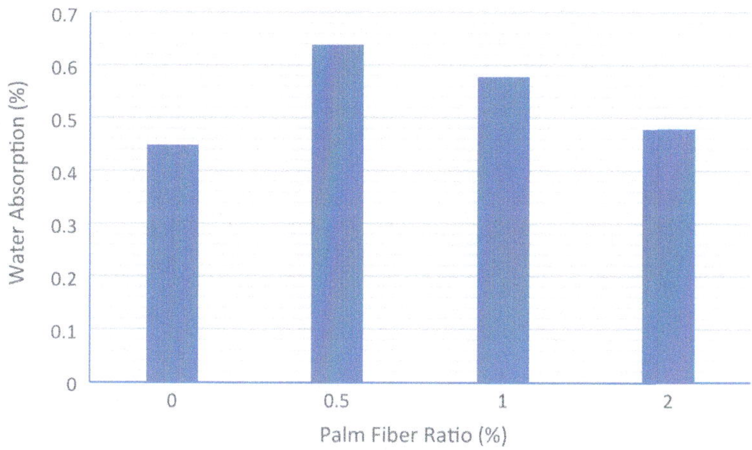

Fig. 6 Effect of palm fiber ratio on water absorption capacity of mortar samples, cited from Ozerkan et al. (2013)

attributed this increase in the water uptake to the Soret effect as the higher the temperature, the higher the progression of moisture. Figure 7 shows a comparison of the water uptake of epoxy composites reinforced with palmyra fibers treated with four different techniques under different temperatures.

- Exposure to combined humidity/temperature conditions

In comparison to the gain in weight during the water immersion, the gain in weight in the exposure to humidity tests is much lower as shown in Fig. 8. The 4% stearic acid

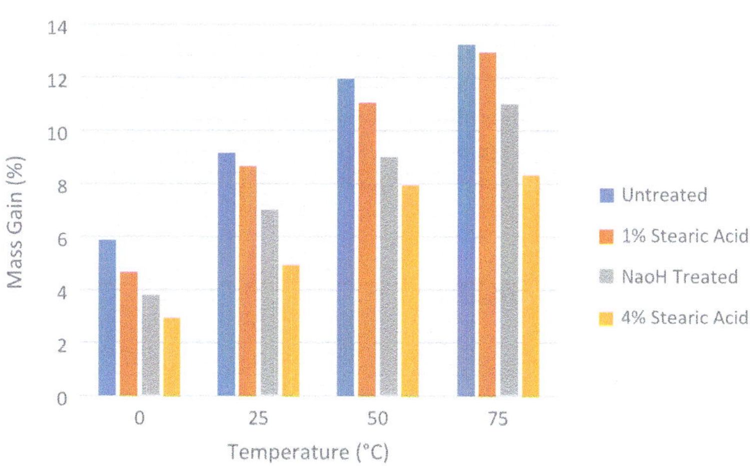

Fig. 7 Weight gain at different temperatures in cryostat/water bath, cited from Jain et al. (2019)

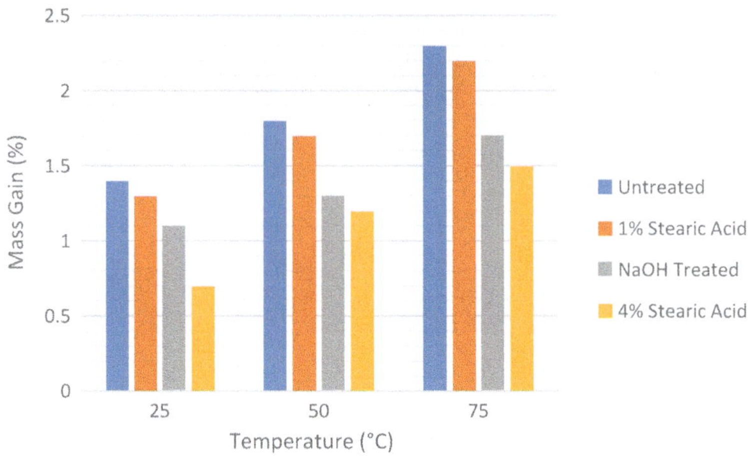

Fig. 8 Weight gain at different temperatures inside humidity environment chamber, cited from Jain et al. (2019)

treatment proved also in these tests to be the optimal treatment technique to increase the hydrophobicity of palmyra fibers. Under the combined temperature/humidity conditions the Soret effect was also clearly visible (Jain et al. 2019).

5 Thermal Effect

5.1 Concept and Scope

The thermal properties of composites reinforced with different palm fibers are of crucial importance as it reveals their heat resistance and thermal stability. Thus, analyzing the thermal properties of the different palm fibers and palm fiber reinforced composites is important to determine the service temperature of these composites (Chollakup et al. 2017; Radzi et al. 2019; Shuhimi et al. 2016).

5.2 Measurement Methods

- Thermogravimetric analysis (TGA)

Thermogravimetric analysis is performed by using a thermal analysis machine. The analysis is mainly done at a temperature range between 30 and 600 °C with a heating rate of 10 °C/min. A schematic representation of the TGA instrument has been illustrated in Fig. 9.

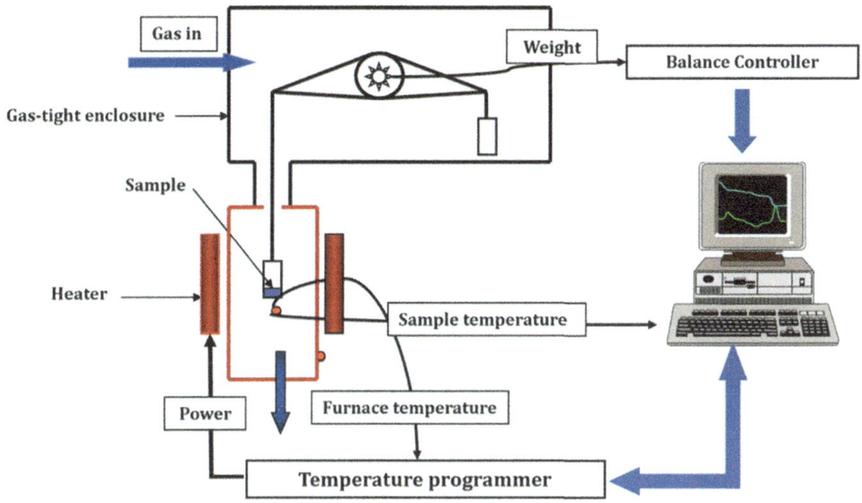

Fig. 9 A schematic representation of the TGA instrument, cited from Akash and Rehman (2019)

- Tribological test under different temperatures

The tribological test under different temperatures is performed by using a pin-on-disc tribometer (shown in Fig. 10) according to ASTM G99-05 standard. The constant applied load is 49.05 N and the speed is 1000 rpm at different temperatures. The specific wear rate (W_s) is calculated by using Eq. 3 (Shuhimi et al. 2016).

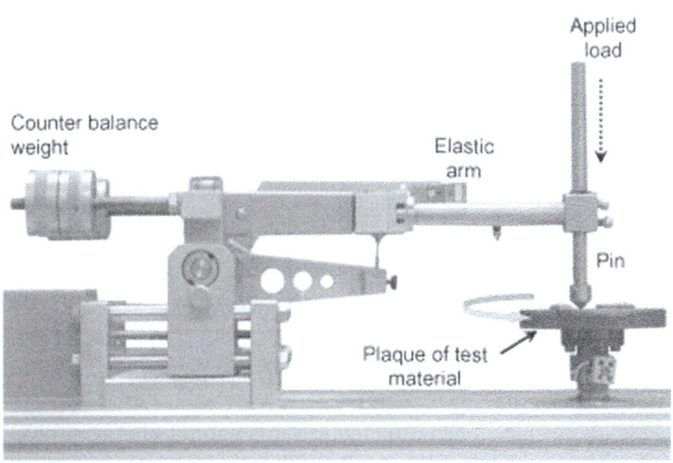

Fig. 10 Pin-on-disk tribometer equipment, cited from McKeen (2016)

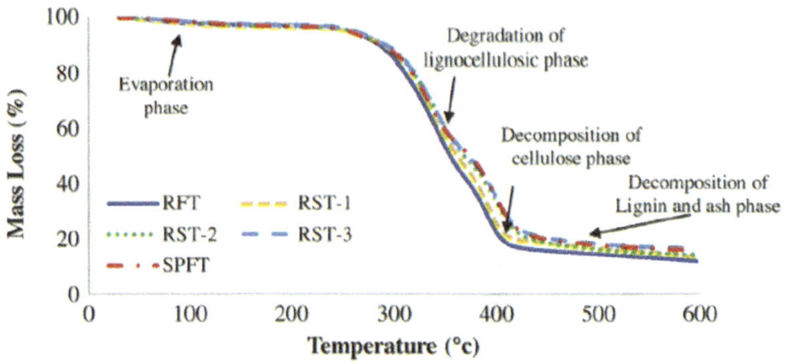

Fig. 11 TGA of RF/SPF hybrid composites, cited from Radzi et al. (2019)

$$W_s = \frac{V_{loss}}{FL} \tag{3}$$

where, W_s is the specific wear rate (mm^3/Nm), V_{loss} is the loss in volume (mm^3), F is the applied load (N), and L is the sliding distance (m).

5.3 *Results and Discussion*

- Thermogravimetric analysis (TGA)

TGA analysis is performed in order to investigate the decomposition and thermal stability of the composites reinforced with palm fibers. In Radzi et al. (2019) TGA was performed on polyurethane composites reinforced with hybrid roselle and Sugar Palm fibers (SPFs). Figures 11 and 12 show the TGA and the differential thermogravimetric (DTG) curves for these composites. Their results showed that the incorporation of the SPFs in the composites improved the thermal properties.

6 Seawater Effect

6.1 *Concept and Scope*

Some research studies have been directed towards natural fiber reinforced composites that can be used in the marine sector for either coastal construction or producing components for fishing boats. Therefore, investigating the effect of exposing natural fiber reinforced composites to the seawater is important (Leman et al. 2008). On the other hand, the major drawbacks of the natural fiber reinforced composites are the

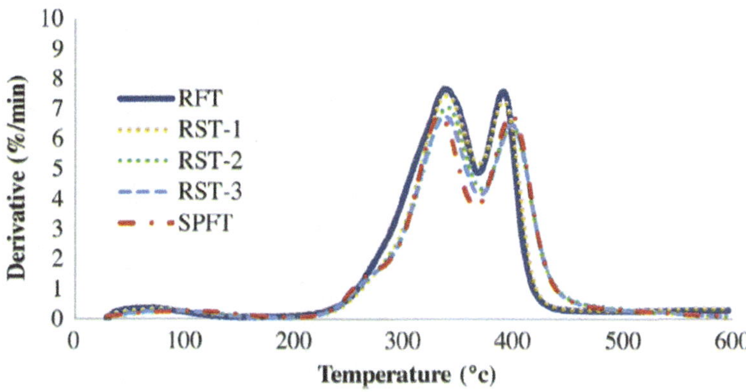

Fig. 12 DTG of RF/SPF hybrid composites, cited from Radzi et al. (2019)

poor interfacial adhesion between fiber and matrix as well as poor fiber dispersion. This could be attributed to the hydrophilicity nature of the lignocellulosic fibers. Therefore, surface modification of natural fibers is important to improve the interfacial bonding and the moisture resistance. Chemical treatments are the most common treatment techniques. The expensiveness, the environmental pollution, and the hazardous effects that threatens life and nature of these chemicals contradict the concept of the sustainable green environment. Thus, the need for a green, sustainable and safe alternative is important. Therefore, recent research studies suggested treating the sugar palm fibers with seawater (Leman et al. 2008; Ishak et al. 2009).

6.2 *Measurement Methods*

- Submersion in actual and simulated seawater

In Machaka and ElKordi (2017), the authors studied the effect of exposing concrete samples reinforced with fan palm fibers (FPF) to five different solutions. These solutions are saturated lime water (control), seawater, solution of 2% concentration of Magnesium sulphate ($MgSO_4$), 2% concentration of sodium sulphate (Na_2SO_4), and solution of sodium hydroxide (NaOH) with 1 N concentration. The durability of concrete/FPF composites was investigated by measuring the change in length of the samples after their exposure to the five solutions. The apparatus shown in Fig. 13 is used to measure the change in length of the concrete composite prisms. In this study, the authors applied practices from ASTM C1012 and ASTM C452. In their proposed measurement procedure, the concrete composite prism length is measured directly after its removal from the mold after 24 h of casting and this reading is recorded as the initial length. Then the measurements of the change in length after submerging the concrete composite prisms in the five solutions were recorded after 7, 14, 30,

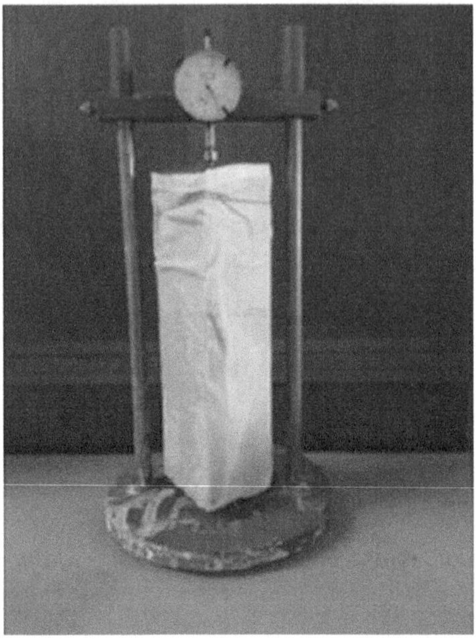

Fig. 13 The apparatus used to measure the change in length, cited from Machaka and ElKordi (2017)

45, 60, 75, 90, 120, 150 days. These measurements were taken within 5 min after the removal of the sample from the solution. Moreover, each sample is covered with a towel submerged in the same solution to continue the exposure even during the measurement period.

- Simulating the seawater surrounding characteristics

The samples in the procedure proposed in Leman et al. (2008) would undergo the same water absorption procedure presented in section "water immersion" except that the used water bath had a relative humidity level of 100% with a constant temperature of 40 °C. These parameters were adjusted to simulate the seawater characteristics though the water used throughout this test is distilled water. The samples were immersed in the water bath for 768 h and data were collected at certain time intervals.

6.3 Results and Discussion

- Submersion in actual and simulated seawater

Adding fan palm fibers (FBFs) to the concretes significantly improves their resistance to seawater and other chemicals. The authors in Machaka and ElKordi (2017) noted that after 150 days of severe exposure to harsh simulated environment, the concrete

Table 7 Length change of concrete prisms with and without FBF after 150 days, cited in Machaka and ElKordi (2017)

Exposure condition	Control concrete (without FBF)	Concrete reinforced with 1% FBF
Water cured	476	393
Sea water	1238	719
Na_2SO_4	581	502
NaOH	621	586
$MgSO_4$	1069	880

samples reinforced with fan palm fibers maintained their durability. Table 7 shows a comparison between the control samples and the 1% FPFs reinforced samples in terms of the expansion in length after exposure to different solutions. Therefore, FPFs proved to improve the volume stability and the performance of different concrete mixtures.

- Simulating the seawater surrounding characteristics

The moisture absorption process of pure epoxy and sugar palm fiber (SPF) reinforced epoxy composites were determined to be reversible with low water uptake saturation levels. The authors in Leman et al. (2008) noted that slightly low water absorption through the thickness direction has occurred. Moreover, the water diffusion in this composite followed the dual sorption-diffusion process which obeys the Fick's law as shown in Fig. 14. Increasing the fiber content up to 20% increased the water absorption of the resultant composites. The authors attributed this to the deposition of the short palm fibers at the lower surface of the composites and thus led to higher water absorption rates by the both fibers and the epoxy resin.

7 Acidic/Alkaline Effect

7.1 Concept and Scope

Recent research studies have been directed towards investigating not only the mechanical and water uptake properties of the date palm fibers reinforced composites but also the durability performance of these composites. Among the several previously performed durability tests, the effects of exposing palm fiber reinforced composites to severe acidic/alkali attack have gained a considerable attention in the recent years. This could be attributed to the importance of studying the feasibility of employing these composites in different applications under severe environmental conditions.

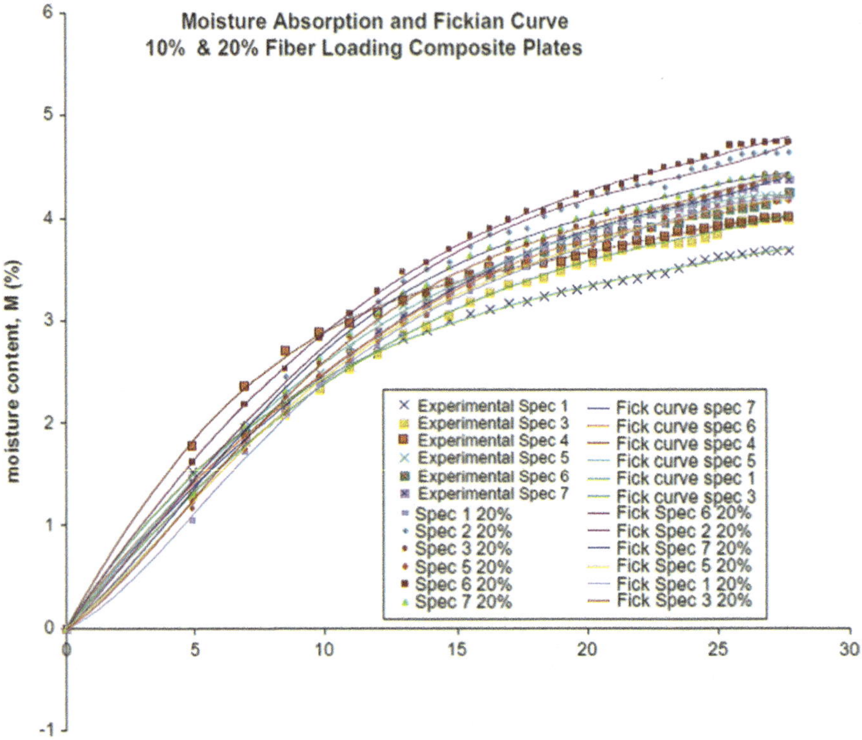

Fig. 14 Moisture absorption and Fickian curves for epoxy composites reinforced with 10% and 20% SPF wt%, cited from Leman et al. (2008)

7.2 Measurement Methods

- Acidic degradation

In Ozerkan et al. (2013), the authors studied the effect of exposing concrete bars reinforced with date palm fibers to acidic attack. The concrete bars were cured in lime water then immersed in sodium sulphate solution. The change in length was recorded at certain time intervals during the 4-month immersion period. The resistance of the mortar bars to sulphate was determined according to ASTM C 1012 (ASTM 2013a, b).

In Kareche et al. (2019), the authors investigated the resistance of the saturated cement mortar/DPF composite samples to the attack by sulphuric acid. The samples were immersed in sulphuric acid solution (5 wt% H_2SO_4) at room temperature for 28 days. The solution was renewed at certain intervals. The weight loss of the composites was calculated according to Eq. 4.

$$\Delta m = \frac{(m_i - m_t)}{m_i} \times 100 \tag{4}$$

where m_i is the initial weight of the sample before immersion in sulphuric acid and m_t is the weight of the sample after immersion in sulphuric acid after certain time period.

- Alkali degradation

In Kareche et al. (2019) the effect of exposing cement/DPF composites to alkali solutions was investigated in terms of compressive strength and weight change. The tests were performed according to ASTMC267-96 standard. The samples were divided and immersed in either sodium chloride solution (5 wt% NaCl) or sodium hydroxide solution (5 wt% NaOH) at room temperature for 28 days.

The durability of male date palm surface fibers (MDPSF) after immersion in three alkaline solutions was investigated in Kriker et al. (2008). These solutions are sodium hydroxide, calcium hydroxide, and Lawrence solution with a pH in the range of 12.5 and 12.95.

7.3 *Results and Discussion*

- Acidic degradation

The corrosive property of sulphuric acid and the high alkalinity of concrete led to the high susceptibly of concrete to acid attack. Due to the complex chemical reactions that occur to concrete after the exposure to acidic attack, the concrete disintegrates and losses its strength. Green corrosion inhibitors have received attention in recent years. In the work proposed by (Kareche et al. 2019), the effect of exposing concrete/DPF composites to sulphuric acid is investigated in terms of weight loss for different exposure times (Fig. 15). Their results showed that the pure mortar exhibited a higher weight loss with increasing the exposure time when compared to DPF reinforced concrete samples. Moreover, increasing the DPFs content reduced the weight loss. The authors attributed this to the existence of organic compounds in natural fibers that are rich in oxygen, nitrogen atoms, and aromatic rings. These natural compounds meet with the fundamental requirements of an inhibitor. Thus, using natural DPFs in cement-based binder improves the durability performance of the cement to resist the attack by sulphuric acid.

Figure 16 shows the effect of immersing mortar sample bars reinforced with different ratios of DPFs in sodium sulfate. The authors in Ozerkan et al. (2013) noted that the incorporation of DPF improved the resistance of the mortar against the sulfate attack and thus improved its durability.

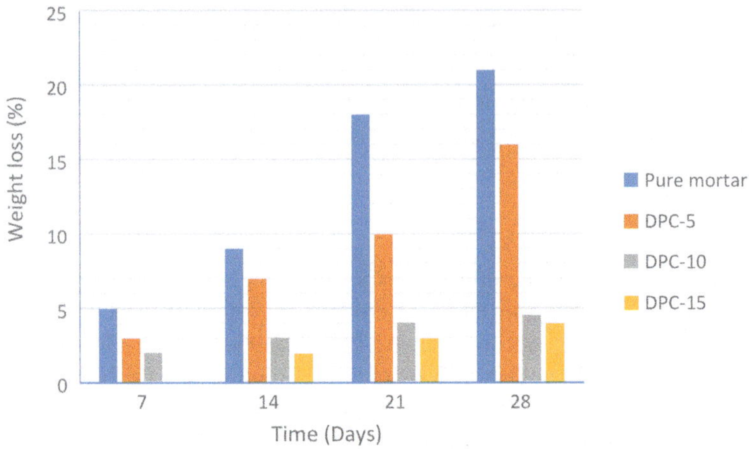

Fig. 15 Weight loss of concrete/DPF composites after immersion in sulphuric acid, cited in Kareche et al. (2019)

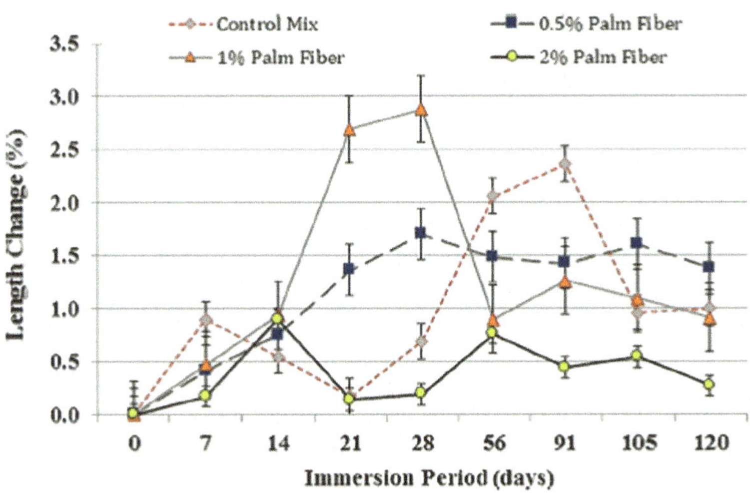

Fig. 16 Length change of mortar/DPFs composites due to sulfate attack, cited in Ozerkan et al. (2013)

- Alkali degradation

In Kareche et al. (2019) exposing the date palm concrete (DPC) to the alkaline solutions for 28 days did not damage its compressive strength as shown in Fig. 17. The authors attributed the increase of the compressive strength of the pure mortar samples and composites after the exposure to salts to the additional hydration that occurred due to the presence of water and to the salt crystallization within the pores

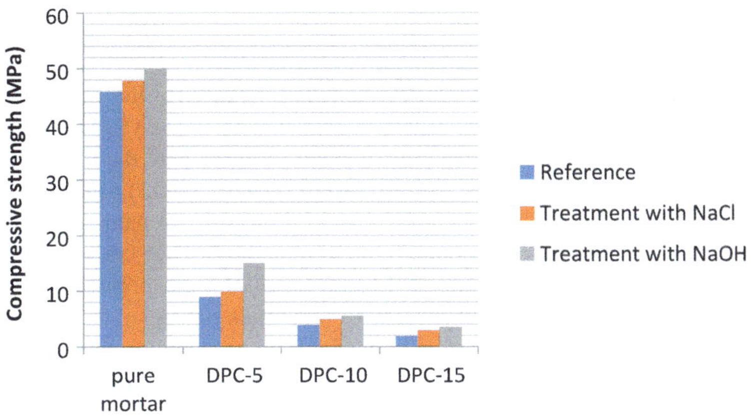

Fig. 17 Compressive strength of DPC samples after immersion in different alkali solutions, cited by Kareche et al. (2019)

of the mortar that made the samples denser. Moreover, the exposure to salts may contribute in the modification of the surface of the fibers and thus led to improving the adhesion with the matrix which in turn improved the compressive strength of the samples.

8 Termites Effect

8.1 Concept and Scope

The capacity of lignocellulosic fibers to be completely assimilated offer a more environmental and sustainable alternative to synthetic polymer. However, this may make them more vulnerable to biodegradation during their use causing a durability problem. Biodegradation can be caused by microorganisms and sometimes by termites (Okada 2002; Mohanty et al. 2000). Palm fibers including date and oil palm fibers are susceptible to termites attack due to their relatively high sugar and starch contents (Bakar et al. 2008). Termites can cause a significant degradation for palm fiber-based materials and may represent a hindrance to their use especially in humid and warm environments. Hence, efficient ways are needed to assess and address this durability problem to make palm fiber-based materials more usable.

8.2 Measurement Methods

The most common method to assess the biodegradation caused by termites is specified in the American Society of Testing Materials Standard Method (ASTM) D-3345-92. Several studies in the literature used this method to investigate the biodegradation of palm fibers caused by termites. For instance, Loh et al. (2011) used this method to evaluate the effectiveness of phenolic resin in protecting oil palm stem plywood against subterranean termites (Coptotermes curvignathus). They cut five specimens with dimensions 25 mm × 25 mm × 13 mm from oil palm plywood boards for both cases treated and untreated fibers. The Coptotermes curvignathus termites were used in this experiment and they were collected from an oil palm plantation. As an essential step in samples preparation for the test, all samples are sterilized in an oven at 130 °C for 2–3 days then conditioned in 65% relative humidity at 25 °C over night. The samples are then weighted and finally sterilized in an autoclave at 110 °C for 15 min just before the test. The test is conducted by placing each sample in a container and adding 1 g of termites containing 90% of workers and 10% of soldiers to each sample. After storing the samples with termites for 28 days at 25.5–27.7 °C, they are removed and cleaned before being dried at 103 °C for 3 days. The final step before weighing the samples is to condition them over night at 25 °C and 65% relative humidity. The biodegradation caused by the termites' attack is evaluated by the percentage of weight loss calculated according to the following Eq. 5:

$$\text{Weight loss}\,\% = (M_i - M_f)/M_i \times 100\% \tag{5}$$

where M_i is the initial weight before exposure to the termites; and M_f is the final weight of the sample after exposure to the termites.

Another study conducted by Bakar et al. (2013) used similar approach to evaluate the resistance of oil palm wood treated by low molecular weight phenol formaldehyde (Lmw-PF) resin against subterranean termites. Similarly, they cut five test blocks (20 mm × 20 mm × 20 mm) prepared at different compression levels (0, 25, and 50%) and conditioned them before measuring their initial weights. Then they placed them in containers and sterilized them at 120 °C for 2 h. The test was conducted using the same composition of the termites and for the same duration and conditions similar to the experiment conducted by Loh et al. At the end of the test, the weight was measured again, and the weight loss was calculated. Some important steps in this test that are recommended by the ASTM standard is to investigate the presence of tunneling, termite mortality, and position of the termites in the containers during the test, mainly at the end of the first and fourth weeks of the test. Tunneling should be high to assure the termites were vigorous and highly active during the test. Termite mortality is a sign for the high material's resistance to termites. This can be attributed to the inability of termites to digest the material (Kajita and Imamura 1991). The position of termites on the surface of the specimen is an indication that the material being tested functions as a repellent.

8.3 Results and Discussions

The results of the termite's resistance test conducted by Loh et al. (2011) showed that the treating oil palm stem with phenolic resin (Low molecular weight phenol formaldehyde) is an effective approach to significantly enhance the biodegradation resistance and protect oil palm stems against termites. For instance, the weight loss of the treated oil palm stem plywood (10.7% for outer veneer and 15.8% for inner veneer) was significantly lower than that for the untreated ones (19.2% for outer veneer and 23.9% for inner veneer). The study also found that the outer layer of oil palm fibers is more resistant to termites' attack due to the presence of more vascular bundles in the outer part of the stem making it difficult for termites to penetrate the structure of the material. Meanwhile, the inner layers are more venerable to the attack of termites due to the presence of high amount of carbohydrates (Lim et al. 1986).

In the other study conducted by Bakar et al. (2013), the Lmw-PF treatment and compression of the oil palm wood was found to significantly enhance its termite resistance which was represented in the reduction in the weight loss after the test. For instance, treated samples showed a significantly higher termite mortality (52–100%) and a lower weight loss (3.22–9.58%) compared to those for the untreated samples; 18% termite mortality and 27.94% weight loss. Figure 18 shows the damage of the oil palm wood samples after the termite test. The investigation of the treated oil palm wood samples showed that 95% of the termites were positioned on the surface of the samples indicating that it behaves as a repellent compared to the untreated oil palm wood samples as the majority of termites were positioned beneath the surface.

9 Microbial (Fungal) Effect

9.1 Concept and Scope

The attack of microorganisms is one of the main reasons for the biodegradation of palm fibers which can be assessed through soil burial. The microorganisms attack results in degradation in the mechanical properties and integrity of composites made of palm fibers which represents a major impediment to their use in various applications (Hill and Abdul 2000). Rot fungi in particular are the most efficient microorganism that is known to degrade lignin and all other components of lignocellulosic materials such as wood (Shary et al. 2007). There are several studies focused on studying the effect of microorganisms on the biodegradation of palm fiber-based composites. This degradation can be represented in weight loss and/or deterioration of mechanical properties. Soil burial is the most common approach to assess the microorganism attack induced mainly by fungi for lignocellulosic fibers.

Fig. 18 Oil palm wood after cleaning at the end of the termite test. Top left: untreated OPW; top right: treated OPWD-0%; bottom left: treated OPWD-25%; and bottom right: treated OPWD-50% (cited from Bakar et al. (2013))

9.2 Measurement Methods

There are different soil-based methods that were used in the literature to study the attack of microorganisms on palm fiber-based composites such as the ASTM D 1413-99 standard (ASTM 1999b), EN ISO 846: 1997 (Plastics-evaluation of the action of microorganisms) standard, and the American Wood e Preserves' Association (AWPA) standard M10-77 (AWPA 1977). Hill and Abdul (2000), studied the effect of soil exposure on the mechanical integrity of polyester matrix composites reinforced with oil palm empty fruit bunch fibers. The test was performed for 12 months according to the EN ISO 846: 1997 standard and using soil with 90% water holding capacity and a 50% moisture content. In this test, the samples are vertically-buried in the soil and kept at 29 °C and relative humidity of 97%. In order to maintain the required humidity, jars containing the samples are suspended above a saturated solution of potassium sulfate. After the required exposure time, samples are taken out and mass loss, moisture content, and tensile properties were determined.

In another study, Loh et al. (2011) used Decay fungi, Pycnoporous sanguinues according to the M10-77 AWPA 1977 standard. In this test, samples are not buried in the soil, but kept on the surface. Soil preparation involves screening, oven drying at 105 °C for 12 h and adding water (about 130% of the water holding capacity of the soil). The amount of water needed in the soil can be calculated according to the following Eq. 6:

$$g = (WHC \times 0.013 \times W_1) + (W_1 - W_2) \tag{6}$$

where W_1 is the weight of the soil in a culture bottle after oven-dry; W_2 is the weight of the soil in a culture bottle before oven-dry; WHC is the water holding capacity.

Fungi was prepared using potato dextrose agar and distilled water with a loop of mycelium then added to the soil. The soil was then placed in an incubator for 3 weeks at 25–27 °C. Finally, the steamed and cooled composite samples were put into the culture bottle.

The D 1413-99 ASTM standard was used by Bakar et al. (2013) to study the microorganism attack on oil palm wood using garden soil. Distilled water was used to maintain the moisture content at 70% in the soil and the white rot fungus (Pycnoporous sanguineus) was generated by placing rubber wood (Hevea brasiliensis) in the soil. Jars containing the soil were then incubated at 27 °C then the sterilized oil fiber wood samples were added, and the test was conducted for four weeks.

Ibrahim et al. (Ibrahim et al. 2014, 2018) used a soil burial testing method to study the effect of microorganism attack on the weight and mechanical properties loss of starch-based composites reinforced with date palm fibers. The method followed was similar to the soil burial methods described in other references found in the literature (Kaith et al. 2010, Di Franco et al. 2004). In this study, Ibrahim et al. buried 80 mm × 8 mm × 2 mm date palm fiber-based composite samples under a mixture of 50 wt% sand and 50 wt% garden soil. The test was conducted at 30 °C and sand moisture content of 30–40%. The test was conducted for 6 weeks and samples were periodically extracted and examined using scanning electron microscopy and tested for their weight loss and mechanical properties.

In another study, Nasser et al. (2017) subjected wood plastic composites made of polypropylene and date palm midrib particles to fungal inoculation by Trichoderma harzianum. The harvested spores from the grown Trichoderma harzianum were spread and filtered through muslin to a final concentration of 1.2×10^6 spores/ml. The samples were incubated for 2 months and the weight loss was determined at the end of the test.

9.3 *Results and Discussions*

The effect of soil exposure on the mechanical integrity of polyester matrix composites reinforced with oil palm empty fruit bunch fibers was studied by Hill and Abdul (2000). The flexural strength and modulus of elasticity of the composites was reduced

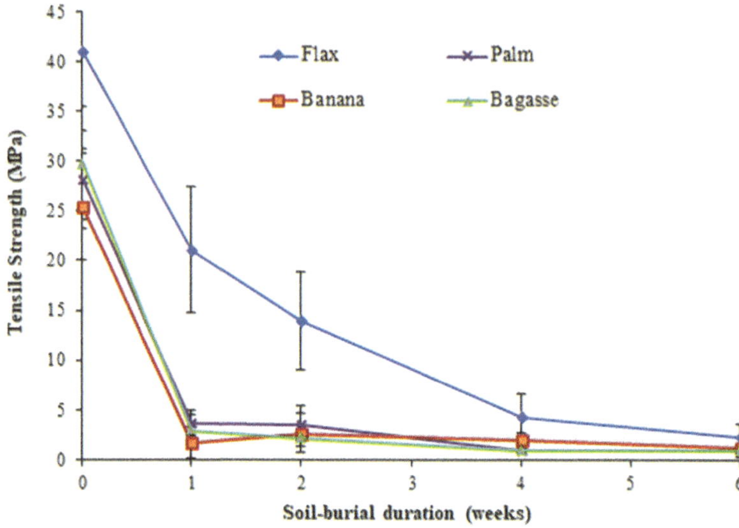

Fig. 19 Effect of biodegradation on the tensile strength of the prepared 50 wt% fiber content composites (cited from Ibrahim et al. (2018))

from 41.6 MPa and 3.85 GPa, respectively, to 27.4 MPa and 2.69 GPa after 12 months of soil exposure. After investigating four types of lignocellulosic fiber starch composites, Ibrahim et al. (2018) found that tensile properties deteriorate drastically, especially in the first week. Palm composite tensile strength decreased from 27 to 4 MPa in week 1, and continued to decrease to 1 MPa in week 6 (see Fig. 19).

The scanning electron microscope (SEM) micrographs of date palm fiber composites after 0, 1, 2, 4 and 6 weeks of soil burial were compared to those of flax-based composites by Ibrahim et al. (2018), see Fig. 20. The SEM micrographs show the erosion of the layer of the matrix after the first week of soil exposure. By the end of the test period (6 weeks), all fibers were exposed from disintegrated matrix. This degradation behavior was associated with a significant reduction in the composites' mechanical strength as shown in Fig. 19.

In the study conducted by Loh et al. (2011), the weight loss of the phenolic-treated oil palm stem plywood specimens showed low weight loss (13%) compared to untreated ones (34%) after exposure to Pycnoporous sanguineus fungi for 12 weeks. Overall, the durability of oil palm stem plywood against white rot fungi attack by 62%. In addition, the outer layer of oil palm stem was more resistance to fungi attack compared to the inner layers.

Similarly, the resistance of treated oil palm wood after 4 weeks of exposure to the white rot fungi was found to be significantly higher than that for the untreated ones (Bakar et al. 2013). For instance, the weight loss of the treated specimens was between 1.71 and 8.99% compared to 16.9% for the untreated specimens. Figure 21 shows that the entire surface of the untreated oil palm wood specimen was covered by the fungi compared to the treated specimens.

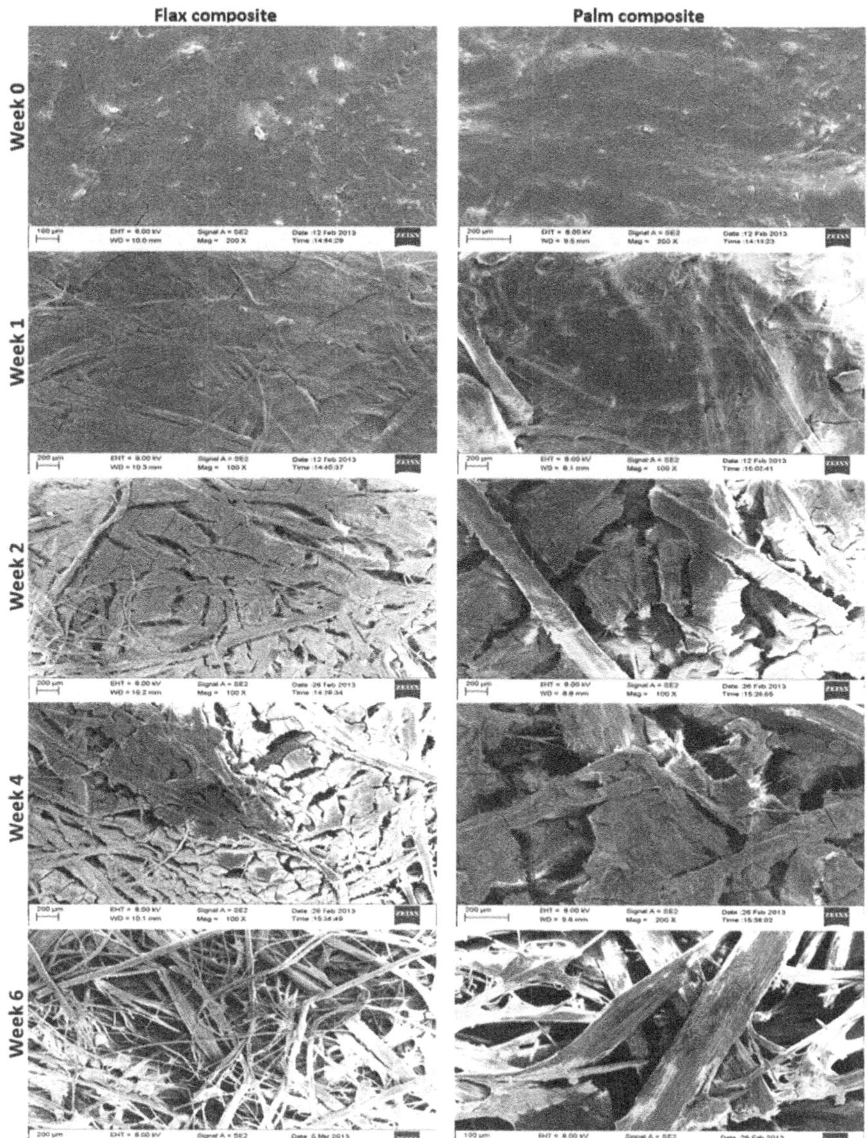

Fig. 20 Surface morphology of the prepared 50 wt% fiber content composites investigated during soil-burial biodegradation test, from week 0 through week 6; flax (left) and palm (right) (cited from Ibrahim et al. (2018))

Fig. 21 OPW after the end of the decay test. Top left: untreated OPW; top right: treated OPWD-0%; bottom left: treated OPWD-25%; and bottom right: treated OPWD (cited from Bakar et al. (2013))

10 Conclusion

Palm residues exhibit promising performance improvement for diverse composite materials. Different physical, chemical, and biological environmental aging factors were explained thoroughly in this chapter. Moreover, their effects on the properties of the palm-based composites were highlighted. Regarding weathering (ultraviolet rays), palm composites were found to be superior to other composites or pure resin (especially PP and epoxy). Aging with humidity exposure decreases tensile (and flexural) strength of palm based composites, contrary to impact strength which increases. In construction industry, palm fibers slightly increased compressive strength of mortar, but substantially improved aging shrinkage, and acidic/alkaline resistance. Termites and fungal attacks are still challenging in palm industry. However, some chemical treatments can double the capacity of palm composites to resist their attacks. Overall, due to its competitive durability, composites reinforced with palm fibers represent versatile material with promising applications, and prospective development.

References

Abu-Sharkh BF, Hamid H (2004) Degradation study of date palm fibre/polypropylene composites in natural and artificial weathering: mechanical and thermal analysis. Polym Degrad Stab 85:967–973

Akash M, Rehman K (2019) Thermo gravimetric analysis. In: Essentials of pharmaceutical analysis, pp 215–222

ATLAS-MTS (2020a) https://www.atlas-mts.com/applications/applications/plastics. Accessed 7 Jan 2020

ATLAS-MTS (2020b) https://www.atlas-mts.com/products/standard-instruments/fluorescent-uv. Accessed 7 Jan 2020

Introduction to the environmental laboratory. http://www.ecs.umass.edu/cee/reckhow/courses/572/572bk11/572BK11.html. Accessed 7 Jan 2020

Bakar AA, Hassan A, Yusof AFM (2005) Effect of accelerated weathering on the mechanical properties of oil palm empty fruit bunch filled UPVC composites. Iran Polym J 14:627–635

Bakar ES, Hao J, Ashaari Z, Yong ACC (2013) Durability of phenolic-resin-treated oil palm wood against subterranean termites a white-rot fungus. Int Biodeterior Biodegradation 85:126–130

Bakar ES, Sahri MH, H'ng PS (2008) Anatomical characteristics and utilization of oil palm wood. Form wood trop for trees—a challenge from the perspective of functional wood anatomy, pp 161–178

Chollakup R, Askanian H, Delor-Jestin F (2017) Initial properties and ageing behaviour of pineapple leaf and palm fibre as reinforcement for polypropylene. J Thermoplast Compos Mater 30:174–195

El-Juhany LI et al (2010) Degradation of date palm trees and date production in Arab countries: causes and potential rehabilitation. Aust J Basic Appl Sci 4:3998–4010

Di Franco CR, Cyras VP, Busalmen JP, Ruseckaite RA, Vázquez A (2004) Degradation of polycaprolactone/starch blends and composites with sisal fibre. Polym Degrad Stab 86:95–103

Hill CAS, Abdul HPS (2000) The effect of environmental exposure upon the mechanical properties of coir or oil palm fiber reinforced composites. J Appl Polym Sci 77:1322–1330

Ibrahim H, Farag M, Megahed H, Mehanny S (2014) Characteristics of starch-based biodegradable composites reinforced with date palm and flax fibers. Carbohydr Polym 101. 10.1016/j.carbpol.2013.08.051

Ibrahim H, Mehanny S, Darwish L, Farag M (2018) A comparative study on the mechanical and biodegradation characteristics of starch-based composites reinforced with different lignocellulosic fibers. J Polym Environ 26:2434–2447

Ishak MR, Leman Z, Sapuan SM, Salleh MY, Misri S (2009) The effect of sea water treatment on the impact and flexural strength of sugar palm fibre reinforced epoxy composites. Int J Mech Mater Eng 4:316–320

Jain D, Sekhon H, Bera TK, Jain R (2019) Comparison of different hydrophobic treatments for the durability improvement of palmyra natural fiber composites under hydrothermal ageing environments. J Nat Fibers: 1–15

Jawaid M, Saba N, Alothman O, Paridah MT (2016) Effect of accelerated environmental aging on tensile properties of oil palm/jute hybrid composites. In: AIP conference proceedings, p 40007

Kaith BS, Jindal R, Jana AK, Maiti M (2010) Development of corn starch based green composites reinforced with Saccharum spontaneum L fiber and graft copolymers—evaluation of thermal, physico-chemical and mechanical properties. Bioresour Technol 101:6843–6851

Kajita H, Imamura Y (1991) Improvement of physical and biological properties of particleboards by impregnation with phenolic resin. Wood Sci Technol 26:63–70

Kareche A, Agoudjil B, Haba B, Boudenne A (2019) Study on the durability of new construction materials based on mortar reinforced with date palm fibers wastes. Waste Biomass Valorization: 1–9

Kriker A, Bali A, Debicki G, Bouziane M, Chabannet M (2008) Durability of date palm fibres and their use as reinforcement in hot dry climates. Cem Concr Compos 30:639–648

Kumar P, Barrett DM, Delwiche MJ, Stroeve P (2009) Methods for pretreatment of lignocellulosic biomass for efficient hydrolysis and biofuel production. Ind Eng Chem Res 48:3713–3729
Kumar R, Singh S, Singh OV (2008) Bioconversion of lignocellulosic biomass: biochemical and molecular perspectives. J Ind Microbiol Biotechnol 35:377–391
Leman Z, Sapuan SM, Azwan M, Ahmad M, Maleque MA (2008a) The effect of environmental treatments on fiber surface properties and tensile strength of sugar palm fiber-reinforced epoxy composites. Polym Plast Technol Eng 47:606–612
Leman Z, Sapuan SM, Saifol AM, Maleque MA, Ahmad M (2008b) Moisture absorption behavior of sugar palm fiber reinforced epoxy composites. Mater Des 29:1666–1670
Lim SC, Khoo KC et al (1986) Characteristics of oil palm trunk and its utilization. Malaysian For 49:3–22
Loh YF, Paridah TM, Hoong YB, Bakar ES, Anis M, Hamdan H (2011) Resistance of phenolic-treated oil palm stem plywood against subterranean termites and white rot decay. Int Biodeterior Biodegradation 65:14–17
Machaka M, ElKordi A (2017) Experimental study of the effect of adding fan palm fibers on concrete durability exposed to severe environments. In: 2nd International conference on bio-based building materials & 1st conference on ECOlogicalvalorisation of GRAnular and FIbrous materials, June 21th–23th, Clermont-Ferrand, France
McKeen LW (2016) Fatigue and tribological properties of plastics and elastomers. William Andrew
Mehanny S, Darwish L, Abd El Haleem M, El-Kashif E, Farag M, Ibrahim H (2019) Effect of glue and temperatures on mechanical properties of starch-based biodegradable composites reinforced with bagasse fibers. Int J Biotechnol Biomater Eng 1
Mehanny S, Darwish L, Ibrahim H, El-Wakad MT, Farag M (2016) High-content lignocellulosic fibers reinforcing starch-based biodegradable composites: properties and applications. In: Composites from renewable and sustainable materials. InTech
Mehanny S, Farag M, Rashad RM, Elsayed H (2012) Fabrication and characterization of starch based bagasse fiber composite. In: ASME International mechanical engineering congress and exposition, Proceedings (IMECE)
Mohanty AK, Misra M, Hinrichsen GI (2000) Biofibres, biodegradable polymers and biocomposites: An overview. Macromol Mater Eng 276:1–24
Nasser R, Mansour MMA, Salem MZM, Ali HM, Aref IM (2017) Mold invasion on the surface of wood/polypropylene composites produced from aqueous pretreated wood particles, Part 1: date palm midrib. BioResources 12:4078–4092
Obi OF (2015) Evaluation of the physical properties of composite briquette of sawdust and palm kernel shell. Biomass Convers Biorefinery 5:271–277
Okada M (2002) Chemical syntheses of biodegradable polymers. Prog Polym Sci 27:87–133
Ozerkan NG, Ahsan B, Mansour S, Iyengar SR (2013) Mechanical performance and durability of treated palm fiber reinforced mortars. Int J Sustain Built Environ 2:131–142
Radzi AM, Sapuan SM, Jawaid M, Mansor MR (2019) Water absorption, thickness swelling and thermal properties of roselle/sugar palm fibre reinforced thermoplastic polyurethane hybrid composites. J Mater Res Technol
Shary S, Ralph SA, Hammel KE (2007) New insights into the ligninolytic capability of a wood decay ascomycete. Appl Environ Microbiol 73:6691–6694
Shuhimi FF, Abdollah MFB, Kalam MA, Masjuki HH, Mustafa A (2016) Amiruddin H (2016) Surface durability of oil palm fiber/epoxy composite at various temperatures. Proc Mech Eng Res Day 2016:108–109
Sreekala MS, Kumaran MG, Geethakumariamma ML, Thomas S (2004) Environmental effects in oil palm fiber reinforced phenol formaldehyde composites: studies on thermal, biological, moisture and high energy radiation effects. Adv Compos Mater 13:171–197
Teo DCL, Mannan MA, Kurian VJ, Ganapathy C (2007) Lightweight concrete made from oil palm shell (OPS): structural bond and durability properties. Build Environ 42:2614–2621
(2015) Digital General Purpose Water Baths. https://www.polyscience.com/products/general-purpose-water-baths/digital-water-baths. Accessed 7 Jan 2020

Date Palm Fiber Composites Applications

Date Palm Fiber Composites in Hot-Dry Construction and Building

A. Kriker, S. Abbani, M. Bouziane, A. Mokhtari, A. Mekhermeche, H. Chaib, and F. Hafsi

Abstract Southern regions of Algeria are characterized by hot and dry climate over period which extends eight months from March to October. Temperatures in summer reach over 45° C in the shade. Actually, modern buildings in these regions are made in cement materials (concrete and mortar) these buildings are weak, not durable and present a lot damages caused by low resistance to cracking and thermal shrinkage. In addition, these cement materials present poor thermal insulation. It is therefore necessary to reinforce them with fibers. Several fibers can be used for that, but for economic and environmental reasons vegetable fibers from the by-products of date palms can fulfill the purpose in a good way. Because their treatment is inexpensive and such valorization contributes favorably to the protection of palm groves against fires, often started by this waste stored in an anarchic manner. The reinforcement of building materials with palm fibers, especially concrete or gypsum, is relatively a newly research topics. In the literatures several researches discuss the reinforcement of building materials with steel, carbon or polymer fibers. On the other hand, only a few scientific works are devoted to vegetable fibers, especially palm fibers. In this chapter it will be investigated the mechanical, physical and thermal properties of certain local building materials (gypsum and earth (clay plus dune sand)) that exist in abundance in the southern regions of Algeria, and their reinforcement by date palm fibers and concretes and mortars reinforced by the same fibers. In addition, it will be presented the mechanical characteristics of the used date palm fibers. This study focuses on the improvement of mechanical and thermal performances of the last materials. It aims the characterization of the newly adopted materials and eventually their use in construction in hot and dry climate. Firstly, the results showed that among the four types of date palm fibers: male palm, Elghers, Deglette Nour, Degla Bida (local names), the male date palm fibers have the best mechanical performances in terms of resistance tensile and deformation at break. Second, for concrete reinforced with date palm fibers and initially kept in a 14-day wet cure, a volume fraction of 3% with fibers 6 cm long was very beneficial for improving mechanical performance. Under the last wet cure conditions, it was noted that beyond 6 months the durability of the fibers was badly affected by the alkaline hydraulic products of Portland cements

---

A. Kriker (✉) · S. Abbani · M. Bouziane · A. Mokhtari · A. Mekhermeche · H. Chaib · F. Hafsi
EVRNZA Laboratory, University KASDI Merbah Ouargla, BP 511, 30000 Ouargla, Algeria
e-mail: kriker.ab@univ-ouargla.dz; a_kriker@yahoo.fr

M. Midani et al. (eds.), *Date Palm Fiber Composites*, Composites Science and Technology,
https://doi.org/10.1007/978-981-15-9339-0_12

(CPJ-CEM II/A 32.5). To resolve the exposed problem, cement with low alkalinity (Portland cement composite (CPJ-CEMII/B 42.5)) plus 10% by mass of fly ash was used. This treatment got better the durability of fiber in cement matrix. Third, it was shown that clay brick with (60% clay + 40% sand + 3% fiber) exhibited the best thermal performance with acceptable mechanical properties. And finally, the reinforcement of the gypsum by date palm fibers (5% by mass of separated fibers and 4% by mass of fibers in mesh layers) have gave the best thermal and mechanical performances. In conclusion, Africo-Asian countries, including Algeria, possess a great wealth of renewable bio-resources. Vegetable fibers especially those obtained from by-products of date palm and the variety of local construction materials can be used to develop industries interested by newly eco-friendly reinforced materials having good mechanical and thermal characteristics. This study wishes to popularize the valorization of date palm by-products in construction materials and proceed to the industrial stages. Consequently, contribute to a sustainable development in these regions of the world.

Keywords Date palm fiber · Reinforcement fiber concrete · Gypsum · Adobe bricks · Hot-dry environment

1 Introduction

Date palm (Phoenix dactylifera L) is one of the most cultivated palms, for its nutritive product (dates) and its good resistant to heat, drought and cold. Generally it is found in the Afro-Asian arid strip which extends from North Africa to the Middle East. Phoenixes are dioecious palms; so there are male palms and female palms. They give multitude varieties of dates. After annually trimming operations, enormous quantities of by products palm are thrown away, except in low scales for artisan products.

In this study, local natural resources (surface date palm fibres) are used for reinforcement the concrete, mortar, gypsum and adobe bricks. They are collected from Ouargla palms city. This city is situated in the south of Algeria. It is characterized by a hot-dry climate during summer. Table 1 provides its average climatic data. Temperature during day is very high (up to 43 °C), the daily amplitude between day and night

Table 1 Monthly mean climatic data of Ouargla Algeria region (Kriker et al. 2005)

	June	July	Aug.	September	October	November
T_{max} (°C)	39.8	43.2	42.2	37.6	30.7	23.6
T (°C)	31.5	34.3	33.6	29.8	23.4	16.6
T_{min} (°C)	23.2	25.3	25.0	22.0	16.0	9.6
RH (%)	28.5	24.4	26.4	34.4	47.6	56.5
WV (Km/h)	16.02	13.43	10.80	10.33	08.86	08.06

With: T_{max}: The maximum temperature, T_{min}: The minimal temperature, RH: The relative humidity, WV: The wind velocity, T: The average temperature regularly reaches about 18 °C, and the humidity level of air stays very low. Furthermore, the site is located in a zone at average wind velocity of 8–16 km/h. In our previous work (Kriker and Bali 1992) we have showed that conventional construction was sensitive and adapted with difficulties to hot-dry climatic conditions. It is therefore necessary to reinforce it by fibers.

According to the literature results of more natural fibres (Savastano et al. 2000; Khenfer et al. 2000; Tolêdo Filho et al. 2000; Bedzki and Gassan 1999; Cook 1980; Swamy 1985), the reinforcement with vegetable fibres in construction materials is good for flexural properties. The aim of this study is then to investigate the mechanical properties of concrete and the mechanical and thermal properties of gypsum and adobe bricks reinforced by male date palm surface fibres in hot-dry climate. It also endeavours to examine the possibility of utilizing these local natural resources in local construction.

2 Properties of Date Palm Fibres

Date palm surface fibres (DPSF) around the trunk, corresponding to four principal palms: Male palm, Deglette-Nour, Degla-Bida and Elghers palms (local manes) were chosen in this study, for it's seemed most suitable for exploitation. Figure 1 present typical date palm fibre. The DPSF are pulled out from trunk in the form of nearly rectangular mesh (length 30–50 cm, wide 20–30 cm) formed with tree superposing layers. It is easy to separate the individual fibre in water.

2.1 Physical and Mechanical Properties of DPSF

Table 2 present the physical mean properties of DPSF. There is no significant deference between the different palm types.

Table 3 present the mean mechanical properties of individual's DPSF. Represented

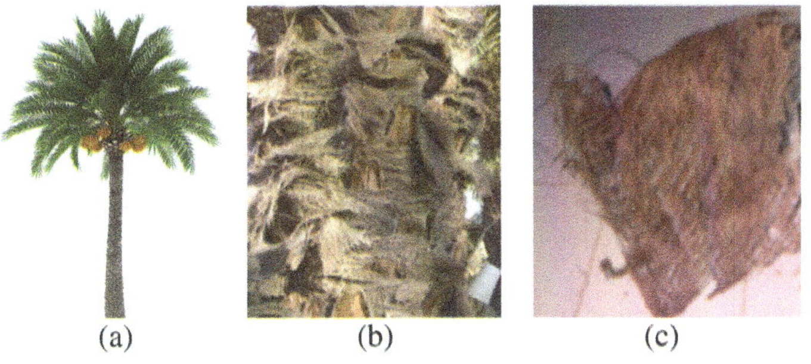

Fig. 1 Typical **a** date palm, **b** trunk, **c** typical surface fibre mesh

Table 2 Physical properties of MDPSF (Kriker et al. 2005)

Property	Lower-upper	Mean-CV (%)
Diameter (mm)	0.1–0.8	0.45–54.43
Bulk density (kg/m^3)	512.21–1088.81	900–17.64
Absolute density (kg/m^3)	1300–1450	1383.33–5.52
Natural moisture content (%)	9.5–10.5	10–5.00
Water absorption after 5 mim (%) under water	60.05–84.12	74 –14.02
Water absorption to saturation (%)	96.83– 202.64	132.5–20.56

Table 3 Mechanical properties of DPSF (Kriker et al. 2005)

Condition		Dry			Wet		
Fibre types	Specimen's length (mm)	tensile strength (MPa)	Elongation (%)	Modulus of elasticity (GPa)	Tensile strength (MPa)	Elongation (%)	Modulus of elasticity (GPa)
Male DPSF	100	170±40	16±3	4.74±2	175±30	17.4±2	3.78±2
	60	240±30	12±2	5.00±2	250±25	13±2	3.25±1.5
	20	290±20	11±2	5.25±3	300±20	12±2	3.55±2
Elghers DPSF	100	88.15±20	11.1±2.5	3.50±1.2	90.10±18	12±3	3.10±1.5
Deglette Nour DPSF	100	72.34±18	8.7±2.2	3.15±1.5	74.34±15	9.5±2.5	2.30±2
Degla Bida DPSF	100	71.15±16	7.5±2.3	2.50±1	73.19±13	8.5±2.7	2.10±1

by 36 fibres the length 10 cm: Indeed, six fibres per diameter (0.8–0.6–0.4 cm) are tested for two types of treatment, first the natural dray fibre and secondly the fibre placed during 24 h in water at this time the fibres are saturate of water. We remark that the natural dray fibres are the greater tensile strength but the smaller elongation. Figure 2 show the set-up fiber tensile strength. The mechanical results show that male date palm surface fibre (MDPSF) has the most tensile strength and higher elongations, for this reason we have select it for reinforcing our concrete. Regarding the mechanical performance of natural fibres used with concrete given by (Cook 1980; Coutts 1983; Lewis and Premalal 1979), MDPSF has an interesting comparatively behaviour.

Like more common vegetable fibre, the MDPSF has porous structures; it's able to absorb the grater quantity of water. Tableau 2 provides the mean percentage in weight of water absorption by MDPSF during the 24 h of immersion. At saturation at 24 h 132.5% in weight of water are then absorbed by fibres. According to resulted reported by from (Tolêdo Filho et al. 2000) relating to fibres of sisal and malva

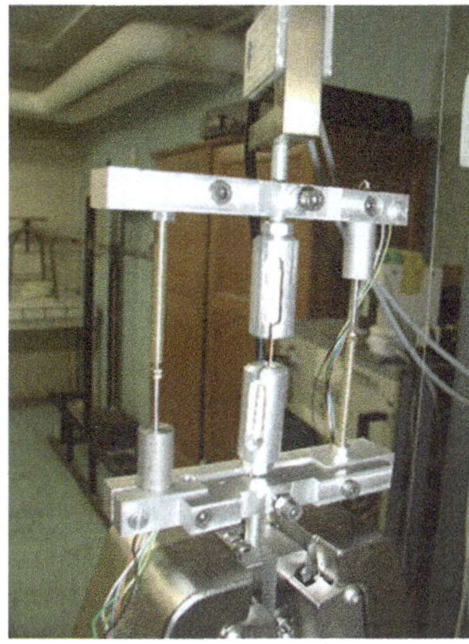

Fig. 2 Set-up fiber tensile strength (Kriker et al. 2005)

the MDPSF absorption percentage to range between 92.2% from sisal and 156.4 to malva. This last propriety affects the water-cement ratio. In mix design proportion, we need add the quantity of water, which will be absorbed by fibers.

2.2 Chemical Properties

Tree major constituents composed the vegetable fibres are: the cellulose (max 75%), the lignin and the hemicelluloses (Bentur and Akers 1989).

The chemical composition of MDPSF in mass percentage is given on following:

- Cellulose: 44.50%
- Hemicelluloses 19%
- Lignin 29%.

2.3 Durability of Date Palm Surface Fibres

The durability of vegetable fibres, in environment where it is exposed, constituted an essential criterion for its use. In reinforced cement the fibres were exposed to alkaline solution $Ca(OH)_2$.

Durability in alkaline solution depends in the type of fibres and its treatments (Lewis and Premalal 1979; Bentur and Akers 1989; Tolêdo Filho et al. 2000). Some studies show that vegetable fibres (sisal, abaca, jute, urena, plantation coir) loosed their strength in alkaline environment (Lewis and Premalal 1979). He reported that all the fibres are attacked by the medium to varying degrees. Elephant grass fibres

Table 4 Tensile strength (MPa) of MDPSF the length 100 mm as function of time in alkaline solution (Kriker et al. 2004)

Time (months)	Solution $Ca(OH)_2$			Solution NaOH		
	Diameter of fibre (mm)					
	0.8	0.6	0.4	0.8	0.6	0.4
0	180±20	160±18	60±15	180±20	160±18	60±15
1	151±11	92±13	39±15	159±14	106±12	58±16
2	146±15	83±15	27±14	153±12	95±15	52±15
3	137±12	77±14	20±07	149±13	88±13	25±13
6	125±12	65±11	–	137±16	74±18	10±06

retained nearly 91% of their original strength at three months, with a constant rate of reduction after.

Water reed fibre retained nearly 57, and 14.28% of their original strength at three and six months respectively. However, we show that the MDPSF fibres retained 43.01, and 37.51% of their original strength at three and six months respectively. According to these results we can see that exception of Elephant grass; all type fibres (MDPSF, water reed, plantain musamba) are adversely affected.

In this chapter, the tensile strength and elongation of MDPSF are also measured as a function of diameter of fibres and time immersion in lime solution with at Ph 12. The results are given in Table 4. The indicated that the MDPSF loosed his mechanical properties in lime solution, a decrease proportionally with diameter of fibres in tensile strength, elasticity modulus and percentage of elongation were unregistered. The fibres getting brittle as function of time immersion in lime solution. We have also distinguished a decrease in the section fibres. This aspect has been discussed in more detail by (Bentur and Akers 1989; Tolêdo Filho et al. 2000) on their study about the sisal, coconut and cellulose fibres. They attributed this mainly to crystallisation of lime in the lumen, walls and voids in the fibre.

2.4 Conclusion

C.1. African-Asian countries, including Algeria have wealth of vegetable bay-product from the date palm that could be valorized in local materials building construction.

C.2. Amongst the four used types of date palm surface fibres, the MDPSF is the most resistant and higher elongations.

C.3. In comparison with most vegetable fibers, the durability of MDPSF is badly affected by alkaline solutions

C.4. Regarding the mechanical performances of common vegetable fibres, the MDPSF has middling tensile strength and feeble elasticity modulus.

Table 5 Chemical properties of used cements (Kriker et al. 2005)

Cement	SiO_2	AL_2O_3	Fe_2O_3	MgO	CaO	Na_2O	SO_3	K_2O	Loss on ignition
CPJ-CEM II/A	21.90	5.73	3.13	1.85	60.18	0.19	2.22	0.83	4.07

3 Properties of Concrete Reinforced by MDPSF

3.1 Materiel Properties

3.1.1 Cement

The cement used was a Portland cement (CPJ-CEM II/A 32.5). Table 5 compiles the main its chemical and physical properties.

3.1.2 Sand

Natural sand was used with the following properties:

- Maximum size: 5 mm
- Finesse modulus: 1.9
- Sand equivalent: 66%
- Absolute density: 2800 kg/m^3.

3.1.3 Aggregate

A natural crushed slice-lime stone aggregate was used with the following physical properties:

- Maximum size: 15 mm
- Minimum size: 5 mm
- Absolute density: 2650 kg/m^3.

3.2 Specimens and Curing Conditions

3.2.1 Specimens Fabrication

A specimen's with the dimension of 7 $\times$ 7 $\times$ 28 cm have been used for the flexural strength concrete test, and cubic specimens (10 cm Sides) have been used for compressive strength concrete test.

Table 6 Mix design proportion (Kriker et al. 2005)

Materials	Mix 1 (0%)	Mix 2 (2%)	Mix 3 (3%)
Cement (kg/m^3)	400	400	400
Aggregate (kg/m^3)	1000	982	973
Sand (kg/m^3)	750	750	750
Water (kg/m^3)	240	270	290
MDPSF (kg/m^3)	0	18	27

Baron-Lesage, adapted by Rossi and Goris methods are used for mix design concrete (Lesage 1974; Rossi et al. 1989; Gorisse 1978). With a constant workability, for all mix concrete, VB test time equal 20 s and slump depth equal 6 cm. The water quantity has been adjusted for the absorption of water properties of the fibres. The percentages of fibres reinforced concrete vary from 0% to 3% in volume. At each mix (each percentage) the length fibres varied from 1.5 to 6 cm. Table 6 give the Mix design proportion.

3.2.2 Curing Conditions

During six months (June-November) the fibre concrete were exposed to two type's environments: Curing 1: in water at temperature varied from 20 to 30 °C. Curing 2: in free atmosphere under uncontrolled severe field conditions and high winds generally oriented northwest and southwest (Table 1). During this period, the middle monthly maximum temperature (MMT_{max}) varied from 42.2 °C in August to 23.6 °C in November, the middle monthly minimum temperature (MMT_{min}) varied from 9.6 °C in November to 25 °C in August. The percentage of relative humidity (RH %) varied between 24.4% in July and 56.5 in November.

The concrete are tested for the intervals of following times: 7, 28, 60, and 90 days, for each test (compressive, flexural strength) three specimens are used.

3.2.3 Test Methods

Standards NF EN 12390-4 and a Maurice Perrier et Cie machine with a rate loading of 3 ± 0.5 kN/s was used to evaluated the compressive strength of concrete.

Standards NFP 18-409 and an Instron universal testing machine by four-point configuration, 210 mm span and cross-head rate of 0.1 mm/min were used to evaluated the concrete flexural properties using an experimental set-up (Fig. 3). The system was equipped with a load-sensor of 50 kN and two LVTD displacement transducers. To determine the deflection at the mid-span of the specimen as function of loading the system is continuously recorded to a numerical data acquisition.

Fig. 3 Set-up for the flexural test

Three parameters were using to evaluated the flexural properties were

1. At the first crack strength (FCS) was determined from the greatest load carried by the composite after the first crack using Eq. (1).

$$FCS = 6M/bd^2 \tag{1}$$

- M is the failure moment of the test specimen
- b and d are the width and depth of the specimen respectively

2. The greatest toughness coefficient (D_{max}):

$$D_{max} = P_{max}/\, P_f \tag{2}$$

With:
(a) P_{max}: The greatest applied load (b) P_f: The load at the first visible crack

3.3 Compressive Strength

As a function of time and reinforced concrete types Fig. 4 present the compressive strength mean, in curing 1 and 2.

In curing 1 (Fig. 4a), the results show that, for each day, the compressive strength decrease with increasing the percentage and the length of fibres. The maximum of compressive strength is obtained by 2% in volume and 1.5 cm in length of fibres. The resistance of this last type remained lower that of the concrete without fibres. This result is on concordance with literatures (Ramaswamy et al. 1983; Lewis and Premalal 1979). In fact vegetable fibres reinforcement concrete was a small benefit effect on compressive strength.

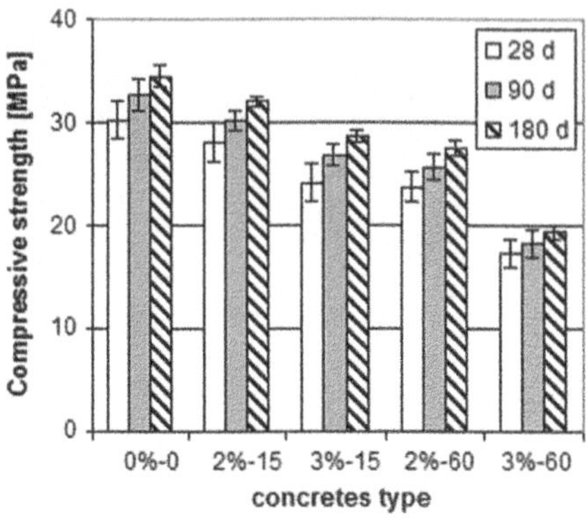

(a) Curing 1

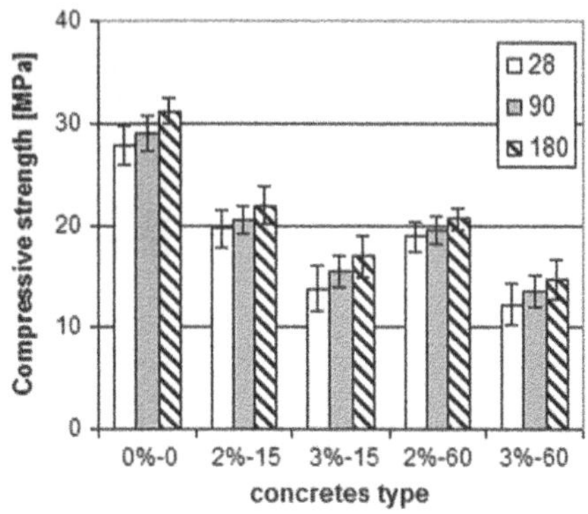

(b) Curing 2

Fig. 4 Compressive strength (Kriker et al. 2005)

However, in curing 2 (Fig. 4b), the maximum of compressive strength is also obtained by 2% in volume and 1.5 cm in length of fibres. The resistance of this last type remained lower that of the concrete without fibres. Just as in the preceding cure the compressive strength decreases with increasing the percentage and the length of fibres for each time. However, the compressive strength of all fibre-concretes in this cure is definitely lower than that in water curing for the on the lack of hydration products caused by the rapid evaporation of water into concrete during early time and with increase of vacuums in concrete.

3.4 Flexural Concrete Properties

Table 7 provides the mean and the coefficient of variation of the FCS and D_{max}, for each period (28, 180 days), in Curing 1or curing 2. Although the FSC continues to increase with time, a reduction in FCS for the concretes reinforced by fibres, compared to the concrete without fibre was observed. For the concretes reinforced by fibres the maximum of FCS was obtained with the low length and percentage fibre (2%-15). Rising the percentage and the length of fibres ameliorates the $D_{max.}$

After the first crack in water curing, examination of the results given in Table 7, and Fig. 5 show that the MDPSF-concretes are relatively weak compared to the more common vegetable fibre concretes of literatures. The D_{max}, for the most ductile MDPSF-concretes (3%-60) remains relatively low compared to that found for the sisal and coconut fibre-concretes by (Tolêdo Filho et al. 2000) That is probably due firstly, to the mediocre mechanical performance of the MDPSF; secondly, to the fibre-matrix interface strength bound. In fact, some researchers (Tolêdo Filho et al. 2003; Savastano et al. 2000) reported that treatment of vegetable fibres by chemical processes improves their mechanical properties and adherence to the matrix and

Table 7 The FCS, P_{max} and D_{max} for MDPSF-concretes as function of time and cures type (Kriker et al. 2008)

			Concretes type				
Flexural properties	Ageing condition	Ageing time (d)	0%-0	2%-15	3%-15	2%-60	3%-60
FCS (CV) (MPa)(%)	Curing 1	28	7.6 (7.11)	6.5 (11.59)	5.5 (14.54)	6.0 (10.82)	5.3 (12.47)
	"	180	8.4 (5.98)	7.4 (8.78)	6.3 (11.11)	6.8 (8.95)	6.2 (12.86)
	Curing 2	28	7.9 (7.28)	6.5 (4.68)	5.8 (5.60)	6.3 (4.92)	5.6 (6.81) 5.6 (6.81)
	"	180	8.7 (3.08)	7.6 (4.15)	6.5 (3.54)	7.2 (6.12)	6.4 (2.66)
P_{max} (CV) (kN)(%)	Curing 1	28	–	0.8 (19.32)	1.7 (11.77)	1.2 (11.06)	3.0 (2.57)
	"	180		0.7 (26.67)	1.4 (11.14)	1.0 (29.69)	2.0 (19.04)
	Curing 2	28		0.6 (25.22)	1.5 (25.62)	1.1 (17.84)	2.2 (20.15) 2.2 (20.15)
	"	180		0.5 (23.19)	1.2 (12.35)	1.0 (17.03)	1.5 (13.37)
D_{max} (CV) (%)	Curing 1	28	–	0.08 (2.85)	0.19 (27.58)	0.10 (12.63)	0.3 (14.57)
	"	180		0.06 (11.19)	0.14 (11.25)	0.09 (24.30)	0.2 (8.36)
	Curing 2	28	–	0.05 (16.27)	0.15 (13.15)	0.11 (9.17)	0.2 (11.38)
	"	180		0.03 (27.14)	0.11 (17.04)	0.07 (15.39)	0.1 (10.22)

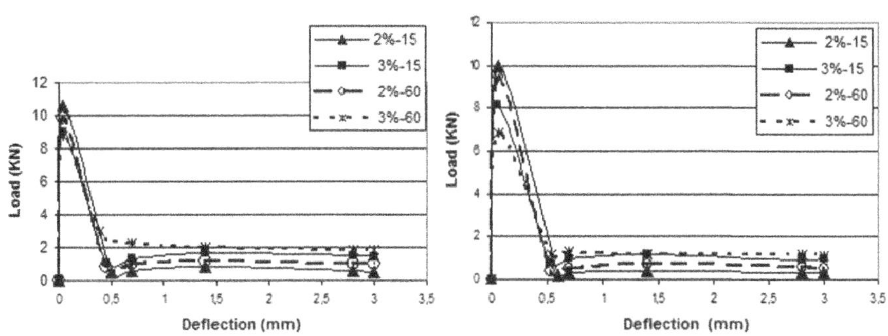

(a) curing1 (b) curing 2

Fig. 5 Load-deflexion curves at 28 days in water curing and hot dry climate (Kriker et al. 2008) hence increases the flexural properties of vegetable fibre-concretes. These two causes shall be investigated in more detail in a following chapter. A suitable treatment is then necessary to retain a moist environment during early time (Kriker et al. 2008). The second cause is the expansion of micro cracking after long time.

3.5 Conclusion

C.1. The compressive strength and the flexural properties of MDPSF fibre reinforced concretes conserved in wet curing is greater than for those conserved in outdoors.
C.2. In wet curing and relatively to the concrete without fibres, the increase in percentage and length of MDPSF in concrete has a beneficial effect on the ductile behavior, but no benefit for the first crack strength (FCS) and the compressive strength.
C.3. In curing 2, the first crack strength of fibre-concretes decreases with time and with rising fibre-percentage and length. But at 28 day rising the percentage and the length of fibres ameliorates the toughness.

4 Durability of Date Palm Fibers Reinforcement Concrete in Hot Dry Climates

In following we will examine two parameters which influence the durability of concrete reinforced by date palm fibers. As we have advanced precisely: the rapid evaporation of the hydration water at early ages and the fall in the durability of the fibers in the alkaline cementation matrix at long term.

4.1 Effect of Curing on Durability of Date Palm Fibers Reinforcement Concrete in Hot Dry Climates

The durability of MDPSF reinforced concrete was investigated according to its flexural properties and some scanning electron micrographs at the fibre-matrix interface zones. Two curing types were used:

1. In water (C1)
2. Initially 14 days in water then in hot-dry climates (C3).

4.1.1 Curing Conditions

After demoulding, the concrete specimens were cured for 24 h in the laboratory with climatic conditions: T = 20 ± 2 °C, and RH = 65%. Thence they were cured until test in two types of environment. For the first type of environment; concrete specimens were placed in water at temperature of 20–25 °C (referenced Curing 1). For the second, the concrete specimens were initially placed for 14 days in water, then they were placed during six months (June–November) in ambient atmosphere under uncontrolled hot-dry climate its referenced Curing 3 (mixed conservation).

Remark: Curing 2: in free atmosphere under uncontrolled severe field conditions and high winds generally oriented northwest and southwest as shown in Table 1.

4.1.2 Flexural Properties

The first crack strength (FCS), the maximum post-crack flexural load (P_{max}) and the greatest toughness coefficient (Dmax), in Curing 1, and 3 were presented in Table 8.

For both Curing 1 and 3 and for each concrete type, the FCS continues to increase with age. However, the FCS decreases when the length and the percentage of fibres increase. Therefore, the 2%-15 concrete type presents the greatest FCS, but remained lower than that of the concrete without fibres. In addition in Curing 1 the FSC is lower than that of concrete conserved in Curing 3. This is due to the thermo activation brought by this last curing 3. In fact, in a former research (Kriker and Bali 1992) we

Table 8 The FCS, P_{max} and D_n for MDPSF-concretes as function of time and curing type (Kriker et al. 2008)

		Concretes type (%-length of fibre)				
Flexural properties	Curing (Time d)	0%-0	2%-15	3%-15	2%-60	3%-60
FCS (CV) (MPa)(%)	C. 1(28)	7.6 (7)	6.5 (12)	5.50(15)	6.02(11)	5.34(13)
	"(180)	8.4 (6)	7.4 (9)	6.30(11)	6.82 (9)	6.21(13)
	C. 3 (28)	7.9 (7)	6.7 (5)	5.80 (6)	6.30 (5)	5.60 (7)
	"(180)	8.7 (3)	7.6 (4)	6.50 (4)	7.22 (6)	6.40 (3)
P_{max} (CV) (kN)(%)	C. 1(28)	–	0.80(19)	1.69(12)	1.18(11)	2.96 (3)
	"(180)		0.70(27)	1.40(11)	0.99(30)	2.02(19)
	C. 3 (28) C. 3 (28)		0.59(25)	1.50(26)	1.15(18)	2.23(20)
	"(180)		0.49(23)	1.20(12)	0.96(17)	1.50(13)
D_{max} (CV) (%)	C. 1(28)	–	0.08 (3)	0.19(27)	0.10(13)	0.34(15)
	"(180)		0.06(11)	0.14(11)	0.09(24)	0.20 (8)
	C. 3 (28)	–	0.0 (16)	0.15(13)	0.10 (9)	0.24(11)
	"(180)		0.03(27)	0.11(17)	0.06(15)	0.14(10)

have noted that the Curing 3 has a beneficial effect on mechanical performance of conventional concrete.

Table 8, show that the P_{max} and D_{max} increase with the increase in the length and percentage of fibres for each term (28 or 180 days), the greatest of P_{max} and D_{max} is obtained with the 3%-60 concrete type for both Curing 1 and 3. But with ageing, the P_{max} and D_{max} decrease for each concrete type. The loss of tensile strength of fibres in alkaline matrix was probably the cause. It is noticed that the P_{max} and the D_{max} of the concrete conserved in Curing 1 exceed slightly those conserved in Curing 3. The relatively losses of P_{max} and the D_{max} between 28 and 180 days of concrete conserved in Curing 1 are slightly lower than that of concrete conserved in Curing 3. For example, for the concrete, which presents largest P_{max} (CE 3%-60 and CE14 3%-60), the loss of its P_{max} between 28 and180 days is respectively 32% and 33%. Comparatively to literature results, about durability of more common vegetable fibre-concrete, according to flexural properties, we can reveal that the MDPSF-concrete, conserved in Curing 1 or 3, has a low durability.

Furthermore, the results of Dmax (Table 8), show that the toughness of MDPSF-concretes losses with ageing. And for each term, (28 or 180 days), the toughness coefficient of MDPSF-concrete remained lower than those of sisal or coconut fibres-concretes (Tolêdo Filho et al. 2003; Bentur and Akers 1989; Aziz et al. 1981). That is probably due for the reasons already exposed in our former research (Kriker et al. 2004), principally to the mediocre mechanical performance of MDPSF. Comparing the toughness properties of the concrete conserved in Curing 1 and 3.

That can be explained by (Gram 1983) observations. In fact, he revealed that the transport of hydration products from the matrix to the fibres had been accelerated by the thermo activation. Therefore, the durability of fibres in the concrete conserved in Curing 3 will be more affected than those conserved in Curing 1.

In addition, when we compare the flexural properties (FSC, P_{max} and D_{max}) of the concrete conserved in Curing 3 with those conserved in (Curing 2) in our preceding research (Kriker et al. 2004), concerning the concrete exposed directly to the hot dry environment. We can see that the conservation in Curing 3 has a beneficial effect on the flexural performances of fibre-concrete at early time (until 28 days). At long term it has a beneficial effect only on the FSC. Thus, some other treatments of MDPSF and MDPSF-concretes are then necessary to improve their mechanical performance, especially after the first cracking. This axis will be investigated later.

Figure 6a, b present the mean bending load-deflection of MDPSF-reinforced concretes at 28 and 180 days in Curing 3. At the beginning of loading the behaviour is elastic until the first crack strength (FCS). Beyond the first crack strength the recorded deflection around 0.05 mm, the initiated crack had an unstable growth leading either to separation of the body into parts when there is no fiber, or to a macro-crack with a deflection around 0.5 mm, when the fibers could stop the crack growth. This crack instability could be connected with MDPSF-fiber elasticity modulus. After that the load became nearly constant with the increasing bending deflection.

Figure 7a, b present some scanning electron micrograph at the broken surface, at the fibre-matrix interface of fibre-concrete conserved during three months respectively in Curing 2 and 3. Observing those microstructures of the fibre-matrix interface,

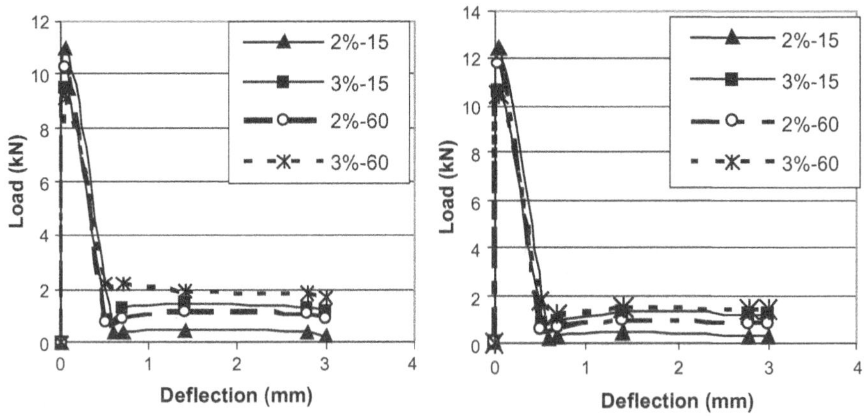

(a) 28 days (b) 180 days

Fig. 6 Load-Deflection in Curing 3 (Kriker et al. 2008)

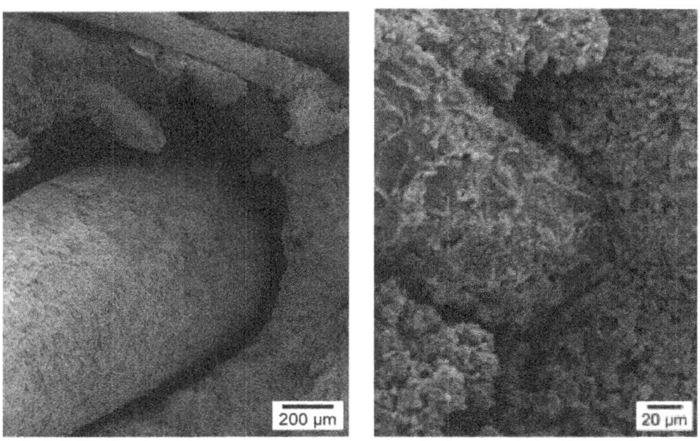

(a): in Curing 2 (b): in Curing 3

Fig. 7 Scanning electron micrograph at the fibre-matrix interface of MDPSF-concrete conserved during three months in various curing. (Kriker et al. 2008)

it can be seen that the voids around the fibres of concretes conserved in Curing 1 are larger than that of the concrete conserved in Curing 3. In addition, Fig. 7b shows the uniform deposition of hydration products on all fibre surfaces. Those observations confirm the fact that the quantity of alkaline hydration products in Curing 2 is lowers that in Curing 3.

Those also explain why the flexural properties of MDPSF-concrete after the first crack at long term in Curing 3, is slightly lower than that in Curing 1. In fact, the durability of MDPSF was more affected by cement alkaline products in Curing 3 than that in Curing 1. That confirms also the findings of (Tolêdo Filho et al. 2003;

Bentur and Akers 1989) regarding mineralization of vegetable fibres due to migration of hydration products, especially calcium hydroxide after natural wet ageing. Like shown it surface of fibre on Fig. 7b.

4.2 Conclusion

The FCS continue to increase with ageing in all Curing 1, 2 and 3 and for each concrete types. But the FCS of concrete without fibre remained greater than those the fibre-concretes.for each term (28 or 180 days),

The FCS of concretes conserved in Curing 3 is slightly greater than that in Curing 1. This can be attributed to thermo activation and hence to higher hydration products in Curing 3. For those reasons, after the first crack, the flexural properties (P_{max} and D_{max}) and the durability of the fibre-concretes in Curing 3 were lower than that in Curing 1.

4.3 Effect on Cement Type on Durability of Date Palm Fibers Reinforcement Concrete in Hot Dry Climates

This section presents the results of experimental investigations of the formulated and characterization of date palm fibers mortar by addition of silica fume. The use of addition mineral is widely used in the production of cements through the world in order to decrease the alkalinity of cement then to improve the durability of the date palm fibers in cement matrix.

Different mortar mixtures were prepared in which the cement was substitute by 10% of silica fume. The durability of MDPSF reinforced concrete was investigated also according to its flexural properties.

4.3.1 Materiel Properties

Composite Portland cement (CPJ-CEMII/B 42.5) from Algeria was used with low alkali content was used. Tables 9, 10, 11, 12 and 13 present its physical and chemical properties of used materials.

Table 9 Physical properties of used cements (Mokhtaria et al. 2018)

Cement	Fineness (m^2/kg)	Setting time (min)	σ_C 28 days (MPa)
CPJ-CEM II/A (C2)	3555	175	45

Table 10 Chemical properties of used cements (Mokhtaria et al. 2018)

Cement	SiO_2	AL_2O_3	Fe_2O_3	MgO	CaO	Na_2O	SO_3	K_2O	Loss on ignition
CPJ-CEM II/A	18.13	4.42	3.03	1.85	60.78	0.64	2.34	0.13	8.07

Table 11 Physical properties of sand (Mokhtaria et al. 2018)

Absolute specific gravity	Apparent specific gravity	Fineness modulus
2.63	1.62	2.74

Table 12 Chemical properties of Silica Fume (SF) (Mokhtaria et al. 2018)

	SiO_2	AL_2O_3	Fe_2O_3	MgO	CaO	Na_2O	K_2O	Loss on ignition
SF	93.07	0.6	0.3	1.85	0.2	0.2	0.5	8.07

Table 13 Properties of super plasticizer (Mokhtaria et al. 2018)

Form	Color	Density	pH	Content of chlorine
93.07	0.6	0.3	1.85	≤0.1 g/l

4.3.2 Specimens Fabrication, Mixing Design and Curing Types

Specimens $(4 \times 4 \times 16)$ cm^3 were used in order to prepare normal mortars, according to standard norm (NFP15-471) with a sand/cement ration equal to 3. The percentage weight of fibers was 0.4 and 0.6%. In fact in our previous studies we were showed that percentage was the optimal (Mokhtari et al. 2015) The rate of mixing water is variable because the use of date palm fibers influence in workability of mortars, for that a super plasticizer was used. Table 14 Present the mix proportion of mortars studied and their notations are presented in Table 7.

Two type of cure are using (air free and tap water).

4.3.3 Flexural Strength

The flexural strength are tested at 7, 14, 28 and 60 days in according with standard EN 196-1. The mechanical properties obtained from tests performed on three specimens testing machine three point test configuration, the maximum flexural strength of mortars determined from the maximum load after the first visible crack using. The evolution of the flexural strength of mortars is represented in Figs. 8 and 9.

According to Fig. 8 an increase in the flexural strength with ages for all various mortars was observed. The RM reinforced by 0.6% of date palm fibers presented the best flexural strength. The use of fibers Improve the flexural strength for all mortars. But a reduction in the flexural strength according with the increase in percentage of silicate fume in the mortar was noticed.

Table 14 Mix proportion of the mortars (kg/m^3) (Mokhtaria et al. 2018)

Notation	Description	Cement (kg)	Water (kg)	Sand (kg)	Silica Fume (kg)	Super plasticizer	Fibres (kg)
MR	Reference mortar	495	218	1487	0	0	0
MSF	Mortar with 10% SF	445.5	218	1487	49.5	0	0
MR-0.4	With 0.4% fibers	495	218	1478	–	4.95	8.8
MR-0.6	With 0.6% fibers	495	218	1487	–	5.95	13.2
MSF-0.4	With 0.4% fibers	445.5	218	1478	49.5	4.95	8.8
MSF-0.6	With 0.6% fibers	445.5	218	1487	49.5	5.95	13.2

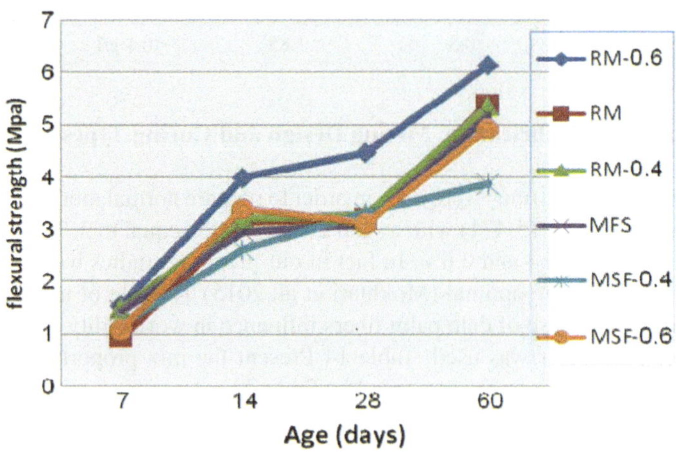

Fig. 8 Flexural strength of mortar keep in tap water cure (Mokhtaria et al. 2018)

According to Fig. 9 we note the crease of the flexural strength for all various mortars according to the ages (7, 14, 28 days) after 28 days until 60 days we remark the increase of the flexural strength for various mortars and the RM reinforced by 0.6% of date palm fibers presented the best flexural strength. The use of fibers Improve the flexural strength for all mortars this result is on concordance with several authors (Aziz et al. 1981; Bentur and Akers 1989; Tolêdo Filho et al. 2003). A diminution in the flexural strength according with the increase in percentage of silicate fume in the mortar was noticed. In hot dry climates an initial wet cure is then necessary.

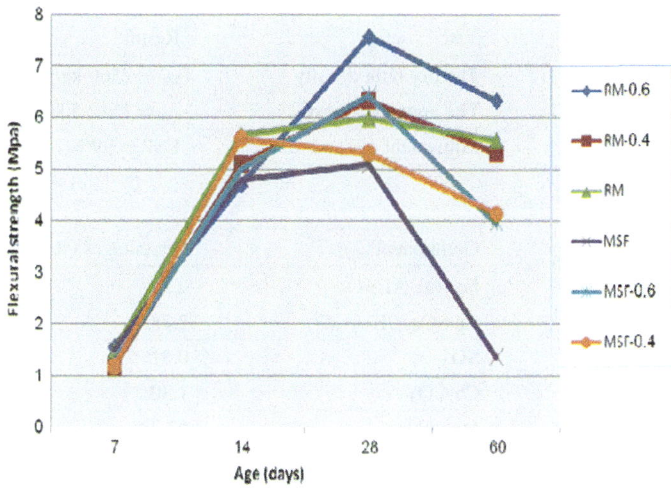

Fig. 9 Flexural strength of mortars keep in air free cure (Mokhtaria et al. 2018)

4.3.4 Conclusion

Reinforcement with date palm fibres has been very beneficial in improving flexural strength in wet treatment; but not in a hot dry curing.

Adding 10% of SF to the mixture did not improve flexural strength compared to conventional mortar; especially in the hot dry curing an initial wet cure with thermal activation is therefore probably necessary.

5 Mechanical and Thermal Properties of Date Palm Reinforced Adobe Bricks

This part presents the results of experimental investigations of the formulated and mechanical, thermal characterization of date palm reinforced adobe clay bricks by addition of sand dunes. The use of addition sand dunes is widely used in the production in order to decrease the fluidity of bricks then to improve the mechanical properties and decrease its shrinkage. The weight percentage of dune sand varied from 0 to 40%, and that of MDPSF varied from 0 to 3%.

Table 15 Physical properties of sand dunes (Chaib et al. 2015)

Test	Result
The dry bulk density	$\rho_s = 2560$ kg/m^3
The apparent density	$\rho_a = 1512.5$ kg/m^3
Equivalent of sand	ESP = 99%

Table 16 Chemical properties of sand (Chaib et al. 2015)

Components	Percentages (%)
Fe $_2O_3$-AL_2 O_3	0.25
Ca SO_4, $2H_2O$	2.78
SO_4	0.51
Ca CO_3	1.30
Insoluble	93.23
Na cl	Trace
Loss to Fire	1.16

5.1 Materiel Properties

5.1.1 The Dunes Sand

Dunes sand was from Sidi-Khouiled (town in Ouargla Algeria). Table 15 present its physical properties. These analyzes were conducted in the laboratory of Civil Engineering of Ouargla University. Three samples for each test were used.

Table 16 present dune sand chemical properties. This analysis was made in the Ouargla L.T.P.S laboratory.

Remark the percentage of the Ca (SO_4), (SO_4) is lower than the recommended limit. Then the sand used is non-aggressive.

5.1.2 The Clay

Table 17 gives the physical properties of used clay. Its from Touggourt (town Ouargla Algeria). These tests are carried out in the Ouargla LT P S laboratory.

The chemical analyses of clay are given in Table 18. This table shows that the

Table 17 Physical properties of clay (Chaib et al. 2015)

Characteristics	Result
The dry bulk density (NF P 94/064)	P = 2.03 g/cm3
Methylene Blue (NF 933-9)	VBS = 8
Limit of Atterberg (NF P 94-051)	WL = 69.58% WP = 24.71% IP = 44.87%

Table 18 Chemical compositions of the clay (Chaib et al. 2015)

Chemical characteristics	Components	Percentages (%)
Insoluble (NF P 15-461)	Insoluble	63.18
Sulphates (BS 1377)	SO_3 $CaSO_4/2H_2O$	0.45 2.46
Carbonates (NF P 15-461)	$CaCO_3$	18.0
Chlorides (MOHR Method's)	Cl^- NaCl	0.42 0.68

insoluble in percentage was of approximately 64%, the levels in sulfates and chlorides are very low.

5.2 *Specimens Fabrication, Mixing Design and Curing Types*

The tests were carried out on bricks of dimensions (240 × 110 × 50 mm) according to (DIN Standard 106). The date palm fibers are separated into individual fibers then washed in bleach to remove these impurities, and then they are cut with lengths: between 1 and 20 cm. We have worked with a fixed water/solid ratio which was to 35%. After demoulding (Fig. 10), the adobe bricks were cured in laboratory during 28 days: temperature of 25 ± 3 C °; Relative humidity HR = 35 ± 5. The compositions were presented as following:

- **Comp. 1**: clay 100% +00% sand (reference composition)
- **Comp. 2**: clay 90% +10% sand
- **Comp. 3**: clay 80% +20% sand
- **Comp. 4**: clay 70% +30% sand
- **Comp. 5**: clay 60% +40% sand

Fig. 10 Date palm reinforced adobe bricks at demoulding

- **Comp. 6-1**: clay 99% + 1% fiber
- **Comp. 6-2**: clay 98% + 2% fiber
- **Comp. 6-3**:clay 97% + 3% fiber
- **Comp. 7**: clay 60% +40% sand + 3% fiber.

5.3 Flexural and Compressive Properties

Using an Instron universal testing machine by four-point configuration, the flexural strength ($\sigma_{f)}$ was determined with an experimental set-up in accordance with standards NFP 18-409. Compression strength ($\sigma_{c)}$ was carried out on cubic specimens (10 × 10 × 10) cm^3, according to the standard NF P 18-406.

Table 19 provide the flexural and compressive adobe bricks as function of percentage of sand and fibbers.

Depending on the increase in the percentage of sand in the mixture the flexural and de compressive strengths increases up to the percentage 20% in consequence it decreases. 20% in dune sand is therefore the optimum percentage.

But, the increase in mass of fibers improves its mechanical performances (Fig. 11).

5.4 Thermal Properties

The method of the hot wire for our experimental program was used, the technique of the hot wire by CT-meter Standard (NF in 993-15).The CT-METER been developed with the aim to assess with precision, the thermal characteristics of a number of homogeneous materials and isotropic.

Table 20 provide the thermal properties of adobe bricks as function of percentage of sand and fibbers represented by their Thermal Conductivity (λ).

A decrease in conductivity with increasing in fraction of dune sand in the mixture was observed. This is probably due to the increased of porosity. In fact, the matrix of the pure clay is very dense because the clay particles are very fine, by adding to the dune sand particles, the melange become more porous which promotes a decrease in thermal conductivity. A decrease in conductivity as a function of the increase in percentage of fibers in the mixture was observed to. That confirms the increase the void in matrix as function of increase de mass fraction of fibers.

Table 19 Flexural and compressive strengths of adobe bricks as function of percentage of sand and fibbers (Mekhermeche et al. 2016)

	Comp. 1	Comp. 2	Comp. 3	Comp. 4	Comp. 5	Comp. 6-1	Comp. 6-2	Comp. 6-3	Comp. 7
σ_f (MPa)	0.30 ± 0.04	0.45 ± 0.03	0.90 ± 0.12	0.06 ± 0.07	0.55 ± 0.06	0.08 ± 0.01	0.85 ± 0.04	1.02 ± 0.06	0.50 ± 0.01
σ_c (MPa)	1.50 ± 0.10	1.68 ± 0.15	2.50 ± 0.17	2.30 ± 0.13	1.90 ± 0.18	2.88 ± 0.1	3.23 ± 0.15	3.55 ± 0.13	3.25 ± 0.15

Fig. 11 Flexural failure of date palm reinforced adobe clay bricks

5.5 Conclusion

As optimization we have chosen composition Comp. 7 which gives a brick having an acceptable mechanical and thermal performance. Indeed, brick C7 will have better mechanical resistance than that of reference brick Comp. 1 but it remains lower than that of Comp. 3 brick with 20% of sand (Comp. 3 is the most resistant brick). According of thermal insulation, brick Comp. 7 is better than brick Comp. 1 but it remains less than brick Comp. 6-3 (Comp. 6-3 has given the best thermal insulation). The choice of this Comp. 7 adobe bricks remains within the acceptable range of standard adobes brick as mentioned by the heights (Bouvenot 1980; Delebecque 1990; Guilaud 1997) as shown in Table 21.

6 Mechanical and Thermal Properties of Date Palm Reinforced Gypsum

This work is part of the evaluation of local materials by improving their performance in the field of thermal insulation, which is considered a first step in the development of new local materials to be used in construction field, the material used in this study is the gypsum reinforced with date palm fiber. In this work physical and thermal conductivity properties will be presented.

Table 20 Thermal conductivity (λ) of adobe bricks as function of percentage of sand and fibbers (Mekhermeche et al. 2016)

	Comp. 1	Comp. 2	Comp. 3	Comp. 4	Comp. 5	Comp. 6-1	Comp. 6-2	Comp. 6-3	Comp. 7
λ (W/m °C)	0.95 ± 0.05	0.87 ± 0.04	0.84 ± 0.03	0.84 ± 0.07	0.83 ± 0.03	0.50 ± 0.02	0.44 ± 0.02	0.41 ± 0.02	0.76 ± 0.03

Table 21 Standard adobe brick properties (Bouvenot 1980; Delebecque 1990; Guilaud 1997)

Properties	Standard Adobe
Compressive strength (MPa)	2–5
Thermal conductivity (W/m °C)	0.46–0.81

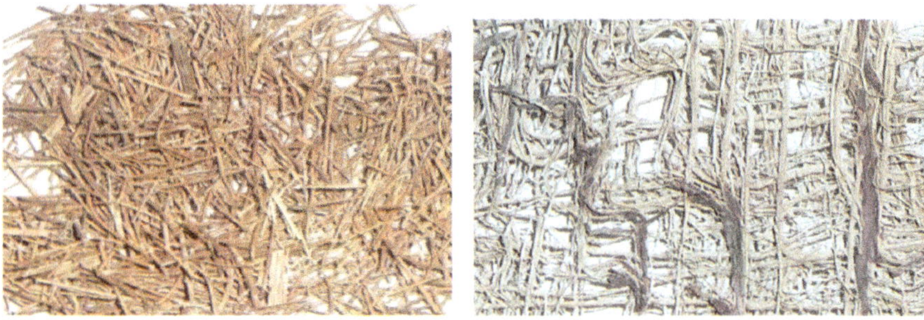

(a) Individual fibers (1cm length) (b) Multi-layered

Fig. 12 Fibers used for reinforcement of gypsum

6.1 *Materiel Properties*

6.1.1 Fibers

Male Date Palm surface fibers were used in two forms:

1. Individual fiber
   The MDPSF was washed and separated in water then cut to length of 1 cm. The choice of 1 cm is practiced to have a good distribution of the fibers in plaster matrix (Fig. 12a).
2. Fibers in rectangular Multi-layered form
   The natural mesh of MDPSF was washed then cut in rectangular Multi-layered form (15 × 8) cm^2 (Fig. 12b).

6.1.2 Gypsum

Gypsum or plaster used in this investigation was from Ouled Djella, Algeria Tables 22 and 23 present its chemical and physical properties.

Table 22 Chemical analysis of gypsum used in the study (Hafsi et al. 2017)

CL	KO	Na_2O	SO_2	MgO	CaO	Fe_2O_3	AL_2O_3	SiO_2
0.002	0.03	0.09	44.95	0.53	32.15	0.08	0.10	0.70

Table 23 Physical and mechanical properties of plaster used in the study (Djoudi 2001)

Bulk density (kg/m^3)	$\rho_a = 840 - 915$
Absolute Density (kg/m^3)	$\rho_s = 1100 - 1300$
Tensile strength (MPa)	After 1 h and 30 d = 3.48 After 24 h = 3.73 After 7 days = 3.99
Compressive strength (MPa)	After 1 h and 30 d = 8.51 After 24 h = 9.27 After 7 days = 10.11

6.1.3 Mortar

The conventional mortar used is a standardized mortar. The confection and mixing were carried out according to standards EN 196-1. The mass composition of a mold consisting of three times

- 03 parts of sand
- 01 part of Cement
- 0.5 part of Water.

Cement used was a Portland cement (CPJ-CEM II/A 32.5). Table 5 present its chemical and physical properties. The same sand seen in Sect. 3.1.2 was used.

6.2 Specimens Fabrication and Curing

A same mould with dimensions of (32 × 27 × 4) cm^3 was used for preparing gypsum and mortar samples. It's separated on six samples with dimensions of 16 × 9 × 4 cm^3.

All samples (mortar or gypsum) were conserved in the laboratory at temperature of 25 ± 3° C and relative humidity (RH) of 35 ± 5%.

6.2.1 Mortar Samples

The mould was filled in two equal layers; each layer was vibrated on a vibrating table during 30 s. Table present the mix fraction of used mortar (Table 24).

Table 24 Mortar mix proportion

Mould volume (l)	Cement (kg)	Sand (kg)	Water (kg)
3.46	1.21	3.63	0.61

6.2.2 Gypsum Samples

1. Conventional gypsum

The mould of conventional gypsum was carried out according to standards NFB 12-401 with Water/plaster (W/P) of 0.6. Table 25 present its mix fraction. The preparation steps in mixer are added as follows:

- Add water then gypsum during 1 min
- Filling mould in a single layer
- Vibration in vibrating table during 20 s.
- Settlement and keeping in the laboratory.

2. MDPSF reinforced gypsum

(a) **Individual fibers**

Table 25 present the individual MDPSF reinforced gypsum proportions. The mass factions of fibers by are (1, 2, 3, 4 and 5%). This mass of fibers for each time is subtracted from the total mass of conventional gypsum (417 kg). The mass fractions are made in mixer as follows.

- Saturate the fibers by adding 132.5% of their weight in water (see Table 2)
- Mix water, gypsum saturated fiber
- Filling mould in a single layer then vibration in vibrating table during 20 s
- Finishing and keeping in the laboratory.

(b) **Multi-layered fibers**

Multi-layered fibers reinforced gypsum is also given in Table 25. The mass fractions of multi-layered fibers are (1, 2, 3 and 4%). This mass of Multi-layered fibers is also for each time subtracted from the total mass of conventional gypsum (417 kg).The mix fractions of Multi-layered fibers reinforced gypsum will be made as following.

- Saturate the fibers by adding 132.5% of their weight in water
- Mix water, gypsum
- Add first gypsum layer (thickness: 1–1.2 cm)
- Add first rectangular multi-layered fibers

Table 25 Mix fractions of gypsum and MDPSF fibers reinforced gypsum

Weight % fibers	0	1	2	3	4	5
Gypsum (kg)	4.175	4.133	4.091	4.050	4.008	3.966
Fibers (kg)	0	0.042	0.083	0.125	0.167	0.209
Water (kg)	2.505	2.505	2.505	2.505	2.505	2.505

(a) With individual fibers (b) With multi-layer mesh fiber

Fig. 13 Date palm reinforced gypsum samples

- Add second gypsum layer (1–1.2 cm thickness)
- Add second rectangular fiber layer
- Add last gypsum layer (1–1.5 cm thickness)
- Vibration in vibrating table during 20 s
- Finishing and keeping in the laboratory.

Figure 13 presents some Date palm reinforced gypsum samples.

6.3 Test and Results

Firstly the following notations will be used:

- **M**: Mortar
- **P**: Gypsum
- **PF**: Gypsum + individual Fiber (1 cm)
- **PFM**: Gypsum+ Multi-layered Fiber.

6.3.1 Bulk Density

Bulk density or apparent volumetric mass of the samples was determined by calculating the weight of the samples and calculating their size by measuring their dimensions and then dividing the obtained mass by their size. The results were provided in Table 26 and Fig. 14.

Table 26 Bulk density of mortar, gypsum and MDPSF reinforced gypsum

% of fibers		0	1	2	3	4	5
Bulk density (kr/m^3)	M	2092.57	–	–	–	–	–
	P	1208	–	–	–	–	–
	PF	–	1201.32	1192.44	0.261179	1165.78	1144
	PFM	–	1162.59	1106.62	1021.47	1027.00	–

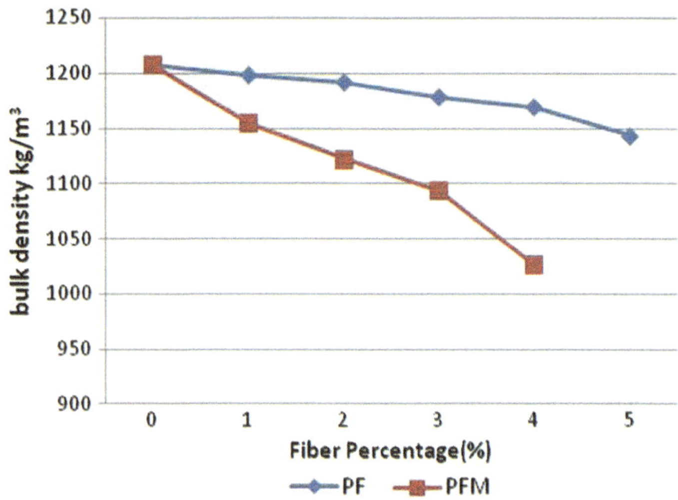

Fig. 14 The bulk density of MDPSF reinforced gypsum as function of mass fraction of fibers (Hafsi et al. 2017)

- Table 26 show that the bulk density of mortar is slightly double of that of gypsum
- The bulk densities of MDPSF are lower than that of conventional gypsum. This is due to fiber textures which provides large voids in the sample which affects the weight of the sample and thus affect the volumetric mass.
- The bulk densities of layer mush MDPSF reinforced gypsum are lower than separated MDPSF reinforced gypsum. Then the mixture multilayer gave amounts volumetric mass lower than in the mixture, despite fibers. With using the same weights of fiber.

6.3.2 Thermal Properties

A thermal conductivity was measured according to the laws of NF EN 933-15 by CT meter device developed by CSTB and which relies on the hot wire method (Fig. 15).

Sample was cut longitudinally into two equal halves to ensure the consistency of the two samples in order to obtain the best measurements. For each measurement six samples of same gypsum- fibers types or mortar were used. Table 27 and Fig. 16 present the thermal conductivity of MDPSF gypsum as function of percentage of fibers.

- Through the results of experiments (Table 27) note the considerable difference between the thermal conductivity of the concrete material grout (1.60 W/m °C) compared textured gypsum (0.58 W/m °C) where conductive grout exceed three times the conductivity of gypsum.
- The results and curves (Table 27 and Fig. 16) showed that the thermal conductivity quite disproportionate inversely with the increase in the proportion of fiber in the PF or PFM samples.

Fig. 15 The thermal conductivity measuring device CT meter

Table 27 Thermal conductivity of mortar, gypsum and MDPSF reinforced gypsum

% of fibers	0	1	2	3	4	5
M	1.60	–	–	–	–	–
P	0.58	–	–	–	–	–
PF	0.60	0.59	0.58	0.55	0.51	0.50
PFM	0.60	0.58	0.54	0.48	0.39	–

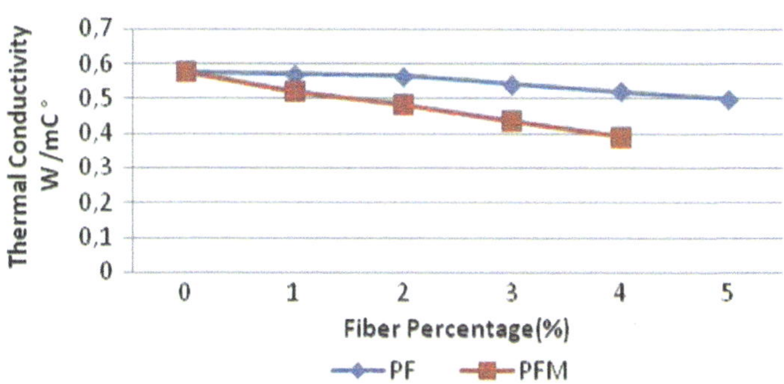

Fig. 16 Thermal conductivity of MDPSF gypsum (Hafsi et al. 2017)

- What also noted from (Fig. 16) that the mixture of multilayer gave amounts of thermal conductivity less than in the individual fiber mixture, by using the same weights of fiber. Those results are in concordance with our previous research (Abbani 2015)

6.3.3 Bulk Density and Conductivity Relationship

Figures 17 and 18 present the relationship between apparent volumetric mass and all of the thermal conductivity of the bulk density of MDPSF reinforced gypsum as function of mass fraction of fibers.

We notice through Figs. 17 and 18 the direct relationship between the volumetric mass and the thermal conductivity, as the greater the volumetric mass the greater the conductivity, and this is due to the fact that the greater the volumetric mass due to the low weight of the fibers And the presence of pores inside, the more voids inside the samples, and thus a decrease in the amount of material, which leads to a decrease in the thermal conductivity and specific heat capacity.

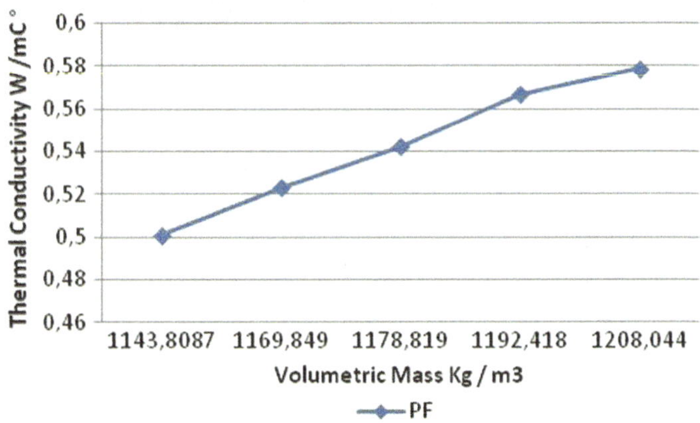

Fig. 17 Thermal conductivity as function of PF Mass sample (Hafsi et al. 2017)

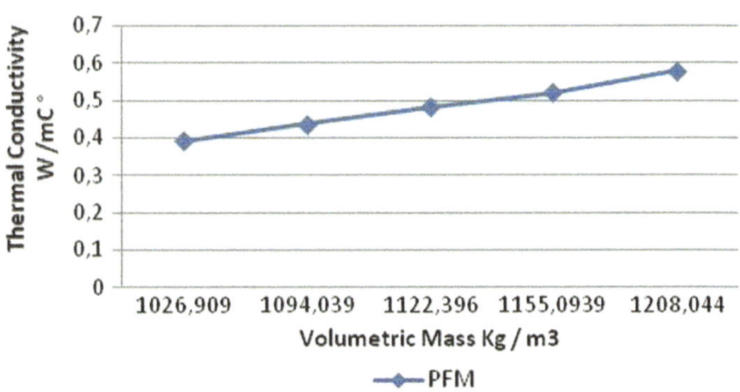

Fig. 18 Thermal conductivity as function of PFM mass sample (Hafsi et al. 2017)

6.4 Conclusion

At the end of this study of the effect of adding surface date palm fiber to gypsum in order to improve its physical and thermal properties, the following points will be concluded.

Through the experiments accomplished on the samples of cement mortar and gypsum, we notice the big difference between the physical properties of these two materials, especially the thermal properties under study, where we found that the thermal conductivity of the cement mortar is three times greater as it is estimated (1.6 W/m °C) than the thermal conductivity of gypsum, which is estimated At (0.58 W/m °C).

The addition of male date palm surface fibers to gypsum has improved the thermal properties of this substance by reducing the volumetric mass of the new compound, due to the low volumetric mass of palm fibers, due to the presence of pores in it in addition to its synthetic structure that allows to leave voids, which led to reducing the mass.

It was also noted through the results of the experiments obtained that the use of palm fibers with gypsum in the form of multi-layer networks led to a decrease in the thermal conductivity of this material is better than the use of separated fibers in the form of a mixture, where for gypsum samples plus cut fibers (PF) was a ratio. The reduction is estimated at 13.33%. As for the samples of gypsum plus multi-layer fibers (PFM), the percentage reached 23.33%. This is due to the reduction of thermal bridges within the samples as the fiber networks separate the gypsum layers from each other

Acknowledgements The authors are grateful to Prof R. N. Swamy of the department of Mechanical Engineering, University of Sheffield for this orientation in the choice of this research axis. They are also indebted to Dr G. Debicki From INSA of Lyon, Prof A. Bali from ENP Algeria, Prof K. M. Khenfer from Laghouat university Algeria for their contribution in experimental programme in these laboratories. They are grateful to Prof. H. El-Mously Founding Father of Date Palm by-products research Ain Shams University, Egypt and Dr. M. Jawaid from University Putra Malaysia (UPM), for their efforts to promote and popularize the vaporization of by-products palm, in particular through the By Palma Word Conference.

References

Abbani S (2015) Contribution to the study of thermal insulation of buildings in desert areas. Doctoral thesis, University KASDI Merbah Ouargla, Algeria

Aziz MA, Paramasivam P, Lee SL (1981) Prospects for natural fibre reinforced concrete in construction. Int J Cem Compos Lightweight Concr 2(3):123–132

Bedzki AK, Gassan J (1999) Composites reinforced with cellulose based fibres. Prog Polym Sci 24:221–274

Bentur A, Akers SAS (1989) The microstructure and ageing of cellulose fibre reinforced cement composites cured in normal environment. Int J Cem Compos Lightweight Concr 11(2):99–109

Bouvenot A (1980) Transferts de chaleur, Edition Masson
Chaib H, Kriker A, Mekhermeche A (2015) Thermal study of earth bricks reinforced by date palm fibers. Energy Procedia 74:919–925
Cook DJ (1980) Concrete and cement composites reinforced with natural fibres. Proc Symp on Fibrous Concrete: 99–114
Coutts RSP (1983) Flax fibre as a reinforcement in cement mortar. Int J Cem Compos Lightweight Concr 4(5):257–262
Delebecque R (1990) Éléments de Construction Bâtiment, Edition Delagrave
Djoudi A (2001) Caractérisation Structurale et rhéologique de béton de plâtre et leur renforcement par les fibres végétales du palmier dattier, Master theses, university of Laghouat, Algeria
Gorisse F (1978) Essais et contrôle des bétons. Edition, Eyrolles
Gram HE (1983) Durability of natural fibres in concrete. Swedish Cement and Concrete Research Fo. 1:83, Stockholm, p 255
Guilaud H (1997) Centre International pour la Construction en Terre, Ecole d'Architecte de Grenoble, Encyclopédie de Bâtiment, Tome 2 Edition Wake, CRATerre
Hafsi F, Kriker A, Abani S (2017) Contribution study to the thermal insulation of the builders in the desert regions of exploiting gypsum fiber reinforced palm. AIP Conf Proc 1814:020029
Khenfer MM, Bali A, Morlier P (2000) The effect of water and fibre length on the fracture resistance of cellulosic fibre cement. Concr Scien Engin 2:56–62
Kriker A, Bali A (1992) Effect of types curing on the mechanical properties of El-Hadjar slag concrete. In: Proceedings international seminary on quality of concrete in hot climate, University of Blida and CTC Sud, Algeria, pp 45–54
Kriker A, Debicki G, Bali A, Khenfer MM, Chabannet M (2005) Mechanical properties of date palm fibres and reinforced date palm fibre concrete in hot-dry climate. Cem Concr Compos 27:554–565
Kriker A, Bali A, Debicki G, Bouziane M, Chabannet M (2008) Durability of date palm fibres and their use as reinforcement in hot dry climates. Cem Concr Compos 30:639–648
Lesage R (1974) Etude expérimentale de la mise en place du béton frais. Rapport de Recherche n° 37, Laboratoire Central des Ponts et Chaussés
Lewis G, Premalal M (1979) Naturel vegetable fibres as reinforcement in cement sheets. Mag of concr Res 107(31):104–108
Mekhermeche A, Kriker A, Dahmani S (2016) Contribution to the study of thermal properties of clay bricks reinforced by date palm fiber. AIP Conf Proc 1758:030004
Mokhtari A, Kriker A, Guemoula Y, Boukrioua A and Khenfer MM (2015) Formulation and characterization of date palm fibers mortar by addition of crushed dune sand. Energy Procedia 74:344–350
Mokhtaria A, Kriker A, Ouaggad H, Merad N (2018) Formulation and characterization of date palm fibers mortar by addition of silica fume. AIP Conf Proc 030078:1–8
Ramaswamy HS, Ahuja BM, Krishnamoorthy S (1983) Behaviour of concrete reinforced with jute, coir and bamboo fibres. Int J Cem Compos Lightweight Concr 1(5):3–13
Rossi P, Harrouche N, Belloc A (1989) Méthode de composition des bétons de fibres métalliques, Annales de l'ITBTP, N° 475, Juin-Juillet
Savastano H Jr, Warden PG, Coutts RSP (2000) Brazilian waste fibres as reinforcement for cement-based composites. Cem Concr Compos 22:379–384
Swamy RN (1985) Natural fibre reinforced cement and concrete. In: Swamy RN (ed) Concrete technology and design, vol 5, Surrey University Press, p 288
Tolêdo Filho RD, Scrivener K, England GL, Ghavami K (2000) Durability of alkali-sensitive sisal and coconut fibres in cement mortar composites. Cem Concr Compos 22:127–143
Tolêdo Filho RD, Ghavami K, England GL, Scrivener K (2003) Development of vegetable fibre-mortar composites of improved durability. Cem Concr Compos 25:185–196

Date Palm Fiber Composites as Wood Substitutes

Hamed El-Mously

Abstract Most of Arab countries are located in an arid zone, very poor in forest coverage. That is why countries like Egypt heavily rely on the importation of lumber for the satisfaction of their needs of furniture and shelter. This represents a heavy burden on the balance of payment of these countries, as well as increasing of pressure on cutting of trees from forests, adding to the causes of climate change. Within this context the idea emerged to use the products of pruning of the date palm to manufacture wood substitutes. Scientific research has been conducted to rediscover the palm midribs as a substitute for wood. Samples of midribs from palms of different species in different localities have been collected. Palm midribs have been characterized in terms of geometry, physical and mechanical properties. The results of research have proven that the palm midribs enjoy mechanical properties compatible with those for soft and hard woods according to ASTM standards. This opens the potentiality of manufacture of tree-free products to substitute imported wood. Palm midribs have been successfully used to replace beech in the manufacture of Mashrabiah (Arabesque) products in the New Valley villages in Egypt. Palm midrib strips and boards have been successfully used as a core material in blockboards satisfying the DIN standards. Single-layer and three-layer particleboards have been successfully manufactured from palm midribs satisfying the international standards. The products of pruning of date palms have been successfully utilized to manufacture medium density fiberboards according to the international standards of the physical and mechanical properties. Samples taken from palm trunks have been tested under static bending. The results show that date palm trunks have strength properties similar to European red pine wood. Thus, there are wide future prospects for the use of date palm byproducts as tree-free wood substitutes. This may have important significance: economically, socially, environmentally and developmentally.

Keywords Date palm · Wood substitutes · Date palm byproducts · Mashrabiah · Blockboards · Particleboards · Medium density fiberboards · Future prospects

---

H. El-Mously (✉)
Department of Design and Production, Faculty of Engineering, Ain Shams University, Cairo, Egypt
e-mail: hamed.elmously@gmail.com

M. Midani et al. (eds.), *Date Palm Fiber Composites*, Composites Science and Technology,
https://doi.org/10.1007/978-981-15-9339-0_13

1 Introduction

The idea of use of the date palm products of pruning to satisfy human needs is a green idea. It reveals an ethical approach to the palm and the environment. Why? Because the existential meaning of the pruning activity is serving the palm by removing the products of pruning (midribs, leaflets, coir, petioles, spadix stems and midribs ends), which have already performed their ecological functions and became redundant to the palm. Left on the palm they may represent a danger of occurrence of fire accidents, facilitate infestation by dangerous insects, such as the red palm weevil and decrease the nourishment of the palm, needed for the future date crop. Meanwhile, during this process of palm service these products of pruning (lignocellulosic crop) are being collected, which may be used to manufacture substitutes for imported wood. Thus a symbiotic relation is developed: between the service of the palm and the obtainment of the products of pruning. This means that the use of palm products of pruning as imported wood substitutes decreases the pressure on cutting of trees from forests contributing to the absorption of CO_2 causing climate change. In addition the decrease of importation of wood will relieve the atmosphere from pollution during the transportation of imported wood for thousands of miles. If the sites of production of wood substitutes from palm pruning products is located near to the rural areas, where the palms are grown, which is quite logic to decrease the cost of transportation, the pollution in transportation will be minimum. **Thus, the use of date palm products of pruning to manufacture tree-free wood substitutes is really a very green idea**.

2 The Arab Region: Shortage in Forest Coverage

The Arab World is located in an arid zone with a forest coverage (% of land area) of 1.6824% in 2016 (Trading Economics 2020). Figure 1. illustrates an estimation of the forest coverage in the Arab World within the period between 1990 and 2015. It is clear that the percentage of forest coverage is decreasing with time, most probably due to urbanization trends, Taking separate countries into consideration, the forest coverage ranges from 0% in Qatar, 0.01% in Oman, 0.07% in Egypt, 0.035% in

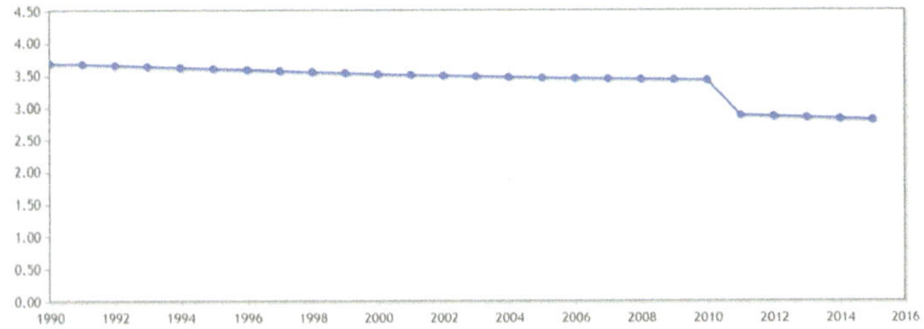

Fig. 1 Forest coverage in the Arab world (Index Mundi 2020)

Kuwait, 1.1% in Jordan, 1.52% in Palestine, 1.9% in Iraq, 2.67% in Syria, 3.83% in United Arab Emirates and 13.41% in Lebanon (Actualitix World Atlas 2016). This situation puts the Arab countries on the top of the list of wood importers in the world.

3 Reliance on the Importation of Wood in the Arab World: Is It Sustainable?

Table 1 illustrates an estimation of wood imports by the Arab countries. It is evident that almost all the Arab countries import wood with United Arab Emirates on the top importing wood annually for 3.44 billion US $, followed by Egypt (3.4 billion US $) and Saudi Arabia (3.16 billion US $). These estimates will increase with time due to the soaring of wood prices on one hand and increase of population and the improvement of the standard of living on the other hand. This represents an increase of pressure on cutting of forests, leading to climate change, increase of pressure on the importing countries, balance of payments in addition to the air pollution due to the transportation of wood thousands of miles from the countries supplying wood (e.g. Romania, Russia, Sweden, Germany, Canada, etc.) to the Arab countries.

Table 1 Arab countries import of wood (WITS 2020)

N	Country	Import (U$ thousand)	Year
1	Algeria	1296837.23	2017
2	Bahrain	329960.03	2018
3	Comoros	3990.81	2013
4	Djibouti	19128.93	2009
5	Egypt	3409824.68	2018
6	Iraq	644335.27	2014
7	Jordan	650620.88	2018
8	Kuwait	809011.57	2018
9	Lebanon	636172.18	2018
10	Libya	489292.41	2010
11	Mauritania	23587.82	2017
12	Morocco	1550497.68	2018
13	Oman	535001.93	2018
14	Qatar	551120.96	2018
15	Saudi Arabia	3156471.48	2018
16	Sudan	316075.83	2017
17	Syria	573348.08	2010
18	Tunisia	558401.68	2017
19	United Arab Emirates	3443407.99	2018
20	Yemen	225141.76	2015

4 Characterization of Palm Midribs: Geometry, Physical and Mechanical Properties

4.1 Date Palm Midrib Geometry

The midrib represents the central part of the palm leaf. Figure 2 illustrates a general view of the palm midrib, divided into three parts: the basal, the middle and the top, as well as the form of the cross-section of its parts. Figure 3 illustrates the variation of the area of midrib cross section, the diameter of the largest circle drawn within the cross section and its area: from the base of the midrib to the top (El-Mously 1995). Table 2 represents a summary of the masses, lengths and dimensions of the cross sections of midribs of palms of different species (Siwi, Amhat and Balady) in El-Fayoum governorate, Egypt (El-Mously 1995). It is clear that there are significant differences among the species, as well as among different localities for the same species.

A study has been conducted on the anatomy of the midrib, (Fig. 4), (Megahed and El-Mously 1995). This study has shown that the palm midrib belongs to the class of monocotyledons, where cross linking is absent in the cross section. The fibro-vascular bundle is the structural unit of the cross section. The cross sectional area of the fibro-vascular bundles, as well as the density of their distribution differ greatly from the epidermal layer to the core of the midrib. It was possible in this study to classify the midrib cross section into three regions: (1) peripheral, (2) transitional and (3) core layers.

It was possible to evaluate the thickness of both the peripheral and transitional layers, where the ratio of fibers increases with adherence to the periphery, by

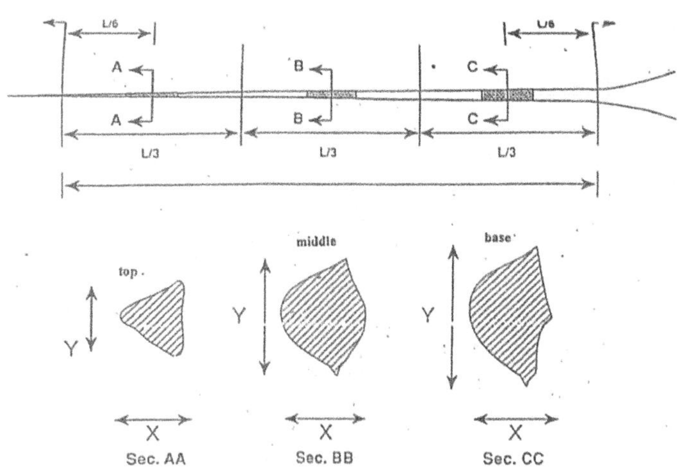

Fig. 2 A diagrammatic sketch showing the palm midrib, as divided to three distinguished parts: base, middle and top (El-Mously 1995)

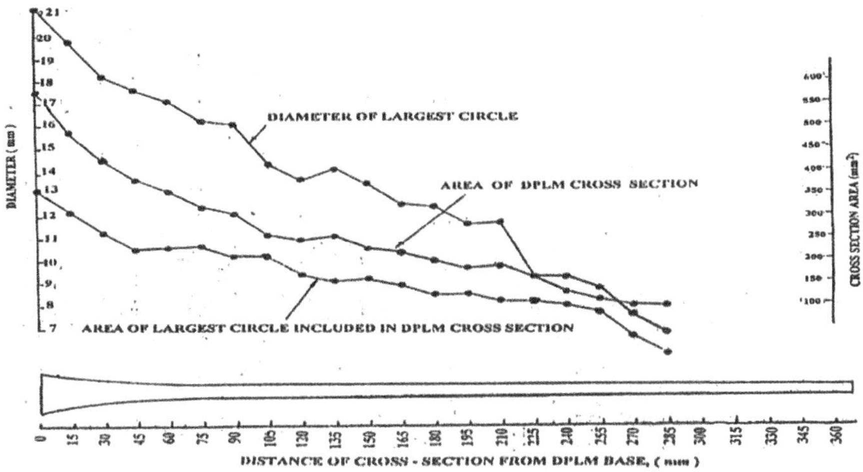

Fig. 3 The variation of cross section, value of the maximum diameter and the area of the largest circle inside the cross section for Siwi palm midribs from the base to the top (El-Mously 1995)

Table 2 Summary of masses, lengths and dimensions of cross sections for data palm midribs (Siwi, Amhat and Baladi species from El Fayoum) (El-Mously 1995)

Species	Locality	n	Mass (gm)	Length (m)		Cross section, dimensions (mm)		
						Base	Middle	Top
Siwi	Garfis	22	362	1.62	X	17.90	15.23	12.19
					Y	14.62	13.20	11.98
Amhat		14	528	2.05	X	20.96	16.70	13.86
					Y	17.20	16.25	14.40
Balady		10	339	1.59	X	19.36	16.75	13.23
					Y	16.38	14.19	13.00
Siwi	El-Kaabi	33	381	2.35	X	31.87	21.48	16.78
					Y	26.15	18.84	17.10
Amhat		22	629	2.59	X	25.77	19.97	15.10
					Y	21.91	17.37	15.72
Balady		19	493	2.56	X	24.51	18.83	14.81
					Y	20.06	16.18	15.35
Siwi	Abou-Ksah	19	657	2.73	X	39.78	23.27	17.07
					Y	25.42	18.84	16.26
Amhat		21	666	2.75	X	32.90	21.90	17.83
					Y	23.02	18.15	15.54
Balady		20	451	2.52	X	27.71	19.41	15.64
					Y	19.26	16.08	14.72

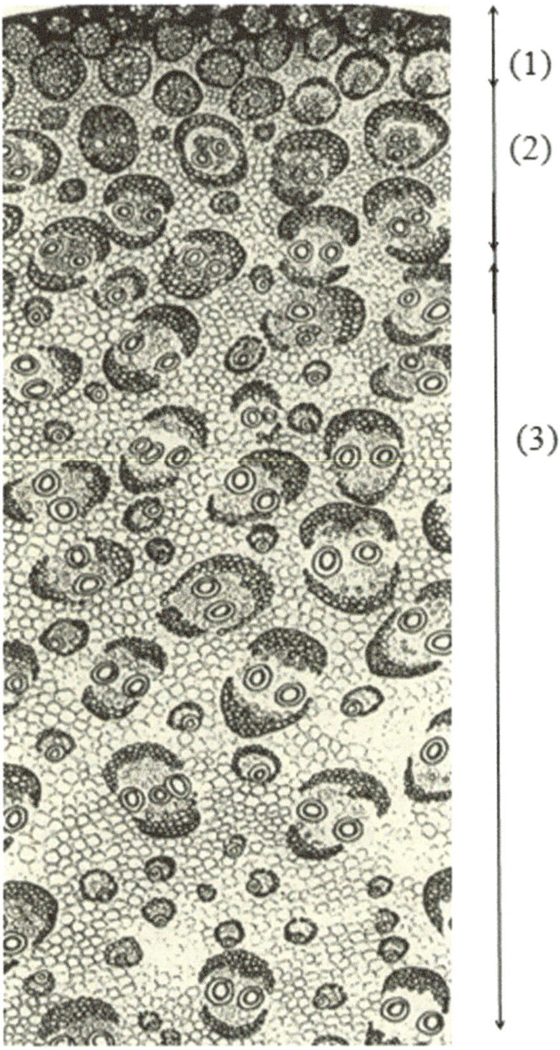

Fig. 4 A diagrammatic illustration of a transverse section of the palm midrib showing the peripheral (1), transitional (2) and core zone (3) (25X) (Megahed and El-Mously 1995)

~1.25 mm. It was also possible to evaluate the average length of fibers by 1.366 mm for Baladi midribs and 1.288 mm for Siwi midribs.

4.2 Date Palm Midribs Dimensional Stability

The research findings have shown the high tendency of palm midribs to swell when immersed in water as compared with wood. The percentages of volumetric swelling of midribs of palms of different species (Siwi, Mantour, Tamr and Males) after immersion in water for 38 h ranged between 980% for Males and 176% for Siwi, whereas the corresponding values for wood of different species (beech, spruce, pitch pine and mahogany) ranged between 13 and 29% (El-Mously 1991). The large values of volumetric swelling of palm midribs can be explained by the absence of radial rays (cross linking) in the palm midribs cross-section as a monocotyledon.

4.3 Mechanical Properties of Palm Midribs

The results of tests, conducted on samples taken from the central parts of midribs obtained from different palm species and geographic locations, indicate that the values of mechanical properties of palm midribs vary with the palm species and geographic locations. Table 3 illustrates the results of tests conducted on midribs of palms of Siwi, Amhat and Baladi species in different localities in El Fayoum governorate. The modulus of rupture (MOR) in bending varied from ~70 for Amhat, ~76 for Baladi to 82 N/mm^2 for Siwi palm species. The compressive strength varied from ~34 for Amhat, 36 for Baladi to ~40 N/mm^2 for Siwi. The tensile strength (UTS) varied from ~66 for Amhat, ~71 for Baladi to ~75 N/mm^2 for Siwi palm species (El-Mously 1995). The corresponding values of the modulus of rupture, compressive strength and tensile strength for beech wood are equal to 104, 46 and 130 N/mm^2 respectively (Forest Products Laboratory 1999).

5 Evaluation of the Annually Available Quantities of Products of Pruning of Date Palms: Egypt as an Example

Taking an example the Siwi species (a dominant palm species in Egypt, it was possible to determine (Table 4) the mass of the annual products of pruning of one palm (El-Mously and Saber 2018).

Egypt possesses 15 million female productive palms (El-Mously and Saber 2018). Consequently, the total annual available mass of the products of pruning of date palms amounts to 810000 tons, which represents a considerable sustainable material base for the establishment of a wide spectrum of industrial activities.

Table 3 Mean values of the main mechanical properties of palm midribs (Siwi, Amhat and Baladi species from El Fayoum Governorate) (El-Mously 1995)

Species	Locality	Static bending, N/mm^2			Comp.//[c] to grain, N/mm^2			Tension//to grain, N/mm^2		
		N	MOR[a]	MOE[b]	N	CS_{max}[d]	MOE	N	UTS[e]	MOE
Siwi	Garfis	13	76.17	5266	28	39.76	20.35	22	69.23	40.61
			(13.40)	(1247)		(5.30)	(4.92)		(8.47)	(5.74)
	El-Kaabi	12	76.60	4349	17	37.66	17.18	18	79.69	–
			(12.24)	(872)		(6.12)	(3.57)		(12.54)	
	Abou-Ksah	9	96.10	5731	10	44.02	19.47	6	83.00	49.91
			(12.66)	(718)		(9.89)	(4.55)		(13.98)	(11.06)
	Species average	34	81.6			39.89		46	75.12	
Amhat	Garfis	35	55.68	3648	51	30.16	13.53	40	52.61	–
			(9.04)	(831)		(11.15)	(4.44)		(9.45)	
	El-Kaabi	50	77.06	5128	64	37.35	18.66	39	76.67	–
			(11.93)	(1085)		(7.24)	(4.56)		(10.19)	
	Abou-Ksah	21	73.09	5042	56	34.08	18.07	30	71.10	–
			(8.50)	(889)		(5.37)	(4.37)		(9.83)	
	Species average	119	69.64		171	34.13		109	66.31	

(continued)

Table 3 (continued)

Species	Locality	Static bending, N/mm^2			Comp.//[c] to grain, N/mm^2			Tension//to grain, N/mm^2		
		N	MOR[a]	MOE[b]	N	CS_{max}[d]	MOE	N	UTS[e]	MOE
Balady	Garfis	26	68.26	4355	43	34.01	16.75	38	64.48	–
			(12.28)	(744)		(5.82)	(3.39)		(9.22)	
	El-Kaabi	21	72.43	4946	32	36.30	19.25	45	73.25	–
			(8.85)	(984)		(5.21)	(4.60)		(13.39)	
	Abou-Ksah	22	88.98	6096	27	38.98	39	39	76.20	
			(11.42)	(955)		(4.67)	(4.53)		(14.65)	
	Species average	69	76.14		102	36.04		122	71.46	

• Values in parenthesis represent standard deviations
N-the number of specimens
[a]Modulus of rupture
[b]Modulus of elasticity
[c]Parallel to the direction of fibro-vascular bundles
[d]Maximum compressive strength
[e]Ultimate tensile strength

Table 4 Masses of different products of pruning per palm (El-Mously and Saber 2018)

N	Secondary products	Mass per palm, Kg (air dry weight)
1	Palm midribs	15
2	Palm leaflets	14.6
3	Spadix stems	9
4	Coir	1.56
5	Midrib end	14
	Total	54.2 kg

6 Modern Achievements in the Use of Date Palm Byproducts as a Substitute of Imported Wood

6.1 Palm Midribs in Mashrabiah (Arabesque) Products

6.1.1 Introduction

The Mashrabiah (Arabesque) (Fig. 5) handicrafts are a part of the cultural heritage in Egypt and the whole Arab region. The word "Mashrabiah" has been used to describe the part of the building, assigned to put the clay pots to cool the drinking water. Its connotation has been widened to include wood partitions, made by turning, to cover the openings of windows or separate the parts of building allocated for men and those for women in houses or mosques. The Mashrabiah design has the advantage of allowing the house dwellers to see people outside the house without being seen by strangers and thus the privacy of dwellers is preserved. In addition, the Mashrabiah

Fig. 5 Historic place of Suhaymi House, Darb al-Asfar, Al-Gamaliyya, Cairo, Egypt

is consonant with the prevailing climate in the Arab region, characterized with a shining sun most of the year seasons. Thus the Mashrabiah offers an appropriate solution to break the high intensity of sun light via the narrow openings between the turned pieces of Mashrabiah, meanwhile allowing the breeze to pass through and cool the house interior. In addition the Mashrabiah (Arabesque) design found wide application as an ornament in different pieces of furniture.

6.1.2 Rationale of Selection of the Mashrabiah Project

The Mashrabiah (Arabesque) products in Egypt and the whole Arab region rely basically on imported beech wood. With the drastic increase of the price of wood the prices of the Mashrabiah products increased considerably and became beyond the financial capabilities of wide strata of population, leading to the collapse of the handicrafts, associated with Mashrabiah. Meanwhile, the drastic change in the style of life in rural areas with the propagation of the Western pattern of consumption has led to the shrinkage of use of palm midribs in different traditional forms of utilization (e.g. in roofing, fencing, crates, etc.). Starting with conducting machinability tests on palm midrib samples on a metal-cutting center lathe, the date palm midrib piece reached a diameter of ~2 mm without breakage, suggesting that the date palm midrib is strong enough to replace wood. Thus, the next thought was to go to remote villages possessing wide palm plantations and teach the public there to use the seemingly useless midribs, treated as waste in the manufacture of Mashrabiah (Arabesque) products that may replace those, made from the expensive imported beech wood. The advantages of such project include reviving the traditional skills of Arabesque manufacture to support Mashrabiah as a distinguishing component of our traditional way of life, while providing cheap sustainable labor opportunities in rural areas, where women could work on lathes in their houses.

6.1.3 Location of Mashrabiah Project

The Dakhla oases, located in the New Valley governorate, Egypt, are highly isolated from the Nile Valley. Their location: distant from the Nile Valley and from each other, together with their geology and natural resources gave them a very strong dimension of subsistence economy. It was thought that the launching of development projects relying on the indigenous raw materials may lead to a peaceful transfer from subsistence to market economy leading to the endogenous development of the local community. Thus, the Gedeida Village in El-Dakhla oases was chosen as a site of the project. The project was launched by the Centre for Development of Small-Scale Industries and financed by a grant from the German Technical Cooperation Agency. Moreover, it was decided to allocate the Arabesque handicrafts in houses to tune better with the specificity of the sociocultural context there and to provide more chances for women to participate in the project.

6.1.4 Project Features

- A training Centre (TC) has been established in the site of an evacuated poultry farm having a total area of 836 m^2. The TC was opened on 2.7.1995. The functions of the TC are as follows:
  - Purchase of date palm leaves' midribs (DPLM) in season to guarantee low prices for the beneficiaries;
  - Preparation of DPLM to be ready for manufacture by the beneficiaries (e.g., drying, treatment against DPLM-borers, etc.);
  - Training of beneficiaries (Fig. 6) in all skills related to Arabesque handicrafts;
  - finishing of products, made by the beneficiaries (e.g., sanding, painting, assembly, etc.);
  - Marketing of products via the permanent exhibition in the TC or exhibitions, organized in EL-Kharga city, Cairo or abroad;
  - Repair of machines and sharpening of tools for the beneficiaries.
- A new multi-purpose machine has been especially designed for this project, tailored to the needs of the beneficiaries to work at home. This machine could perform turning, drilling, saw cutting of DPLM, as well as its conversion into rectangular strips. The same machine could be used for turning of any available solid wood in the oases (casuarina, tamarisc, etc.). Forty machines have been locally manufactured: 10 for the TC and 30 for the beneficiaries in houses.

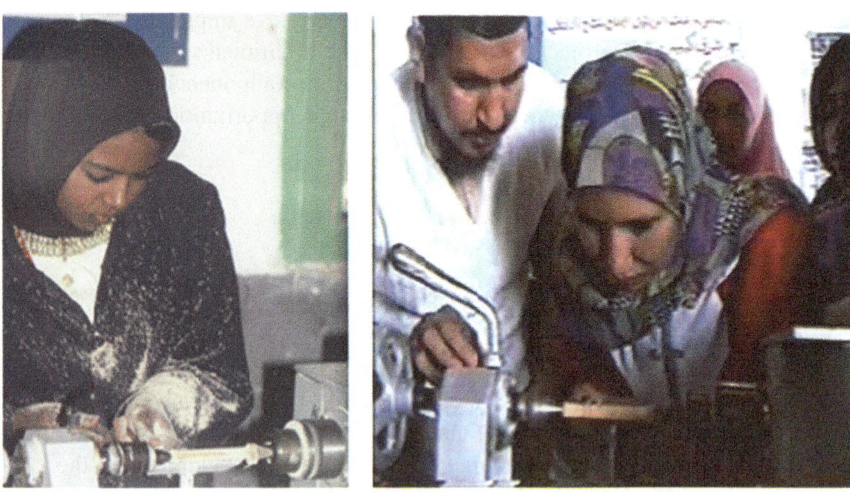

Fig. 6 Mashrabiah (Arabesque) products from palm midribs in the New Valley, Dakhla villages

6.1.5 General Comments

- The beneficiaries liked work on lathes. After 3–4 months women and men previously having no idea about Arabesque were able to attain a high level of craftsmanship. They liked the lathe itself (we called it a machine, so that they may get familiar to it like the sewing machine: being a necessity in each house) and tended it carefully. They asked us to have training on sharpening of tools and on machining of other Arabesque models. They began to dispense with the DPLM, prepared for them in the TC, and relied more on DPLM they obtained from the pruning of palms of their fathers and relatives. We felt that many of them want to be independent from the TC and we encouraged them. Some of them were able to make contracts by their own with customers: visitors to -or inhabitants of -the oases. Two girls and a young man, whom we trained, wanted to make a joint project: but they were careful to find how this may fit with their traditions. Actually we felt that the culture of industry or the industry as a culture is easily and smoothly permeating among oases inhabitants, via the introduction of a new form of utilization (Arabesque) relying on a raw material they are familiar with: DPLM.
- It is interesting here to talk about the response of the oases carpenters. They preferred the DPLM Arabesque to the beech Arabesque coming to them from Cairo. First, the quality of the DPLM-Arabesque was decisively better. the artistic capabilities of especially girls (art being a part of their traditional way of life) and the training built on scientific approach guaranteed better quality. Secondly, it was a cheaper product. Thirdly, the carpenters could tailor the Arabesque item according to their specific designs of the furniture pieces and ask the girls to produce it, whereas they had to adapt their designs to the ready-made Arabesque product coming from Cairo. Fourthly, the carpenters felt that they are supporting local beneficiaries, when purchasing DPLM products. Gradually DPLM-Arabesque items began to permeate through the furniture pieces in El-Dakhla (chairs, cupboards, etc.) and as decorations of the houses of the newly married couples: as if the local community is glad and proud of the new forms of utilization of its indigenous resource: DPLM.
- The selection of the house as a site for the Arabesque handicrafts proved to be very appropriate. The Arabesque lathe (so called machine) became a part of the social life in the houses of the beneficiaries. These are large mud brick houses, where extended families live and are the main place for social life in the oases. We discovered during our field visits that beneficiaries train their relatives spontaneously on their Arabesque machines: sisters, brothers, aunts, etc. Therefore, we coped with this new feature of the project and declared that whoever gets trained on Arabseque by any beneficiary could be given a lathe after passing an examination in the TC. In addition, through beneficiaries working in houses, other new applicants came to the TC asking for training on Arabesque handicrafts.
- One of the most interesting developments that had happened is that beneficiaries (especially women) began to use products of pruning of other fruit or wood trees in their vicinity. Totally independent of the TC we found that they are acquiring new

Fig. 7 Mashrabiah (Arabesque) products from palm midribs in the New Valley villages

habits: collection of the products of pruning (olive trees, lemon, etc.), drying of the raw material in the open courts of their houses and their use for the production of a wide variety of products. This meant that their work with the palm midrib on lathes has inspired them to rediscover other byproducts of fruit trees as substitutes of imported wood that could be used to produce new products.

- In order to widen the scope of maketing of the Mashrabiah (Arabesque) project, beyond the needs of furniture, new designs of modern products (e.g. pen holders and handkerchief case covers (Fig. 7) have been made. These products provided a dimension of independence to benefiaries being able to market their products by themselves apart from the carpenters!

6.2 Strips of Uniform Cross-Section from Palm Midribs

One of the big challenges often met in order to use palm midribs in different applications (e.g. parquet boards, blockboards and lumber substitutes) was how to convert the palm midrib irregular cross-section into strips of uniform cross-section (Fig. 8). For this purpose two families of machines have been designed and manufactured:

(a) Machines producing date palm strips by cutting using carbide-tipped disc saws;
(b) Machines producing date palm strips by applying the principle of skinning (peeling).

Thus, it was possible to produce date palm strips with cross-sections ranging from 4 × 4, 6 × 6, 8 × 8, 10 × 10, 12 × 12 and 14 × 14 mm with different lengths according to the palm species and the required final product.

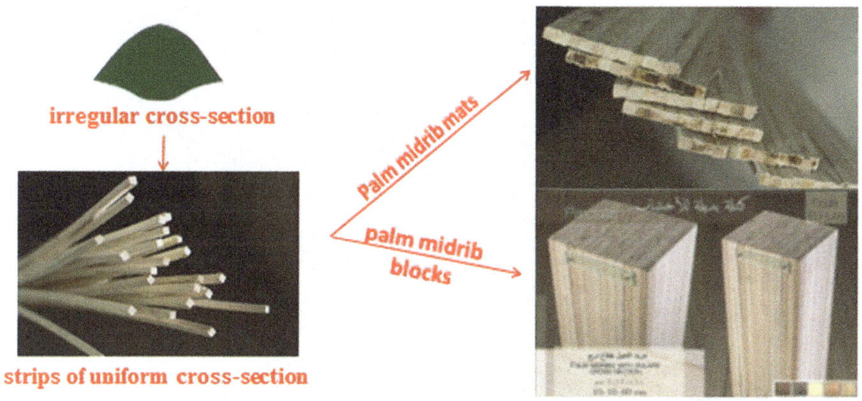

Fig. 8 Strips and boards of uniform cross-section from palm midribs

6.3 Boards of Uniform Dimensions from Palm Midribs Strips

In order to manufacture different products from palm midribs (e.g. parquet boards, lumber substitutes and furniture pieces) it was necessary to assemble date palm strips into a board of uniform dimensions. For this purpose special equipment has been designed and manufactured. Thus the glue (urea formaldehyde) was applied on the side surface of the strips and laid in the aforementioned equipment and the necessary pressure was applied until the resin got cured. Thus, it was possible to produce palm midrib strips boards (Fig. 8) with dimensions 100 mm × 600 mm and varying thickness according to the strips thickness.

6.4 Parquet Boards from Palm Midribs

The idea of manufacture of parquet boards emerged from the necessity of finding a local alternative for manufacture of flooring products basically relying on imported oak, beech and red European pine, the prices of which are soaring with the success of the environmental movements in decreasing the rate of cutting of wood from forests. The parquet boards have been manufactured using for each piece 3 palm midrib boards (mats) of dimensions 700 × 60 mm and thickness 8 mm. After exit from the press the parquet pieces were sent to a parquet factory for preparing the groove-and-tongue assembly and for finishing (Fig. 9). Parquet specimens were tested in the laboratories of the Egyptian Organization for Standardization and Quality (ESO) to measure the changes in dimensions due to change of temperature (30–35–40 °C) and humidity (55, 60 and 80%). The results indicate that the changes in longitudinal dimensions at different temperatures and relative humidity values, as well as

Fig. 9 Parquet pieces from palm midribs before assembly to satisfy a demand of flooring 120 m^2

the changes in thickness and width with increase of relative humidity to 80% are comparable for those for oak and beech. **Thus palm midrib parquet could be used in flooring purposes.**

6.5 Super Strong Material from Palm Midribs

Specimens for tensile tests were prepared from the palm midribs of siwi palm species (one of the most dominant species in Egypt) (El-Shabasy and El-Mously 1997). Four successive layers with thickness of 2.5 mm were cut parallel to the outer surface. Tensile test specimens were shaped from these layers according to ASTM D143/1988. Tensile test was carried out on date palm specimens, as well as specimens from beach, red European pine wood for comparison. It was noticed that the palm midrib outer layer exhibited superior values of tensile strength of value 248 N/mm^2 as compared with the inner layers (69 N/mm^2) (Fig. 10). The density values of the outer and inner layers were found 1.14 and 0.8 gm/cm^3 respectively. This variation can be related with microstructure. The average number of fibro-vascular bundles per square millimeter was about 6 bundles in the surface layer and decreased inward to become less than one bundle in the center. Also the percentage of fibers near the surface reached about 55% and decreased inward to reach about 10% in the central zone. The outer layer of the palm midrib enjoys a specific tensile strength about 1.5 of beech and red pine and more than 4 times that of mild steel. **Thus the palm midrib outer layer is a superstrong material that should be used in special purposes, as for example a reinforcement element in new biocomposites.**

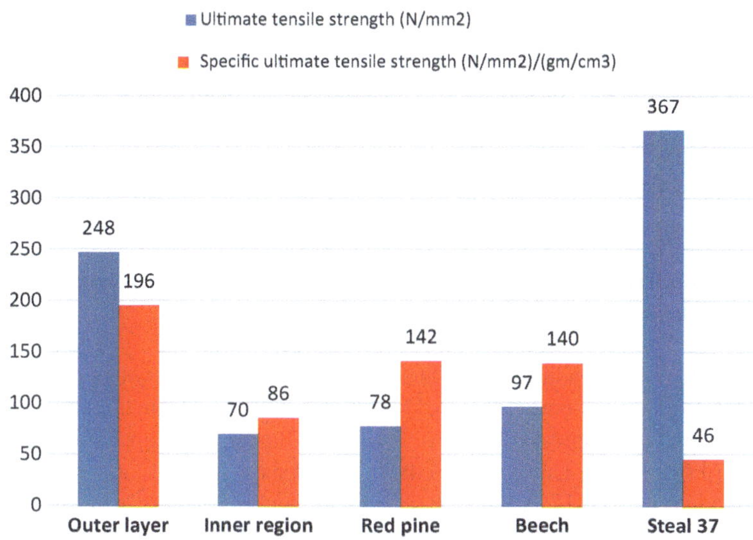

Fig. 10 Comparison of the values of ultimate tensile strength and the specific tensile strength for the palm midrib outer layer and transitional region and the corresponding values for the palm midrib inner region, beech, red European pine, and mild steel (El-Shabasy and El-Mously 1997)

6.6 Palm Midribs as a Core Material in Blockboards

6.6.1 History of Emergence of the Blockboard

The Blockboard as a wood-based panel has emerged for the first time in Germany within the First World War 1914–1918. This innovation was a response to the difficulty of importing of hardwoods, such as beech, oak and mahogany etc., from Africa, India and South America. The main idea was to use local and cheap woods, such as spruce, and cover it from both sides with a thin veneer layer, made from hardwoods. Besides satisfying the European market demands of hardwoods, the blockboard is distinguished with high strength, ease of working and production, as well as low cost compared to solid hardwoods (El-Mously 1993).

6.6.2 Rationale of Use of Palm Midribs as a Core Material in Blockboards

The lumber-core blockboard is a high quality product, used in durable furniture and other applications (e.g. construction industry, containers and wall paneling). People in Egypt prefer to have durable furniture that may not need to be changed quickly. This conservative attitude to furniture has led to the consistent increase of the demand on blockboard 66% of which is being satisfied by importation (El-Mously 1993). Due to its reliance on imported spruce wood, Egypt's blockboard industry is in a very uncertain position. With the advent of GATT and freedom of trade, its position was expected to be worse. Thus, it was found necessary to search for a local substitute for imported spruce wood as a core material for blockboard industry in Egypt. Thus the idea emerged to use the date palm midribs as a potentially cheaper and more sustainably available substitute for spruce wood.

6.6.3 Comparison Between Palm Midribs-Core and Spruce-Core Blockboards

A research has been conducted (El-Kinawy 1997) to compare between palm midrib-core (PMC) and spruce-core (SC) blockboard specimens of 12 mm thickness in terms of physical and mechanical properties. Figure 11 illustrates the appearance of PMC blockboard specimens. The moisture content and density for PMC and SC specimens were found equal to 8.17%, 9.1%, 0.68 gm/cm^3 and 0.57 g/cm^3 respectively. The thickness swelling after 2 h for PMC specimens was found 2%: much lower than that for SC (6.6%). The values of the modulus of rupture, and modulus of elasticity in bending for PMC specimens was found 77% and 85% of the corresponding values of SC respectively. The value of glue bond strength for PMC specimens (0.692 N/mm^2)

Fig. 11 A photograph showing the appearance of DPLM blockboard

was found higher than that for SC (0.558 N/mm^2). **These results prove that it is feasible to use the palm midrib as a core material to substitute the imported spruce wood.** It was also possible in this research to determine the optimum process conditions (8 kg/cm^2 pressure, 120 °C temperature and 5 min pressing time).

6.6.4 Establishment of a Pilot Unit for the Manufacture of Palm Midrib-Core Blockboards

This project has been supported by a grant from the German Technical Cooperation Agency, as well as a grant from the Ministry of local Authority in 1993. Samples of palm midrib-core Blockboards have been sent for testing to Munich Institute for wood Research in Germany. A report has been issued by Munich Institute emphasizing that PMC blockboards enjoy high quality according to German standards DIN and could be thus used as a substitute for traditional spruce-core blockboards: in furniture, wall and ceiling paneling, containers, equipment, etc. During the production running of the pilot unit, PMC blockboards have been purchased by UNICEF for the manufacture of furniture for 100 community schools, established in Asiut, Souhag and Kena governorates. A certificate has been issued and signed by the UNICEF representative in Egypt stating that the production of palm midrib-core blockboard was of good quality and that UNICEF has been very happy to support this initiate as it is in line with the development of appropriate technology.

6.7 Lumber-Substitutes from Palm Midribs

The value of Egypt's imports of wood and wood products in 2014 could be estimated by ~2 billion US$ (El-Mously and Saber 2018). Taking into consideration an annual growth rate of demand on wood of 6.68% the cost of Egypt's import of wood and wood products in 2050 will attain ~22.8 billion US$. Therefore, it is necessary to search for alternatives for solid wood, made from locally available lignocellulosic resources, such as date palm midribs.

Air-dried DPLM of known species, (Tayssier 1996), were converted into strips of 10 × 10 mm cross-section. Lumber-like blocks with cross-section 70 × 70 mm and length 420 mm have been produced in a test rig under 4 different levels of pressure and 3 levels of pressing time. The gluing process was conducted at room temperature using urea–formaldehyde as a resin and citric acid as a hardener. Mechanical properties (Fig. 12) in static bending (MOR and MOE), compressive strength and shear strength parallel to grain, nail-pull and hardness were determined.

The results showed also that, there is a specific set of pressing conditions at which the strength properties attain their highest level. The conditions are 9 kg/cm^2 pressure and 1 h pressing time. For this set of conditions the value of modulus of rupture and modulus of elasticity were found to be 70 N/mm^2 and 5915 N/mm^2 respectively. This

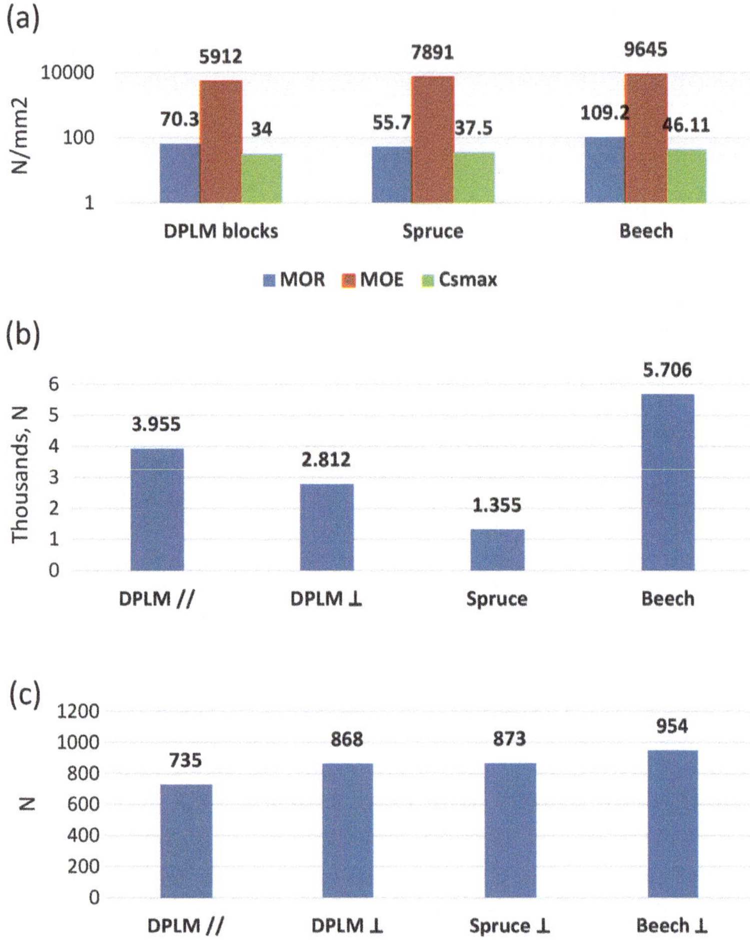

Fig. 12 A comparison between the mechanical properties of palm midirb blocks, spruce and beech wood (Tayssier 1996): **a** Modulus of rupture, modulus of elasticity and maximum crushing strength. **b** Nail withdrawal test. **c** Hardness

research has been published in EUROMAT conference, 1997. It has been granted the HOUWINK-PRIZE of the best poster, presented in EUROMAT 97 conference.

The aforementioned findings open the possibility of manufacture of lumber-like products of any desired dimensions from DPLM as a substitute of solid wood.

6.8 Particleboards from Palm Midribs

In one of the pioneer researches (El-Mousty et al. 1993), one-layer boards were manufactured at El-Nasser Company for the Production of Particleboards and Resins, El-Mansoura from palm leaves' midribs and flax shives. The raw material was produced using Pullman Double Stream Mill. The boards were nominally 122 by 244 by 1.6 cm in size. The screen analysis of palm leaves' midribs and flax shives are listed in Table 5.

The manufacturing conditions were as follows:

1. Binder: commercial urea–formaldehyde, 60% resin.
2. Resin content: 8.2% (based on oven dry weight of particles).
3. Catalyst: 6% ammonium chloride.
4. Mat moisture content: 10%
5. Pressing conditions:

    5.1 Press temperature: 150 °C.
    5.2 Press time: 11 min including 2 min closure time.
    5.3 Pressure: 25 kg/cm^2.

Specimens representing both types of particleboards were prepared from each board for determining the average modulus of elasticity (MOE), modulus of rupture (MOR), internal bond (IB) strength, specific gravity, moisture content and dimensional stability properties. The mechanical properties test specimens were conditioned at 20 °C and 65 RH prior to testing. All tests were conducted in accordance with ASTM D 1037 – 89 (1989), except that the static bending specimens had a span of 30 cm and width 5 cm.

The results of strength properties indicated that MOR of flaxboard (125 kg/cm^2) was higher than that of palm board (106 kg/cm^2). MOE of flax board was lower (17831 kg/cm^2) than obtained for palm board (18513 kg/cm^2). However, there is no significant differences between stiffness (MOE) in both species of particleboard. Internal bond (IB) strength of flaxboard (4.1 kg/cm^2) statistically equals that obtained for palm board (4.3 kg/cm^2). The lower MOR value of palm board is probably due to the higher raw material specific gravity and particle geometry (Vital et al. 1974; Shuler and Kelly 1976; Gertjejansen et al. 1978).

The results of dimensional stability properties indicated that thickness swelling after 2 and 24 h soaking in water of palm boards (22.2 and 25.3% respectively) were higher than obtained for flaxboards (12.7 and 19.1% respectively). Water absorption after 2 h soaking was higher for palm boards (78%) than for flaxboards

Table 5 Screen analysis of palm leaves, midribs and flax shives

Size	Palm leaves' midribs	Flax shives
Coarse %	15–30	15–20
Medium %	50–70	60–80
Fine %	15–35	10–15

(64.9%). However, water absorption after 24 h soaking of flaxboards (104.8%) was statistically higher than palm board (89.6%). Regarding the linear expansion after 2 h soaking, palm boards (0.217%) was higher than that obtained for flaxboards (0.107%). However there is no significant difference between both species in linear expansion after 24 h soaking. The greater thickness swelling recorded by palm boards could be attributed to the larger internal voids in the boards, because of their large particle size (Gertjejabsen et al. 1978). **It could be concluded from the results of this study that palm leaves' midribs is suitable for production of medium density particleboards in the light of increasing demands for inexpensive panel products in Egypt**.

In 1998 two industrial experiments have been conducted on the manufacture of particleboards from date palm midribs (El-Mously 1993). The first was conducted in the Mansourah Particle Board Nasr Company using ~1.15 tons of palm midribs of Siwi species, obtained from the New Valley governorate, to manufacture single layer board of dimensions 2240 × 1220 × 16 mm. The second experiment has been conducted in MATIN Factory of particleboard industry in Alexandria. A quantity of 20 tons of Siwi palm midribs has been brought from Siwa oasis and used to manufacture 3 layer boards of dimensions 4300 × 1830 × 8 mm. Table 6 illustrates a comparison between the results of both experiments. **The results of both experiments show that palm midrib particleboards satisfy the Egyptian Standards for Particleboards No. 0.906, 1991, which opens a wide potentiality for industrial utilization of date palm midribs**.

In a research (Ferrandez-Garcia et al. 2018) rachis of the fronds of three different palm species—date, canary, and Washingtonia palms in a municipality in Spain were obtained by trimming the leaflets. They had an initial moisture content of 74%. They were air dried for 8 months to an approximate moisture content of 8%. Particles were obtained using a laboratory-scale ring-knife chipper, after which they were sieved using sieves of 8, 4, 2, 1 and 0.25 mm. A combination of the fractions, retained on each sieve was used for the panel manufacture: 0.25–1 mm, 1–2 mm and 2–4 mm.

Table 6 Comparison between 2 industrial experiments to manufacture one layer and 3 layer particleboards from palm midribs (El-Mously 1993)

Item	1st experiment	2nd experiment
Palm midrib quantity (ton)	1.15	20
Board type	One layer	Three layers
Board dimensions (mm)	2240 × 1220 × 16	4300 × 1830 × 8
Modulus of rupture (N/mm^2)	20.3	21.9
Density (gm/cm^3)	0.647	0.844
Face strength (N/cm^2)		1.07
Internal bond (N/mm^2)		0.91

Particles were mixed by injection with 8% UF with a 65% solid content in a resin blender for 5 min. The mixture was placed into iron molds 600 × 400 mm to form the mat. Mats were then pressed in a hot plate press under 2.6 MPa pressure for 5 min at 130 °C. After pressing the particleboards were conditioned at 20 °C and 65% relative humidity for one week in a vertical position.

The boards with highest density were achieved with the smallest particle size (0.25–1 mm) for the three species. Concerning the thickness swelling after 2 h the date palm showed the lowest value, whereas the canary showed the highest value. As far as the modulus of rupture is concerned, the canary gave the highest value, whereas the palm gave the lowest. Concerning the thermal conductivity no significant dependence was observed between the particle sizes or palm species. In general using a higher processing temperature and a higher density resulted in the improved mechanical behavior of the particleboard. The addition of higher amounts of UF was found to improve the mechanical properties and to reduce the thickness swelling. This study suggests the possibility of using particleboards as thermal insulation materials, since they have better thermal properties than commonly-used woods with similar densities. The results of this study show, as well, that the physical and mechanical properties of particleboards were influenced with the palm species in addition to other parameters such as the particle size, pressing temperature, pressure, pressing time, raw material, amount of adhesive used, particle moisture and particleboard density.

In another research (Hegazy and Aref 2010), it was, stated that particleboards form about 57% of the total production of wood-based panels and that particleboard consumption is constantly increasing. It was mentioned in the same source that particleboard industry suffers a shortage of raw materials due to deforestation, forest degradation and increasing demand for wood-based panels and that there is a necessity of search for alternative lignocellulosic resources, such as fast-growing tree plantations and agricultural residues. Three fast-growing trees namely Acacia Saligna, Conocarpus erectus and Malia azedarach were taken, as a source of material for particleboard, as well as palm (Phoenix dactylifera L.) midribs. Panels were pressed at two target density levels of 750 and 650 kg/m^3 and a thickness of 13 mm using 10 percent urea–formaldehyde resin.

Based on CEN standards, particleboards should have a maximum thickness swelling (TS) value of 8 and 15% for 2 and 24-h immersions. The average value of the tested specimens after 2-h immersion exceeded this level, except for date palm boards, compressed at the low value level, which attained an average value of 6.3 percent. The same was found for 24-h immersion: the date palm boards, compressed at the low level were the only boards that passed the TS requirement with an average value of 13.7%. As far as the bending stiffness is concerned the studied specimen with a board density 650 kg/m^3 from all the tested species didn't meet the requirements of particleboard standards. However, increasing board density up to 750 kg/m^3 has led to attaining acceptable bending stiffness. It can be thus concluded that at a density of 750 kg/m^3 or higher all tested species, including date palm midribs can be used in the particleboard industry.

In another work (Said and Ahmed 2013) fronds of three palm cultivars, namely: Barhi, Saqie and Sukkari, were collected, and leaflets stripped. After converting into particles they were sieved and the particle sizes: 0.64 mm < D ≤ 0.25 mm and 0.25 < E < 0.12 mm used. These particles were oven-dried to 3% MC and then blended by using UF (50% solid content) at a level of 18% oven dry particle weight). The study focused on the evaluation of the strength properties of modulus of rupture (MOR), modulus of elasticity (MOE) and internal board strength (IB). From the analysis of the results, it is clear that the palm cultivar has a significant effect on all the mechanical (Saqie cultivar has significantly recorded the highest mean values for MOR, MOE and IB) and dimensional stability properties, while particle size had a significant influence only on MOR, IB and linear expansion. Hot water extraction treatment has also a significant effect on all dimensional stability properties and MOE, while panel density has a significant effect on all mechanical properties, as well as linear expansion.

6.9 Medium Density Fiberboards from the Products of Pruning of Date Palms

The medium density fiber boards (MDF) are classified among the lignocellulosic composition panels. This industry relies basically on wood resources. There is a growing trend at the present time to utilize agricultural residues as an alternative to wood in view of the environmental concerns of cutting of wood trees in forests.

Within a project, conducted in Bahariah oases, Egypt (El-Mously and Saber 2018), samples of products of pruning of date palms have been collected with quantities, proportional to the real masses of palm midribs, palm leaflets, spadix stems, coir and palm midrib ends per palm and sent to the laboratories of Nagga-Hammady MDF Company. Tables 7 and 8 illustrate the results of testing of the physical and mechanical properties. The results of testing are as follows.

Table 7 Summary of results of tests of the physical properties of MDF specimens, made from the products of pruning of date palms

The test	Thickness	Density	Moisture content	Thickness swelling	Volumetric Swelling	Formaldehyde emission
Unit	Mm	Kg/m^3	%	%	%	m gm./100 gm
Result	**12**	**752**	**5.2**	**12.7**	**58.04**	**22.54**
Standard deviation	0.04	24.28	11	1.3	2.5	0.87
Followed standard	EN323	EN323	EN322	EN317	BS1142	EN120
Acceptance limits	–	–	4–11	15		≥30

Table 8 Summary of results of tests of the mechanical properties of MDF specimens, made from the products of pruning of date palms

The test	Thickness	Modulus of rupture	Modulus of elasticity	Internal bond	Surface soundness
Unit	mm	N/mm^2	N/mm^2	N/mm^2	Newton
Result	**12**	**24.40**	**2911**	**0.9**	**1.35**
Standard deviation	00.04	3.49	280.52	0.2	0.30
Following standard	EN323	EN310	EN310	EN319	EN311
Acceptance limits	–	20.00	2200	0.55	1.2

Physical and chemical properties

- Humidity (5.2%, which falls within the limits of EN 322 standards: 4–11%);
- Water absorption (12.7%, which is less than the corresponding value in EN 317: 15);
- Formalin emission (22.54 mg/100 mg, which is less than the corresponding value in EN 120: 30).

Mechanical properties

- Modulus of rupture (24.4 N/mm^2, which is higher than the corresponding value in EN 310: 20);
- Modulus of elasticity (2911 N/mm^2, which is higher than the corresponding value in EN 310: 2200);
- Internal bond (0.9 N/mm^2, which is much higher than the corresponding value in EN 319; 0.55);
- Surface strength (1.35 N, which is higher than the corresponding value in EN 311: 1.2).

It can be concluded from the aforementioned results that there is a great potentiality for the establishment of MDF industry in Egypt and the whole Arab region relying as a base material on the products of pruning of date palms.

In another research (Hosseinkhani et al. 2015) Dry process was employed to produce MDF in pilot plant scale using three categories of manufacturing parameters, i.e., two resin types, two resin content levels and three pressing times. Due to the availability of the materials in the pilot plant, date palm and also soft wood fibers (Pinus silvestries L.) as reference were used for the MDF production under the same parameters. Three boards were manufactured with combination of variables, and samples of all boards were tested according to the methods of European (EU) standard. As main technological properties modulus of rupture (MOR), modulus of elasticity (MOE), internal bond strength (IB), and formaldehyde emission were determined. Also, samples of each board were tested for thickness swelling and water absorption after soaking in water for 2 h and 24 h respectively. In addition all the mechanical and physical properties of boards were derived from analysis procedures. The results of this study showed that boards made of date palm pruning residues fibers featured better properties than the MDF property requirements which was recommended by ASTM and EN standards particularly the mechanical properties. In addition, the formaldehyde (HCHO) emission of the panels featured that almost all the boards met the minimum requirement according to EN 120 especially for the panel bonded with MUF resin. **Therefore, Date palm could be proposed as an alternative material for the manufacture of MDF**.

6.10 Oriented Strand Board from Palm Midribs

Strands (Hegazy et al. 2015) with thickness 0.07 cm and a width ranging from 1.25 to 2 cm were prepared from palm midribs of four cultivars: barhi, saqie, khalas and sukkari. Strands were washed by soaking in water at a temperature of 80 °C for 20 h. The bending characteristic, internal bond strength, thickness swelling, water absorption and linear expansion along and across the grain orientation were tested. The internal bond strength values were found acceptable. But the samples manufactured from water-soaked strands had lower mechanical and physical properties as compared with the unwashed samples. The bending properties need to be improved by using different approaches, better resin distribution or modification of the strand size. No significant differences were found between the mechanical properties of panels manufactured from different date palm cultivars.

6.11 Lumber Substitutes from Date Palm Trunks

Three points bending test has been performed on air dried specimens of size 50 × 20 × 450 mm from palm trunk, red European pine and beech (similar specimens were taken for comparison). The span length was 400 mm. The test was performed on a universal testing machine at a crosshead speed of 5 mm/min in compliance with ASTM D1037-72.

The results of tests are shown in Table 9. It is clear from this table that the palm trunk has an overage bending strength amounting to 82.4 MPa, which is 15% lower than red European pine and 36% lower than the beech wood. The value of the Young's modulus in bending of the palm trunk is 6311.6 MPa, which is about 19% lower than red European pine and 31% lower than beech wood. **This opens the potentiality of using date palm trunks as a substitute for imported wood**. This may help in satisfying the basic needs of millions of people in Southern countries of furniture, doors, windows, etc. at a much lower cost and without cutting trees from forests leading to the green house phenomenon and global warming.

Table 9 The average bending strength, and Young's modulus in bending for the date palm trunk and other wood species

No	Sample Name	σ_b (MPa)	E_b (MPa)
1	Trunk	82.412 (13.59)	6311.638 (974.34)
2	RED European pine	97.144 (16.32)	7807.751 (1524.57)
3	Beech	128.52 (29.27)	9176.618 (1212.68)

7 Future Prospects of Use of Date Palm Byproducts for Sustainable Development

Taking Egypt as an example of Arab countries, located in an arid zone with very poor forest coverage and relying on importation of wood to satisfy the population need in furniture and shelter, Egypt has imported in 2014 wood and wood products for ~2 billion US $ (El-Mously and Saber 2018). Taking into consideration an annual growth rate of demand on wood of 6.68% the cost of importation of wood and wood products will reach ~22.8 billion US $in 2050, which represents an intolerable burden on future generations. Thus there is a necessity to find local lignocellulosic alternatives for wood in Egypt and the whole Arab region. There is a strong evidence (see Sect. 6) that the date palm byproducts could represent a sustainable alternative of imported wood. This will relieve the country's balance of payment from the pressure of importation of wood and wood products. Meanwhile the economic utilization of the date palm byproducts will represent a strong motive for regular pruning of date palms, which will increase the production of date, secure the palm plantations from fire accidents and facilitate the combat with the red palm weevil threating the life of date palms. In addition, the economic utilization of date palm byproducts in a wide spectrum of industries (e.g. Mashrabiah products, blockboards, particleboards, MDF, parquet boards and lumber substitutes, etc.) will secure sustainable labor opportunities beginning from rural areas and reaching urban areas. Over and above this measure will stimulate a wave of innovation and create the necessity of building of scientific and technological capabilities nation-wise to rediscover palm byproducts for industrial utilization.

Let us quantitatively compare the date palm products of pruning with traditional aspens forests and short rotation trees. The annual yield of the products of pruning of date palm per hectare is ~8.5 tons[1] (air-dry weight at ~10% moisture content). The corresponding estimation for the traditional aspens and short rotation trees are correspondingly equal to 2.5 ton (Youngquist et al. 1993) and 9.0 ton (Forest Products Society 1999). This means that the lignocellulosic yield of date palm is higher than that for red European pine and near to that for short rotation trees correspondingly. This points to the big economic significance of the products of pruning of date palms.

[1]Taking as an example the siwi species in Egypt and 65 palm per acre the annual yield of products of pruning is equal to: $54.2 \times 65 \times 2.4 \approx 8.5$ tons.

In other words this means that the date palms may be grown in future to obtain two crops: the date palm crop and the lignocellulosic crop (products of pruning of the date palm).

In March, 2013 EU has issued a low prohibiting the importation of any furniture or other timber product made from illegally logged timber to the EU market (Hontelez 2014). This law, together with the increasing concern for the preservation of natural forests world-wide provide a strong competitive advantage-and create a market niche-for tree-free products, manufactured from the products of pruning of date palm.

Demand for timber products (World Bank Group Action Plan 2016) is growing rapidly, with the demand for global industrial round wood predicted to quadruple by 2050 (Indufor 2012). This increase surpasses by a large amount the supply growth, deepening the projected yearly supply deficit from one billion cubic meters in 2012 to 4.5 billion cubic meters in 2050 (World Bank Group Action Plan 2016). This points to the necessity of finding alternatives to wood lignocellulosic resources. The significant advantage of the products of pruning of date palm is that they need no extra land or water or agricultural inputs. They are simply byproducts of date palm, cultivated for obtainment of the date crop. These, products of pruning should be looked at as an additional lignocellulosic crop, obtained for the same investment in land, water, labor and agricultural inputs, made to obtain date.

8 Conclusion

It is clear from the foregoing passages that the date palm byproducts represent a wealth of renewable materials resources, which are sustainably available with huge quantities in many counties in the world (e.g. Egypt and the Arab region).They enjoy (e.g. date palm midribs) mechanical properties compatible with those for known wood spaces. This opens a huge potentiality of production of wood substitutes (e.g. blockboards, particleboards species, MDF, as well as lumber-like products) from date palm byproducts. This will help many countries in the South build their own endogenous scientific and technological capabilities in a wide spectrum of industries providing sustainable labor opportunities in rural and urban areas. Meanwhile, the rational. And economic utilization of dates palm byproducts, will save the environmental in many countries from the open-field burning of these byproducts, decrease the pressure on their balance of payment due to importation of wood and decrease the pressure, as well, on the cutting of trees from forests being one of the main causes of climate change.

References

Actualitix World Atlas—Statistics by Countries (2016) https://en.actualitix.com/country/moor/middle-east-and-north-africa/forest-area-percent-of-land-area-wb-data-btml. Accessed 2020

El-Kinawy MS (1997) A study on the appropriate conditions of the press-cycle for the manufacture soft palm midrib-core blockboards. Dissertation, Ain Shams University

El-Mously HI (1993) The third stage report of the date palm leaves' midribs utilization project as an industrial material. Center for Development of Small-Scale Industries and Local Technologies, Ain Shams University

El-Mously HI (1991) Utilization of date palm leaves midribs as an industrial material. Joint research between Faculty of Engineering and Academy of Scientific Research and Technology

El-Mously HI (1995) Date palm utilization project. Centre for Development of Small-Scale Industries, Ain Shams University

El-Mously HI, El-Morshedy MM, Megahed MM, Abdel-Hai Y (1993) Evaluation of particleboard made of palm leaves' midribs as compared with flaxboard. In: El-Mously H (ed) Proceedings of the symposium industrial utilization of local renewable materials for development, the 4th international conferences on production engineering and design for development, Ain Shams University, Cairo, 28 Dec 1993

El-Mously HI, Saber M (2018) Medium density fiberboards from date palm residues: a strategic industry in the Arab world. In: 1st international conference for palm byproducts (Bypalma), Aswan, 15–17 Dec 2018

El-Shabasy AB, El-Mously HI (1997) Study of the variation of tensile strength across the cross section of date palm leaves' midrib. In: Proceeding of the 5th European conference on advanced materials and applications, Maastricht, 21–23 Apr 1997

Ferrandez-Garcia C-E et al (2018) Physical and mechanical properties of particleboard made from palm tree pruning. Forests 9(12):755

Forest Products Laboratory (1999) Wood handbook—wood as an engineering material. USDA Forest Service, Madison, WI

Gertjejansen RO, Haavik DW, Okoro SPA, Carino HF, Hall HJ (1978) Properties of particleboards from mixture of Philippine hardwoods. For Prod J 28(12):21–27

Hegazy SS, Aref M (2010) Suitability of some fast-growing trees and date palm fronds for particleboard production. For Prod J 60(7–8):599–604

Hegazy S, Ahmed K, Hizairoglu S (2015) Oriented strand board production from water-treated date palm fronds. BioResources 10(1):448–456

Hontelez J (2014) Implementation guide for companies trading FSC-certified materials in the European Union, EU timber regulation

Hosseinkhani H et al (2015) Utilization of date palm (phoenix dactylifera L.), pruning residues as raw material for MDF manufacturing. J Mater Sci Res 4(1):46–62

Index Mundi (2020) https://www.indexmundi-com/Factsarab-world/indicator/AG.LND.FRST.ZS. Accessed 2020

Megahed MM, El-Mously H (1995) Anatomical structure of date palm leaves' midrib and its variation across and along the midrib. In: IUFRO XX world congress, Tampere, 6–12 Aug 1995

Proceedings of the Meeting of the Eastern Canadian Section on the Forest Products Society (1999) Winnipeg, Manitova, Canada

Said SH, Ahmed K (2013) Effect of date palm cultivar, particle size, panel density and hot water extraction on particleboards, manufactured from date palm fronds. Agriculture 5:267–285

Shuler CE, Kelly RA (1976) Effect of flake geometry on mechanical properties of eastern spruce flaketype particleboards. For Prod J 26(6):24–28

Tayssier AM (1996) An investigation into the conditions of manufacture of lumber-like blocks from date palm leaves' midribs. Dissertation, Ain Shams University

Trading Economics (2020) https://tradingeconomics.com/arab-world/Forest-area-percent-of-Land-area-wb.data.html. Accessed 2020

Vital BR, Lehmann WF, Boone RS (1974) How species and board densities affect properties of exotic hardwood particleboards. For Prod J 24(12):37–45

World Bank Group. Action Plan FY 16-20 (2016) International Bank for Reconstruction and Development, The World Bank

World Integrated Trade Solutions (WITS) (2020) https://wits.worldbank.org/. Accessed 2020

Youngquist JA et al (1993) Agricultural fibers in composition panels. In: Proceedings of the 27th international particleboard composite materials symposium, Pullman, Washington State University, 30 Mar–1 Apr 1993

Date Palm Fiber Composites for Automotive Applications

Tamer Hamouda and Nermin M. Aly

Abstract Natural fiber composites are being targeted in several industries particularly the automotive industry because of their environmental and economic benefits, driven by the increasing demands for lightweight, low emissions, high fuel efficiency and recyclability in vehicles. The proper utilization of the available natural renewable resources become crucial for producing lightweight and low-cost applications in the automotive industry. Natural fibers like flax, jute, kenaf, coir, hemp, and sisal have achieved great success in reinforcing composites with different types of polymer matrices. They are widely used in automotive interior components such as door panels, seat backs, dashboards, truck liners, headliners, decking. Recently, there is a growing interest in the uses of the date palm tree, which is one of the most essential agriculture crops found abundantly around the world, mainly in the Middle East and North Africa. Huge amounts of date palm trees biomass are accumulated annually without appropriate use. These quantities are of potential interest to be used in many industries to produce alternative cheap sustainable materials. Date palm fibers and the secondary products of the tress such as palm midrib, leaflets, spadix, stem and mesh have been utilized as promising reinforcing fibers in various composites industrial applications with competitive performance. In this regard, this chapter presents the importance of natural fibers based composites in automotive applications and draws attention to adopting date palm fibers composites in the automotive industry, owing to their superior properties that qualify them to produce sustainable added-value products in this major industry.

Keywords Date palm · Date palm fiber composites · Biocomposites · Automotive applications · Interior and exterior components · Sustainability

T. Hamouda (✉) · N. M. Aly
Spinning and Weaving Engineering Department, Textile Industries Research Division, National Research Centre, 33 El Bohouth St., Dokki, Cairo 12622, Egypt
e-mail: tehamoud@ncsu.edu

M. Midani et al. (eds.), *Date Palm Fiber Composites*, Composites Science and Technology,
https://doi.org/10.1007/978-981-15-9339-0_14

1 Introduction

In the last decades, the natural resources of wood have been decreasing considerably, although the needs for raw wood products are still rising. The environmental concerns about the high consumption of wood sources lead to searching for other resources of wood supplies (Jonoobi et al. 2019). On the other hand, the increased awareness of the negative impacts of using synthetic fibers in products had resulted in finding new efficient green solutions that meet the market's demands, and are coping with the global requirements towards the industrial sustainability. Natural fibers derived from renewable resources are offering a promising alternative to synthetic materials in polymer composites reinforcement, and are attracting much attention as wood substitutes due to their remarkable environmental and economic benefits and recyclability (Alawar et al. 2008). The superior properties exhibited by natural fibers such as abundant availability, low density, high specific strength and stiffness, biodegradability, high acoustic and thermal insulation, lower Co_2 emissions, and low cost are offering great potential to engineers in designing composite materials structures for various applications with enhanced performance (Ghori et al. 2018; Elbadry 2014; Aly 2017).

Natural fibers such as flax, jute, kenaf, coir, date palm, hemp, abaca, sisal, bamboo and palmyra leaf had showed great success in reinforcing polymer composites with different types of polymer matrices (Ghori et al. 2018; Mochane et al. 2019). These natural fibers are used in developing sustainable composite materials and assist in reducing the amount of waste disposal problems and environmental pollution (Al-Oqla et al. 2014). Furthermore, the usage of natural fibers as reinforced bio-fillers in polymer composites had gained much interest lately, as they afford significant advantages such as low density per unit volume, minimal tool wear, low cost and degradable characteristics (Jawaid and Abdul Khalil 2011). Consequently, natural fibers-based composites are extensively used in diverse construction applications such as walls and roofs, furniture, automotive interior components, packaging (Ashori 2008; https://www.grandviewresearch.com/industry-analysis/natural-fiber-composites-market). Also, in several products such as mobile cases, laptop cases, tennis rackets, and snowboards. The global natural fiber composites market size was valued at USD 4.46 billion in 2016, and it is expected to record a compound annual growth rate (CAGR) of 11.8% from 2016 to 2024. The increasing requirements for lightweight products, reduced Co_2 emissions and energy consumption principally in the automotive sector are among the key trends rising natural fiber composites market growth. Figure 1 illustrates the global natural fiber composites market revenue (%) by application in 2016 (https://www.grandviewresearch.com/industry-analysis/natural-fiber-composites-market).

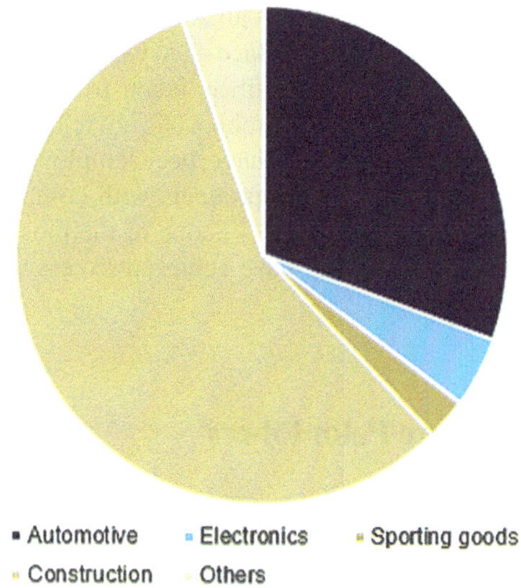

Fig. 1 The global natural fiber composites market revenue by application in 2016 (https://www.grandviewresearch.com/industry-analysis/natural-fiber-composites-market)

1.1 Natural Fibers Composites Performance Concerns

The performance of natural fibers-based composites is depending on several parameters such as the reinforcement fibers, the matrix, and their interfacial compatibility. Nevertheless, there are some drawbacks associated with the use of natural fibers in composites, regarding their behavior in the polymeric matrix and processing (Alawar et al. 2008). The mechanical behavior of natural fibers is highly related to the harvesting conditions which vary according to the climate conditions, seasons and soil conditions. Additionally, their properties are influenced by their degree of maturity, the part of the plant they are taken from (leaf or stem), and the fibers extraction methods such as retting and disintegration (Alawar et al. 2008; Elbadry 2014; Kalia et al. 2009). All of these factors result in considerable variation in the fibers properties compared to their synthetic counterparts such as glass and carbon fibers. Many issues occurred at the interface phase between the fibers and the matrix like; low moisture resistance, poor compatibility with the hydrophobic polymeric matrices caused by the hydrophilic nature of natural fibers, the propensity to form aggregates through processing, quality variations and, low thermal stability (Alawar et al. 2008; Elbadry 2014; Aly 2017; Osoka et al. 2018). This weak fiber/matrix interfacial bonding reduces the mechanical properties and can cause internal strains, porosity, moisture absorption, and environmental degradation (Alawar et al. 2008).

Several studies have investigated the use of various surface modifications methods of natural fibers by different treatments such as alkalization, bleaching, grafting of monomers, acetylation, etc. (Saheb and Jog 1999). To improve fibers surface wettability, enhance compatibility and strengthen the fiber/matrix interfacial bonding, which in turn affords effective stress transfer between the fibers and the matrix (Alawar et al. 2008; Elbadry 2014). The most common process for natural fibers surface modification is alkalization, it enhances fiber/matrix adhesion by the formation of a high fiber surface area, which is needed for fiber/resin optimization (Alawar et al. 2008). Natural fibers taken from agriculture biomass like date palm fibers, bagasse, rice straw, bamboo, etc. are released annually with massive amounts around the world, and they have been employed in various composites applications to produce added-value products with competitive performance as wood substitutes. Date palm fibers (DPF) is one of the most cost-effective available waste materials in agriculture that have achieved excessive success in many industries in the last decades.

2 Date Palm Fibers

Date palm (*Phoenix dactylifera*) is a multipurpose tree due to ancient cultivation, processing and utilization, they are found abundantly in the Arabian Gulf in Saudi Arabia, Northern Africa, Pakistan, India and California in the United States (Ghori et al. 2018). Date palm trees annual global production is about 42%, which is 20% and 10% higher than coir and sisal/hemp fibers production, respectively and cheaper (Alshammari et al. 2019). Every year after date palm cultivation and fruit harvesting, huge quantities of residues are accumulated in the agricultural lands without proper utilization (Ashori 2008). It was found according to El Mously study in 1995 that, a date palm tree produces annually a dry weight of 9.75 kg of midribs, 7 kg of spadix stems, 8 kg of leaflets, and 1.25 kg of mesh (El-Mously 2005). Accordingly, the worldwide date palm residues can be estimated to be about 4200 ton of raw fibers yearly (AL-Oqla and Sapuan 2014). These wastes are mostly left in the agricultural lands or burned, which results in many health and environmental issues, as well leaving these wastes for a long time expose them to risk of being highly flammable. So, the utilization of this natural resource in bio-applications can meet the increasing demands for producing low-cost biodegradable composite materials (Ashori 2008). Each part of the date palm tree can be used for one or more uses. Traditionally, date palm leaves are used in making baskets, mats and ropes in several areas of the world (Ghori et al. 2018).

Date palm leaf is compound and it is divided into; midrib, midrib base and leaflets. The midrib is the center part that carries the leaflets and is referred to as rachis or petiole. The mesh is a naturally woven mat of crossed fibers of different diameters surrounding the stem and is also referred to as tree surface, bark and sheath (Ibrahim 2011; Bourmaud et al. 2017). Date palm leaves are considered as waste materials of date trees and are widely used in different applications (Ghori et al. 2018). Fibers can be extracted from different parts of the date palm tree such as the mesh, the spadix stems, the midribs and the leaflets, Fig. 2 (Ibrahim 2011). The extraction technique of fibers is considered one of the most significant factors to obtain high performance fibers. There are several methods used to extract the fibers either mechanically, chemically or by using biological treatments, each process has its own advantages and drawbacks. Some studies concerned with fibers extractions,

Fig. 2 Date palm trees (left) and their structural parts such as; **a** the spadix stems, **b** the leaflets, **c** the midribs, and **d** the mesh (Elseify et al. 2019) (reproduced)

whereas others worked on separation of cellulose fibers from other constituents such as lignin and hemicellulose. The most common process used is the mechanical extraction and alkaline treatment with NaOH (Ibrahim 2011; Elseify et al. 2019). These secondary products of the date palm tree can be employed and developed in various composites industries.

2.1 *Properties of Date Palm Fibers*

The physical and mechanical properties of date palm fibers are crucial in determining their suitability for different industrial applications. Fibers' length, diameter, aspect ratio (length/diameter), and density, as well as their thermal conductivity and cost are key criteria to determine their potential use and compatibility for different composites applications (Hakeem et al. 2014). Table 1 summarizes some of the physical and mechanical properties of date palm fibers compared with other natural fibers such as coir, sisal and hemp (Ghori et al. 2018; AL-Oqla et al. 2015). Date palm fibers density is one of the significant properties that contribute in reducing composite materials weight to be used in different applications such as automotive and aerospace industries. It was found from the data in Table 1 that, date palm fibers density is lower compared to the other natural fibers (AL-Oqla and Sapuan 2014; Hakeem et al. 2014). The fibers aspect ratio (length/diameter) is closer to coir and slightly greater than hemp, while sisal has a larger aspect ratio compared to all the other types. Thus, date

Table 1 The physical and mechanical properties of date palm fibers compared with other natural fibers (Ghori et al. 2018; AL-Oqla et al. 2015)

Properties	Date palm	Coir	Sisal	Hemp
Cellulose (wt%)	46	32–43.8	60–78	68–74.4
Lignin (wt%)	20	40–45	8–14	3.7–10
Hemicellulose (wt%)	18	0.15–20	10–14.2	15–22.4
Density (g/cm^3)	0.9–1.2	1.15–1.46	1.33–1.5	1.4–1.5
Length (mm)	20–250	20–150	900	5–55
Diameter (μm)	100–1000	10–460	8–200	25–500
Moisture content (%)	5–12.1	8	10–22	6.2–12
Thermal conductivity (W/mK)	0.083	0.047	0.07	0.115
Tensile strength (MPa)	97–275	95–230	363–700	270–900
Elongation at break (%)	2–19	15–51.4	2–7	1–3.5
Tensile modulus (GPa)	2.5–12	2.8–6	9–38	23.5–90
Cost per weight (USD/kg)	0.02	0.3	1	1.2

palm fibers are classified as one of the appropriate fibers for the automotive industry, owing to the ability of producing both continuous and short fibers that are suitable for manufacturing complex parts with an isotropic nature which in turn will enhance the productivity of this industry (AL-Oqla and Sapuan 2014).

On the other hand, the mechanical properties of fibers are influenced by their structure, chemical composition, microfibrillar angle, cell dimensions and defects (Jonoobi et al. 2019). The chemical composition of date palm fibers concerning cellulose, lignin and hemicellulose contents compared with the other fibers is presented in Table 1. The presented data shows that, sisal and hemp fibers have better mechanical properties including high values of specific modulus and strength than coir and date palm fibers because of their higher cellulose contents (Jonoobi et al. 2019; AL-Oqla et al. 2015). Nevertheless, due to the presence of higher cellulose content than lignin in date palm fibers, they offer desired mechanical properties and low moisture absorption compared to hemp and sisal which have an effective role in the automotive industry (Ghori et al. 2018). The microfibrillar angle is found between the microfibrils and the fiber axis which is responsible for the fibrils' mechanical properties. The fibers having longer cell length, higher degree of cellulose polymerization and smaller microfibrillar angle show higher strength and stiffness properties (Jonoobi et al. 2019). Date palm fibers have a moderate value of elongation at break property as it shows an acceptable value of elongation which is three times less than coir fiber and two times greater than sisal (AL-Oqla and Sapuan 2014).

The specific modulus of elasticity with respect to cost ratio has a great influence on choosing the suitable type of natural fibers for various composites applications. It is clarified from Table 1 that, date palm fibers modulus of elasticity is approximately twice that of hemp and more than three times of sisal, which reveals that these fibers are highly preferable in most of the industrial applications mostly automotive industry as a result of having the best stiffness compared to its cost. Automotive applications had shown great attention to the acoustic and thermal insulation properties of the fibers used in the interior design of vehicles. The hollow tubular structure of natural fibers can afford better insulation properties against heat and noise. Moreover, their lower thermal conductivity is a highly desired property for the natural fibers used in automotive applications. It can be noticed from Table 1 that, the thermal conductivity of date palm fibers is lower than that of hemp and still close to sisal, which makes date palm fibers suitable for the insulation purposes in the automotive sector. This assures the potential and feasibility of using date palm fibers in the automotive industry from technical and economic standpoints (AL-Oqla and Sapuan 2014). Furthermore, date palm fibers are employed in other diverse applications such as textile, building and construction materials, sports items, baggage, cabinets, and mats. Also, surface modified date palm fibers are used in producing machinery parts like transmission cloth, air-bag tying cords, conveyor belt cord and some cloths for industry uses (Ghori et al. 2018).

2.2 *Date Palm Fiber Composites*

The development of composite materials reinforced with date palm fibers obtained from different parts of the palm tree using various types of polymer matrices have been adopted in several researches (Ghori et al. 2018). Mahdavi et al. (2010), had examined the mechanical and morphological properties of date palm fiber composites fabricated using fibers from trunk, rachis and petiole of the tree in a high-density polyethylene matrix. Other studies had demonstrated that, the performance of date palm fibers can be enhanced by chemical modifications to reduce the limitations and to improve the compatibility between the fibers and the matrix in date palm composites (Ghori et al. 2018). The influence of different chemical treatments on date palm fibers surrounding the stems of the tree was studied (Mirmehdi et al. 2014; Sbiai et al. 2010). Alawar et al. (Alawar et al. 2009), had investigated the behavior of date palm fibers using different alkali treatments with a range from (0.5–5%) and acid treatment with (0.3, 0.9 and 1.6 N). The results indicated that, chemically treated fibers showed an increase in their tensile strength and considerable morphology improvement.

The usage of date palm fibers as filler in polymer composites was studied in few researches (Almi et al. 2015a, b). The behavior of date palm fibers as reinforcing filler compared with wood flour in polyester/flax woven sandwich biocomposites in a polyester matrix and their influence on the physical and mechanical properties were investigated by Aly and El Nashar (2016, 2019). Virgin and recycled thermoplastics are widely used as matrices for date palm fibers. AlMaadeed et al. (2014), had reported an improvement in the mechanical properties of recycled low-density polyethylene composites filled with date palm wood powder. Also, the effect of natural and artificial weathering degradation on date palm fibers composites in polypropylene (Abu-Sharkh and Hamid 2004) and polyester matrices (Al-Kaabi et al. 2005) were studied. Alodan et al. (2018), had studied the influence of date palm fibers loading with different concentrations on the physical, mechanical and thermal properties of low-density polyethylene composites.

The environmental awareness has led to developing new hybrid composites with more than one type of reinforcements from natural resources. Hybridization comprises a combination of natural fibers and reinforcing filler in a single matrix, which assists in enhancing the mechanical properties of the produced composites (Girijappa et al. 2019). In a study by Rafeeq et al. (2013), date palm fibers and coconut shell particles were used as reinforcing fillers for epoxy matrix composites. The produced hybrid composites exhibited reasonable properties and can be used for low-cost applications. Megahed et al. (2019), examined the properties of starch-based hybrid composite reinforced with chopped randomly oriented flax, sisal and date palm fibers. It was found that, the hybrid composites containing 20 vf % sisal, 5vf % flax and 25 vf % date palm fibers, as well as 35 vf % sisal, 5 vf % flax and 10 vf % date palm fibers showed the optimum mechanical properties and thus can be used as competitive eco-friendly composites for diverse applications.

3 Automotive Industry

The global automotive composites market is valued at 8.37 USD billion in 2018 and it is estimated to reach 18.33 USD billion by 2025 with a compound annual growth rate (CAGR) of 11.85% over the forecast period (https://www.marketwatch.com/press-release/automotive-composites-market-size-2019-share-demand-growth-and-forecast-by-2025-2019-07-04). This was driven by the increasing critical requirements for lightweight, high performance, low emissions, high fuel efficiency, and recyclability in automotive applications (Hakeem et al. 2014). Weight reduction is the most significant cost-effective approach to reduce fuel consumption and greenhouse gases from the transportation sector. It is influenced by the need to comply with European legislation to reduce emissions (from <130 g CO_2/km in 2015 to <95 g CO_2/km by 2021). It was assessed that, for every 10% reduction in vehicle's total weight, the fuel consumption improves by 7% (Ghassemieh 2011). As well as, every 1 kg reduction in vehicle's weight causes about 20 kg reduction of CO_2. Metals account about (30–65%) of the vehicle's total weight, so by optimizing the vehicle's design and materials replacement, weight reduction could be enhanced (Ashori 2008; Ghassemieh 2011). Materials experts from various automakers had estimated that, an all advanced composite auto-body could be (50–67%) lighter compared to a similarly sized steel auto-body with a mass reduction about (40–55%) for an aluminum auto-body and a (25–30%) for an optimized steel auto-body (Koronis et al. 2013). The automotive industry requires suitable compatibility between the material and products to fulfill performance criteria such as ultimate breaking force and elongation, flexural properties, impact strength, fire retardancy, crashworthiness, acoustic and thermal insulation, dimensional stability, suitability for processing temperature and dwell time (Faruk

2006). Also, to cope with the regulation and legislation of environmental and safety issues (Ghassemieh 2011). According to the EU guideline 2000/53/EG issued by the European Commission, 85% of the vehicle's weight had to be recyclable by 2005, and the recyclable percentage should be increased to be 95% by 2015 (Koronis et al. 2013).

Hence, the searching for new sustainable materials with desirable distinctive characteristics can increase the prospects for novel designs in the automotive industry (Hakeem et al. 2014). Safety concepts in the automotive industry are very essential factors to be considered during manufacturing such as crash behavior and penetration resistance. Crashworthiness is the ability to absorb impact energy through controlled failure modes and mechanisms which provides a gradual decline in the load profile during absorption. Whereas penetration resistance is related to the total energy absorbed without allowing projectile or fragment penetration (Jacob et al. 2002). The legislation in automotive design requires that in case of an impact at speeds up to 15.5 m/s (35 mph) with a solid, immovable object, the occupants of the passenger compartment should not experience a resulting force that produces a net deceleration greater than 20 g. The behavior of composite materials differs significantly compared to metals when exposed to crashes. Metal structures are collapsing under crush or impact by buckling and/or folding in accordion (concertina) type including wide plastic deformation, while composites fail through a sequence of fracture mechanisms including fiber fracture, matrix cracking, fiber/matrix de-bonding, delamination, and interplay separation. The damage mechanisms are highly dependent on the composite structure geometry, lamina orientation and type of trigger and crush speed, all of which can be suitably designed to develop high energy absorbing mechanisms (Hakeem et al. 2014).

The automotive industry is continuously working on optimizing the cost versus quality to remain competitive in the market, since cost is a significant consumer driven factor. There is always a comparison between the cost of new material and that presently used in the product. In general, costs comprise of raw materials cost, manufacturing value-added, and cost of designing and testing the products. The increasing incorporation of natural fibers in the automotive industry driven by the environmental issues posed by the non-degradable and non-recyclable contents of salvaged vehicles, had attracted manufacturers attention towards the greening of automotive industry (Ghassemieh 2011; Faruk 2006).

3.1 Natural Fibers Reinforced Composites Applications in Automotive

Natural fibers reinforced composites are the materials of choice for the automotive industry, although their usage is limited to semi-structural and non-structural components in vehicles, due to their load-bearing strength compared to synthetic fibers (Ul-Islam and Butola 2019). Even though, there are important aspects concerning using natural fibers instead of glass fibers in reinforcing composites for vehicle's interior components, they are much safer than glass fibers since there are no-sharp edges fracture surfaces occur in case of crash, the high absorptivity of fibers makes excellent acoustics and air cleaning effect and also the fibers don't cause allergic reactions or skin irritations (Ashori 2008; Mueller 2004). Natural fiber-based composites are used to produce automotive lightweight parts with good mechanical properties to improve fuel efficiency and reduce CO_2 emissions. They contribute in weight reduction by 30% and cost reduction by 20% during vehicle manufacturing. This segment accounted for a revenue share off over 30% in 2015 (https://www.grandviewresearch.com/industry-analysis/natural-fiber-composites-market). In 2012, the total volume of 80,000 tons from different wood and natural fibers are used in 150,000 tons of composites in passenger cars and lorries produced in Europe (90,000 tons of natural fibers composites and 60,000 tons wood-plastic composites). Wood plastic composites represent 38% of total biocomposites used in the automotive industry with hemp 25% and flax 19% (Industry News 2015). Natural fiber composites such as jute, kenaf, hemp, flax, sisal, wood fiber are widely used in the manufacturing of automotive interior components such as door panels, seat backs, dashboards, truck liners, headliners, decking and frames. Moreover, they are employed in the exterior body parts (https://www.grandviewresearch.com/industry-analysis/natural-fiber-composites-market). The growth outlook for bio-fibers in automotive components is expected to increase by 54% annually. Table 2 presents the typical weights of natural fibers used in the current automotive applications (Koronis et al. 2013).

The automotive composites market manufacturing process is classified into compression molding, injection molding, resin transfer molding (RTM), and others. On the other hand, thermoplastic composites offer numerous advantages to the industry such as zero solvent emissions, reduced material scrap, improved work safety conditions, elimination of paintings steps through using high molecular weight polymer surface films, getting rid of tedious production steps via automation, and highly improved recyclability compared to thermoset based composites (https://www.marketwatch.com/press-release/automotive-composites-market-size-2019-share-demand-growth-and-forecast-by-2025-2019-07-04). Several automotive manufacturers are using biocomposites in their components such as Audi, Opel, Daimler-Chrysler, Fiat, Ford, Mercedes-Benz, Peugeot, Volvo, Volkswagen,

Table 2 Typical of natural fiber weights' used in automotive application (Faruk 2006)

Automotive component	Typical weight of fibers (kg)
Front door liners	1.2–1.8
Rear door liners	0.8–1.5
Boot liners	1.5–2.5
Parcel shelves	2.0
Seat backs	1.6–2.0
Sunroof sliders	0.4
Headliners	2.5

and BMW (Ashori 2008). Table 3 presents automotive manufacturers, models, and components using natural fibers (Faruk 2006).

For instance, Mercedes E-class door panels were produced using sisal reinforced epoxy composites. Abaca fibers were used in the reinforcement of the spare wheel pan covers of Mercedes A-class (W169). Figure 3 shows wood fibers that were used in 2006 Mercedes S-class front door lining and driver's seat back rest, while flax fibers were used in the parcel shelves and trunk covers. Toyota used biocomposites

Table 3 Automotive manufacturers, models, and components using natural fibers (Faruk 2006)

Automotive manufactures	Model and applications
Audi	A2, A3, A4, A4 Avant, A6, A8, Roadstar, Coupe: Seat back, side and back door panel, boot lining, hat rack, spare tire lining
BMW	3, 5 and 7 series and others: Door panels, headliner panel, boot lining, seat back
Daimler-Chrysler	A, C, E, S class: Door panels, windshield/dashboard, business table, piller cover panel A class, Travego bus: Exterior underbody protection trim M class: Instrumental panel. (Now in S class: 27 parts manufactured from bio-fibers, weight 43 kg)
Fiat	Punto, Brava, Marea, Alfa Romeo 146, 156
Ford	Mondeo CD 162, Focus: Door panels, B-Piller, boot liner
Opel	Astra, Vectra, Zafira: Headliner panel, door panels, pillar cover panel, instrumental panel
Peugeot	New model 406
Renault	Clio
Rover	Rover 2000 and others: Insulation, rear storage shelf/panel
Seat	Door panels, seat back
Volkswagen	Golf A4, Passat Variant, Bora: Door panel, seat back, boot lid finish panel, boot liner
Volvo	C70, V70
Mitsubishi	Space star: Door panels, Colt: Instrumental panels

Fig. 3 Under floor protection trim of Mercedes, A-class made from abaca fiber reinforced composites (Faruk 2006)

Fig. 4 Spare tire cover made from kenaf fiber reinforced PLA composites for Toyota RAUM (https://www.slideserve.com/vincent/man-made-cellulose-fibres-as-reinforcement-for-polylactic-acid-pla-composites)

made of Polylactic acid (PLA) matrix reinforced with kenaf fibers in the spare tire cover of the RAUM 2003 model, Fig. 4. Biocomposites were used in BMW structural parts like bumpers, fender liners, shields, and suspension system components (Faruk 2006; Akampumuza et al. 2017; Fields 2015). Figure 5 shows door panels of BMW i3 made of kenaf fiber composites (Schmiedel et al. 2014).

Mitsubishi motor used bamboo fibers and a plant-based resin polybutylene succinate (PBS), and the floor mats made from PLA and nylon fibers for its interior components (Matsuda 2008). Hemp fibers are widely used in the door panels and also used in the exterior parts of vehicles. In the ECO Elise concept car launched in July 2008, hemp fibers were used instead of its typical glass fibers in the reinforcement of the composite body panels, the double-curvature fixed hard top and the spoiler, Fig. 6a, b. The visible hemp fibers in the unpainted stripe from the bumper to the spoiler made a striking eco-contrast to the metallic finish which indicates that this car is different. Also, the seats, door panels, shifter boot, horn pad and other interior surfaces were upholstered with special undyed eco-wool and the carpet is woven from sisal fiber (Malnati 2009).

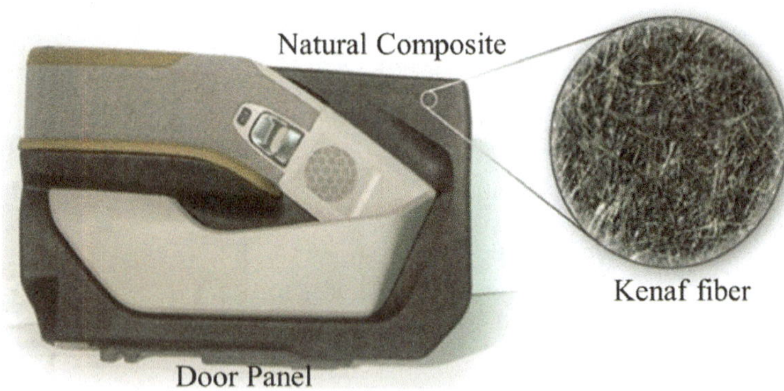

Fig. 5 BMW i3 door panel (left) with enlarged view of kenaf fiber structure (right) reproduced image (Schmiedel et al. 2014)

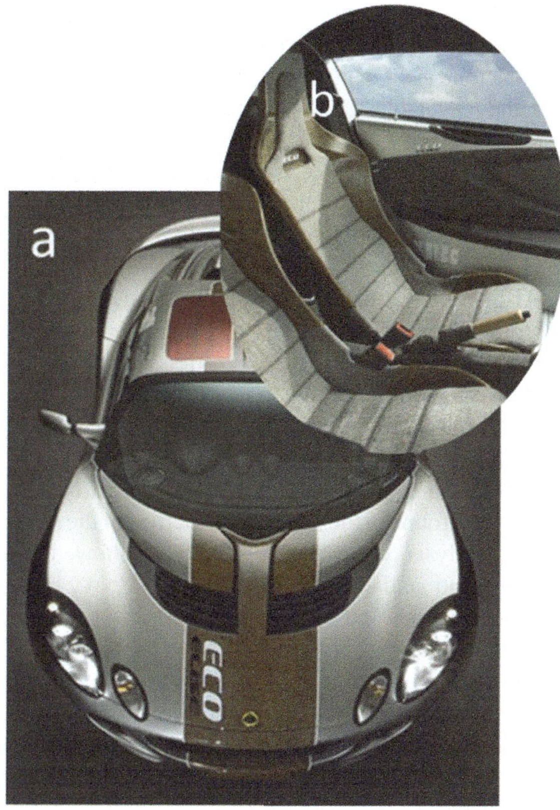

Fig. 6 The ECO Elise concept car sports; hemp fiber reinforced polyester composites are visible in the bold, unpainted bumper-to-spoiler stripe (**a**) and also in parts of the car seats (**b**) (Malnati 2009)

3.2 Date Palm Fiber Composites Applications in Automotive Industry

The employment of date palm fiber composites in the automotive sector will have a significant environmental impact through achieving efficient sustainable waste management practice (Ghori et al. 2018). Several studies had presented the industrial applications of date palm fibers composites based on their physical and mechanical properties suitable for automotive components manufacturing (Al-Oqla et al. 2014). The tremendous properties of date palm fibers such as eco-friendly, performance and cost-effectiveness have attracted the automotive industry to adopt date palm fibers owing to their improved productivity and high sustainability (Ghori et al. 2018). In the university of Portsmouth, a team of researches has developed a new biocomposite material using date palm fibers biomass that can be employed in non-structural parts in automotive, such as bumpers and door linings. Also, the research team included researchers from the University of Cambridge, INRA (Institut national de la recherché Agronomique, a French public research institute dedicated to agricultural science) and the University of Britanny, South. The date palm fiber polycaprolactone (PCL) biocomposite is completely biodegradable, renewable, sustainable and recyclable. The mechanical properties of the DPF/PCL biocomposites were tested and it was found that, they achieved increased tensile strength and better low-velocity impact resistance compared to the traditional man-made composites. This study provided a comprehensive assessment of the improved mechanical properties of DPF/PCL biocomposites (http://www.jeccomposites.com/knowledge/international-composites-news/bio-composite-material-using-date-palm-fibre-biomass).

Alshammari et al. (2019), had investigated the performance of epoxy composites reinforced using 50 wt% of date palm fibers (DPF) extracted from different parts of the tree, leaf sheath (G), palm tree trunk (L), fruit bunch stalk (AA) and leaf stalk (A) as reinforcing fillers. Their mechanical strength, density, water absorption, and morphological properties were investigated. The results demonstrated that, date palm fibers mainly fruit bunch stalk (AA) filler improved the strength and moduli of the fabricated composite considerably compared to the epoxy composites. Similar trends were found in the physical properties of the proposed composites. The morphological analysis of the fractured surfaces showed relatively less fiber pull-out, debonding, and void content within the AA/epoxy biocomposites, which reveals their high performance and improved properties compared to the other composites. Thus, these DPF/epoxy composites can be used as potential cost-effective candidates for lightweight structural interior and certain exterior vehicle parts like package trays, trunk liners, seat backs, racks, spare tire linings, headliner panels, dashboards, back cushions, boot linings, windshields, noise insulation panels, and door trim panels, taking in consideration better mechanical strength and minimum energy absorption tendencies.

Mehanny et al. (2016), had studied the properties of six types of natural fiber reinforced starch-based composites produced using flax, banana, bagasse, date palm fiber, bamboo and hemp fibers as the reinforcements. The biocomposites were fabricated by using compression molding process. The results showed that, the biocomposites exhibited higher tensile strength and modulus of elasticity at fiber weight content from (50–70%). The tensile strength was increased from 2–12 MPa for thermoplastic starch, to reach 55, 45, 32, 44, 365 MPa for flax, bagasse, date palm fiber, banana, bamboo, and hemp composites when fiber content increased to the optimum content (50–70%). In addition, increasing the fiber content and choosing a fiber with high cellulose content improved significantly the biocomposites moisture resistance and thermal stability. So, the properties exhibited by the starch-based high-content natural fiber composites are promising for many paneling applications in automotive.

In another study done by Ibrahim (2011), the properties of date palm fibers composites based on glycerol plasticized potato starch were investigated. Various compositions of date palm fibers (3.5, 6.9, 10, 12.9, 15.6, 18.1, and 22.8% by volume) were used with the same matrix formulation. Tensile strength and 3-points bending tests were carried out to examine the mechanical properties of the proposed biocomposites, also X-ray diffraction was done to determine the change in the crystallinity of potato starch before and after processing, as well as its crystallinity before and after adding date palm fibers. The morphology of the thermoplastic matrix before and after adding date palm fibers was also studied. It was found that, the tensile strength of the biocomposites increased with increasing fiber content up to 22.8% by volume. The tensile strength and young's modulus increased from 0.975 MPa and 40 MPa, respectively for the neat matrix and recorded a value of 6.67 MPa and 800 MPa for the biocomposites with 22.8% by volume of DPF. The crystallinity increased from 27.16% for thermoplastic starch to 36.03% for the 22.8% biocomposites. Also, the electrical resistivity showed an increase from 0.48 MÎ© for the plasticized matrix to reach 3.26 MÎ© for the biocomposites at 22.8% by volume. So, because of these biocomposites improved properties, they could be used in automotive applications.

Recently natural fibers have been used in reinforcing automotive friction composites to replace asbestos fibers, as a result of the health hazards caused by asbestos such as asbestosis, mesothelioma and lung cancer. The gradual reduction of using asbestos in brake friction materials is widely increasing worldwide seeking for the development of safer alternatives. Natural fibers like corn, palm and sugar bars showed relatively high friction coefficient suitable for friction materials. The automotive braking friction pads composed of more than 10 metallic organic filling and binding constituents to achieve maximum value and stability of friction coefficient, solid state lubricity, wear resistance, vibration damping, long life and low maintenance costs (Bakry et al. 2013; Ramadan et al. 2011). Bakry et al. (2013), had investigated the behavior of biocomposites reinforced by corn, palm, and sugar bars fibers (10, 15, 20 and 25 wt%.) to be used as friction materials to replace the conventional friction materials based on asbestos. These fibers were mixed by carbon, barium sulfate, silica, metallic powders and phenol formaldehyde. The new biocomposites could be used as environmentally friendly friction materials for brake lining and clutch facings. Experiments were done to measure both friction coefficient and wear. As well, SEM was used to characterize the wear mechanisms of the biocomposites, their tribological properties were compared with three commercial brake linings.

The results showed that, the addition of corn, sugar bars, and date palm fibers to the composites had increased their friction coefficient and decreased wear. The maximum friction value (0.58) was achieved by composites composed of 30 wt% iron and 25 wt% sugar bar fibers. Corn fibers showed more compatibility with aluminum powder, it gave the highest friction coefficient and relatively lower wear compared to the other biocomposites containing corn. The wear resistance of the biocomposites composed of date palm fibers and aluminum recorded a lower values compared to the biocomposites containing corn and date palm fibers. The lowest wear values were found in the composites composed of 25 wt% corn fibers and 30 wt% aluminum and the composites containing 20–25 wt% sugar bar fibers. The biocomposites composed of date palm fibers and aluminum displayed the highest friction values compared to the other specimens due to the high adhesion of date palm fibers with aluminum. The best wear resistance was exhibited by the biocomposites containing high content of date palm fibers and low content of iron.

The concept of utilization of natural fibers composites in automotive exterior parts is gaining much interest, taking in consideration that the exterior parts are more complex and must be able to withstand extreme weather conditions such as exposure to wetness and chipping (Faruk 2006). Osoka et al. (2018), had investigated the mechanical properties of new biocomposites prepared from three natural fiber sources (Empty plantain bunch fiber, Empty palm bunch fiber and Rattan palm fiber) with two resins polyester and epoxy to be used in automotive applications. It was found that, the biocomposites produced with polyester resin showed the lowest impact strength, about half of that obtained with epoxy. All of the biocomposites produced using epoxy resin achieved impact strength higher than 20 kgfm/cm^2, which is about nine times the impact strength of mild steel used in auto-body parts, with the maximum impact strength recorded for the rattan/epoxy biocomposites. Hence, the produced biocomposites can be used as a substitute for low carbon steel in auto-body parts because of their good impact strength, whereas empty plantain bunch/polyester biocomposites will serve better for structural purposes.

4 Conclusion

Natural fibers are tremendously used in several composites industries due to the increased awareness of the negative impacts of using synthetic fibers. They are being widely employed in the automotive industry in the manufacturing of interior components for a number of passenger and commercial vehicles. Biocomposites reinforced using agricultural wastes are offering great potential to be used in producing sustainable eco-friendly, lightweight, cost-effective and recyclable parts in the automotive sector. Recently, the utilization of date palm fibers in different composites applications had contributed in enhancing the industrial sustainability and wise management of environmentally waste disposable issues. Fibers can be extracted from different parts of the date palm tree such as the midribs, spadix stems, mesh, and leaflets with different physical and mechanical properties. They can be used in reinforcing composites with various types of matrices which in turn will widen the performance properties of the produced biocomposites. As well, their performance can be improved by applying chemical modifications to the fibers. Accordingly, date palm fibers composites had attracted the attention to be incorporated in automotive industry, owing to their versatility, wide availability, price advantage and competitive properties. These biocomposites are promising for many automotive applications such as door panels, trunk liners, seat backs, spare tire linings, friction pads, headliner, etc. Furthermore, the development of new hybrid biocomposites produced using date palm fibers and other types of natural fibers will assist in expanding their applications in the automotive industry.

References

A bio-composite material using date palm fiber biomass (2018). http://www.jeccomposites.com/knowledge/international-composites-news/bio-composite-material-using-date-palm-fibre-biomass. Accessed 5 Jan 2020

Abu-Sharkh B, Hamid H (2004) Degradation study of date palm fiber/polypropylene composites in natural and artificial weathering: mechanical and thermal analysis. Polym Degrad Stab 85:967–973

Akampumuza O, Wambua P, Ahmed A et al (2017) Review of the applications of biocomposites in the automotive industry. Polym Compos 38:2553–2569

Alawar A, Hamed AM, Al-Kaabi K (2008) Date palm tree fiber as polymeric matrix reinforcement, DPF polypropylene composite characterization. Adv Mater Res 47:193–196

Alawar A, Hamed AM, Al-Kaabi K (2009) Characterization of treated date palm tree fiber as composite reinforcement. Compos Part B: Eng 40:601–606

Al-Kaabi K, Al-Khanbashi A, Hammami A (2005) Date palm fibers as polymeric matrix reinforcement: DPF/polyester composite properties. Polym Compos 26:604–613

AlMaadeed MA, Nógellová Z, Janigová I et al (2014) Improved mechanical properties of recycled linear low-density polyethylene composites filled with date palm wood powder. Mater Des 58:209–216

Almi K, Lakel S, Benchabane A et al (2015a) Characterization of date palm wood used as composites reinforcement. Acta Phys Pol A 127:1072–1074

Almi K, Benchabane A, Lakel S et al (2015b) Potential utilization of date palm wood as composite reinforcement. J Reinf Plast Compos 34:1231–1240

Alodan HA, Alsuhybani MS, Alshammari BA et al (2018) Effect of fiber loading on physical, mechanical and thermal properties of low density polyethylene/palm tree waste fiber composites. Sci Adv Mater 10:1341–1350

AL-Oqla FM, Sapuan SM (2014) Natural fiber reinforced polymer composites in industrial applications: feasibility of date palm fibers for sustainable automotive industry. J Clean Prod 66:347–354

Al-Oqla FM, Alothman OY, Jawaid M et al (2014) Processing and properties of date palm fibers and its composites. In: Biomass and bioenergy. Springer

AL-Oqla FM, Salit MS, Ishak MR et al (2015) Selecting natural fibers for bio-based materials with conflicting criteria. Am J Appl Sci 12:64–71

Alshammari BA, Saba N, Alotaibi MD et al (2019) Evaluation of mechanical, physical, and morphological properties of epoxy composites reinforced with different date palm fillers. Mater 12:2145

Aly NM (2017) A review on utilization of textile composites in transportation towards sustainability. IOP Conf Ser: Mater Sci Eng 254:042002

Aly NM, El Nashar DE (2016) Polyester/flax biocomposites reinforced using date palm leaves and wood flour as fillers. Int J Eng Technol 8:1579–1588

Aly NM, El Nashar DE (2019) The influence of using agriculture wastes as reinforcing fillers on hybrid biocomposites properties. Paper presented at the 19th world textile conference (AUTEX), Ghent, Belgium, 11–15 June 2019

Ashori A (2008) Wood-plastic composites as promising green-composites for automotive industries. Bioresour Technol 99:4661–4667

Automotive composites market size 2019 share, demand, growth and forecast by 2025 (2019). https://www.marketwatch.com/press-release/automotive-composites-market-size-2019-share-demand-growth-and-forecast-by-2025-2019-07-04. Accessed 20 January 2020

Bakry M, Mousa MO, Ali WY (2013) Friction and wear of friction composites reinforced by natural fibers. Mat-wiss u Werkstofftech 44:21–28

Bourmaud A, Dhakal H, Habrant A et al (2017) Exploring the potential of waste leaf sheath date palm fibers for composite reinforcement through a structural and mechanical analysis. Compos Part A: Appl Sci Manuf 103:292–303

Elbadry EA (2014) Agro-residues: surface treatment and characterization of date palm tree fiber as composite reinforcement. J Compos 2014:1–8

El-Mously H (2005) The palm fibers for the reinforcement of polymer composites: prospects and challenge. In: The first Ain Shams conference on environmental engineering

Elseify LA, Midani M, Shihata LA et al (2019) Review on cellulosic fibers extracted from date palms (*Phoenix Dactylifera L.*) and their applications. Cellulose 26:2209–2232

Faruk O (2006) Cars from jute and other bio-fibers. Macromol Mater Eng 29:449–457

Fields M (2015) Sustainability at ford, transforming ford into an automotive and a mobility company. http://Corporate.Ford.Com/Microsites/Sustainability-Report-2015-16/Products-Greener. Accessed 10 Jan 2020

Ghassemieh E (2011) Materials in automotive application, state of the art and prospects. In: Chiaberge M (ed) New trends and developments in automotive industry. InTech open

Ghori W, Saba N, Jawaid M et al (2018) A review on date palm (*Phoenix Dactylifera*) fibers and its polymer composites. IOP Conf Ser: Mater Sci Eng 368:012009

Girijappa YGT, Rangappa SM, Parameswaranpillai J et al (2019) Natural fibers as sustainable and renewable resource for development of eco-friendly composites: a comprehensive review. Front Mater 6:1–14

Hakeem KR, Jawaid M, Umer R (2014) Biomass and bioenergy—processing and properties, vol 1. Springer

Ibrahim A (2011) Preparation and characterization of natural composite based on plasticized potato starch and date palm fibers. Master Thesis, American University in Cairo, Department of Mechanical Engineering, Egypt

Industry News (2015) Nova-institute analyzes growth potential of WPC & NFC in European market. https://omnexus.specialchem.com/news/industry-news/nova-institute-analyzes-growth-potential-of-wpc-and-nfc-in-european-market-000176482. Accessed 20 Jan 2020

Jacob GC, Fellers JF, Simunovic S et al (2002) Energy absorption in polymer composites for automotive crashworthiness. J Compos Mater 36:813–850

Jawaid M, Abdul Khalil HPS (2011) Cellulosic/synthetic fiber reinforced polymer hybrid composites: a review. Carbohydr Polym 86(86):1–18

Jonoobi M, Shafie M, Shirmohammadli Y et al (2019) A review on date palm tree: properties, characterization and its potential applications. J Renew Mater 7:1055–1075

Kalia S, Kaith BS, Kaur I (2009) Pretreatments of natural fibers and their application as reinforcing material in polymer composites—a review. Polym Eng Sci 49:1253–1272

Koronis G, Silva A, Fontul M (2013) Green composites: a review of adequate materials for automotive applications. Compos Part B: Eng 44:120–127

Mahdavi S, Kermanian H, Varshoei A (2010) Comparison of mechanical properties of date palm fiber-polyethylene composite. BioResources 5:2391–2403

Malnati P (2009) ECO Elise concept: lean, speedy and green composites technology. http://www.compositesworld.com/articles/eco-elise-concept-lean-speedy-and-green. Accessed 30 Dec 2019

Man-made cellulose fibers as reinforcement for poly(lactic acid) (PLA) composites (2009). https://www.slideserve.com/vincent/man-made-cellulose-fibres-as-reinforcement-for-polylactic-acid-pla-composites. Accessed 20 Feb 2020

Matsuda M (2008) Expectations for biomass-plastics from the perspective of customers. Paper presented at BioJapan 2008, Yokohama, Japan, 15–17 Oct 2008

Megahed H, Emara M, Farag M et al (2019) Enhancement of the mechanical behavior of starch-palm fiber composites. Mater Res Proc 11:201–210

Mehanny S, Darwish L, Ibrahim H et al (2016) High-content lignocellulosic fibers reinforcing starch-based biodegradable composites: properties and applications. In: Poletto M (ed) Composites from renewable and sustainable materials. InTech Open

Mirmehdi M, Zeinaly F, Dabbagh F (2014) Date palm wood flour as filler of linear low-density polyethylene. Compos Part B: Eng 56:137–141

Mochane MJ, Mokhena TC, Mokhothu TH et al (2019) Recent progress on natural fiber hybrid composites for advanced applications: a review. EXPRESS Polym lett 13:159–198

Mueller DH (2004) Improving the impact strength of natural fiber reinforced composites by specifically designed material and process parameters. Int Nonwovens J 13:31–38
Natural Fiber Composites (NFC) market size, share & trends analysis report by raw material, by matrix, by technology, by application, and segment forecasts (2018). https://www.grandviewresearch.com/industry-analysis/natural-fiber-composites-market. Accessed 20 Jan 2020
Osoka E, Onukwuli OD, Kamalu C (2018) Mechanical properties of selected natural fiber reinforced composites for automobile application. Am J Eng Res 7:384–388
Rafeeq SN, Abdulmajeed IM, Saeed AR (2013) Mechanical and thermal properties of date palm fiber and coconut shell particulate filler reinforced epoxy composite. Indian J Appl Res 3:89–92
Ramadan MA, Khashaba MI, Ali WY (2011) Fade and temperature rise of automotive friction materials. J Egypt Tribol 8:53–66
Saheb DN, Jog JP (1999) Natural fiber polymer composites: a review. Adv Polym Technol 18:351–363
Sbiai A, Maazouz A, Fleury E et al (2010) Short date palm tree fibers/polyepoxy composites prepared using RTM process: effect of tempo mediated oxidation of the fibers. BioResources 5:672–689
Schmiedel I, Barfuss GS, Nickel T et al (2014) Use of visible natural fibers in vehicle interiors. Dev Inter 116:20–23
Ul-Islam S, Butola BS (2019) Advanced functional textiles and polymers: fabrication, processing and applications. Wiley

Potential Future Applications of Date Palm Fiber Composites

Anuradha Gupta

Abstract Although, the existence of natural fiber dates back to millions of years, the application in the high performance areas is fairly new and in early adoption stage. The composite market still remains an untapped area and has potential worth billions of dollars. Composite materials characteristically provide enhanced strength or durability over a wide spectrum of high performance technical products. Application areas like automotive, consumer goods, sporting goods, aerospace, construction, and biomedical applications are a few examples where market potential is enormous. This chapter provides an overview of the mechanical properties and performance along with its environmental, economic and social implications of date palm fiber in high performance composite applications. This chapter discusses how adopting date palm fiber rallies sustainability and promotes efficient waste management practices with detailed SWOT (Strength, Weakness, Opportunity, Threats) analysis to highlight different aspects of date palm fiber. Nevertheless, date palm fiber is devoured with limitations which restricts composite applications.

Keywords Date palm fiber · Applications · Sustainability · SWOT analysis · Fiber selection

1 Introduction

The significance of natural fiber in composite application is gaining momentum in the current materials engineering market. This has given boost to the acceptance of plant-based material for composites to replace synthetic fiber (like glass, carbon, aramid). There has been a lot of work published to improve the shortcomings of natural fibers, especially in the last two decade. Natural fibers demonstrate many expedient characteristics like low density which contributes to the lightweight composite material with relatively higher specific strength and stiffness. Additionally, renewable source

A. Gupta (✉)
College of Textiles, North Carolina State University, Raleigh 27606, USA
e-mail: agupta20@ncsu.edu

M. Midani et al. (eds.), *Date Palm Fiber Composites*, Composites Science and Technology,
https://doi.org/10.1007/978-981-15-9339-0_15

and ease of processing with less hazardous manufacturing are some of the advantages inherent to natural fibers. The natural fiber also helps in reducing petroleum dependency in composite manufacturing (Francucci et al. 2014; Khot et al. 2001).

2 Natural Fiber Reinforced Composites—Properties and Applications

Natural fiber reinforcements are a prevalent option for use in composite manufacturing, aiming to reduce the overall carbon footprint. In order to provide application specific and tailorable properties to meet performance requirements. Hemp, flax and date palm fibers have been used as reinforcement in composites due to high specific strength and modulus. They possess a high vibration dampening, high fiber yields, and pest and drought resistance. The production of date palm fiber production is in increasing trend. Egypt along with many Middle Eastern countries like Saudi Arabia and Iraq, produce majority of date palm owing to their hot and arid conditions, which boosts the production.

The comparative performance of the bio-composites is lower than the composites from synthetic materials due to the drawbacks or inconsistencies inherent to the natural fiber. There are some challenges of using natural fiber in composite due to its hydrophilic nature including date palm fiber. Date palm fiber tends to absorb water which can result in swelling of the fiber which lowers the dimensional stability and mechanical properties of their composites. This deterioration of the physical and chemical properties are reflected in the composites as water presence makes the interfacial bonding between fiber and matrix weak. This overall impacts the performance of composites negatively however the moisture absorption can be curtailed to an extent by chemical and physical treatments. These surface modification methods help to improve interfacial bonding of fiber and matrix which is extensively discussed by Bledzki et al. (1999).

In order to produce uniform and optimized fibers for specific sectors like automotive, construction, etc., improved treatment methods can be embraced (Craig 2005). Since date palm fibers are prone to water absorption, suitable low-cost coating and encapsulation methods like acetylation of the hydroxyl groups present in the fiber should be developed to reduce its hydrophilicity. This will minimize fiber swelling and water absorption rate and also improve matrix–matrix bonding (Bledzki and Gassan 1999; John and Thomas 2008; Mohanty et al. 2001, 2002; Netravali and Chabba 2003).

Composite materials are used in various sectors namely automotive, heavy truck, aerospace, civil infrastructure, marine, and durable goods. The main purpose of using composites instead of metal are durability, strength, lightweight that qualify them for a wide range of applications. There has been a tremendous development in the applications of fiber reinforcement composites. The bifurcation of composites usage is shown in Fig. 1, which depicts sector-wise distribution. However, the advent of

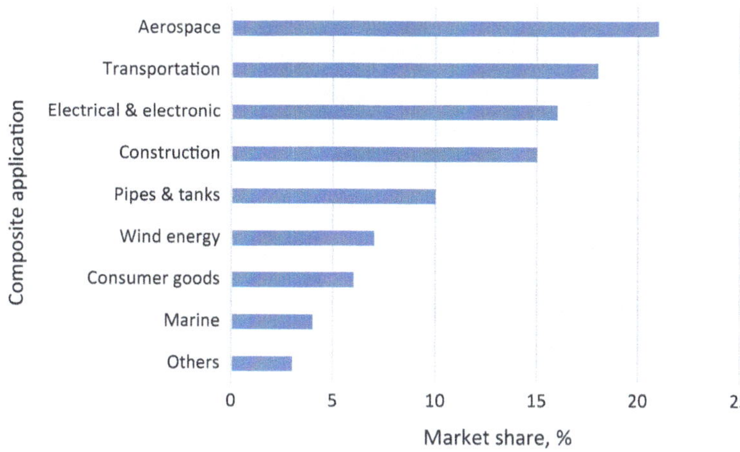

Fig. 1 Sector-wise distribution of the composite market (adapted from Zini and Scandola 2011)

natural fiber composites gives a whole new dimension to the market as they are eco-friendly and easily available. Date palm fiber has the potential to become a major contributor to this market.

Date palm fiber is gaining popularity due to the availability of multiple fiber types and advantages like easy processing, high specific strength and a wide variety of usage. Fiber reinforced composites often aim to improve the strength and stiffness to weight ratios. These properties will reduce the weight of the components produced by the fibers and therefore fibers used for composite materials will have high strength, high flexibility and it is most widely used for textiles and other major fields.

However, natural fiber composites are now replacing many interiors as well as exterior parts of an automobile. The list of interiors and exterior parts are delineated in Table 1.

Advanced composites in aircraft and helicopters are around 20–30% lighter compared to conventional metals. They are used in fairings, landing gears, engine cowls, rudder, fin boxes, doors, floor boards, etc. where metallic and non-metallic

Table 1 Example of interior and exterior automotive parts from natural fiber (adapted from Singh et al. 1998, Mohanty et al. 2002)

Vehicle parts	Natural material
Glove box	Wood/cotton fiber molded, flax/sisal
Door panels	Flax/sisal with thermoset resin
Seat coverings	Leather/wool
Seat/backrest	Coconut fiber/rubber
Trunk panel	Cotton fiber
Trunk door	Cotton with PP/PET fibers
Insulation	Cotton fiber
Floor panels	Flax mat with PP

Fig. 2 Natural fiber composites applications

materials are combined to develop an advanced composite (Prasad and Ramakrishnan 2000). The Advanced Light Helicopter (ALH) which is made up of around 60% composite structure is a good example of PMCs application. However, the most recent one in the field of aircraft is the use of hemp fibers in rotorcraft interiors instead of glass fibers which prove that hemp is comparable in terms of strength, weight, and cost. Hemp being an example of natural fiber, is biodegradable and has a low environmental impact but characteristics like durability, fire retardant, and hydrophilic nature should also be factored in when used in aeronautics (Van Vuure et al. 2015). There are a lot of modern applications of natural fiber composites, especially flax and hemp as shown in Fig. 2.

3 Natural Fiber Composites—Past and Present

The automotive industry is one of the sectors where composites, both polymer-based and natural fiber-based, are used extensively. Nowadays, composites from synthetic fibers are replaced with natural fiber composites due to the regulations being imposed on the manufacturing firm. It started decades back in 1941 when Henry Ford used hemp-based plastic in the body of an automotive (Fig. 3). Replacing metal with hemp significantly reduced its weight and increased impact strength without denting.

Porsche revolutionized the sports cars market by launching a GTS Club sport model of a sports car in 2019 which was made out of natural fiber body parts, see Fig. 4. Hemp and flax fiber reinforced composites were used to manufacture driver and co-driver doors and rear wing replacing previously used carbon fiber. The main advantages of using natural fiber are its low density and high specific stiffness which make lighter composites giving a better fuel efficiency. These natural fibers have many commonalities with carbon fiber and allowed the vehicle a lighter weight of 2910 lb.

Fig. 3 Henry Ford demonstrating the hemp car

Fig. 4 Porsche 718 GT4 sports car made from natural-fiber (hemp & flax) composites

These natural fibers have many commonalities with carbon fiber and allowed the vehicle to be lighter. The motivation for using natural fiber in a sports car was to further improve drivability with faster lap times and usage of sustainable material. Since date palm fiber share similar chemical and mechanical properties like hemp and flax, there is a huge potential of using date palm fiber for sustainable automotive industry. Date palm fiber has higher specific modulus of elasticity to cost ratio compared to hemp, coir and sisal as shown in Fig. 5. This property of date palm has a very high influence in selecting natural fiber for applications. Specific elastic

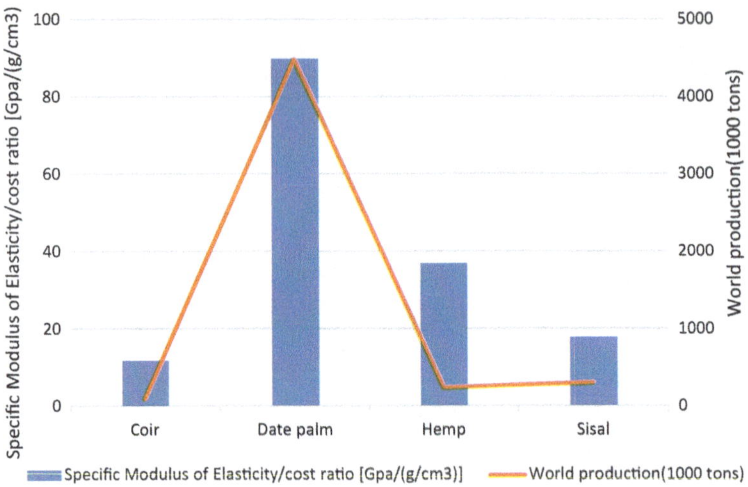

Fig. 5 Specific modulus of elasticity/cost ratio along with world production comparison of natural fiber (adapted from AL-Oqla and Sapuan 2014)

modulus is derived using average values of stiffness over average value of fiber density in (GPa * cm^3)/g. The ratio is twice the ratio of hemp and approximately three times that of sisal. This property can be used in various application especially for industry and automotive sector.

The cost of date palm is low because of the abundance availability of date palm. This will ensure continuous supply of date palm in the long run. Figure 3 shows that the production of date palm is way higher than sisal (about 10 times). This will reduce the dependency of raw material or ensure unhindered supply. As date palm fiber is derived as a byproduct of date palm, it significance is enhanced as it also supports sustainable waste management practice.

4 SWOT Analysis of Date Palm Fiber Composites

The SWOT analysis gives an overview of the strengths, weaknesses, opportunities and threats of natural fiber composites. The purpose of SWOT analysis is not only to gauge the advantages and disadvantages of natural fiber composites but to understand the future prospects and threats. By recognizing all these aspects, it will give a fair understanding of the applications.

Strengths

- Biodegradable
- Global concerns towards sustainability
- Demand for natural fiber and its byproducts

- Low density and high specific strength and stiffness
- Less expensive if produced in abundance
- Low hazardous processing and manufacturing
- Increased flexibility
- Renewable resource and recyclability.

Weaknesses

- Variation in inherent properties
- Poor interfacial bonding
- Low temperature processing
- Low production technologies.

Opportunities

- Demand for light and sustainable materials
- Growth in composite market
- Research focus on natural fiber
- Technological advancement for natural fiber processing
- More sustainable approach by companies..

Threats

- Climate change
- High performance fibers for specialized application
- Pressure to keep the continuous flow of material.

The SWOT analysis shows there are more opportunities and strengths of date palm fiber which can be used in various application like automotive, consumer goods, sporting goods, constructions, etc. Although, there are a lot of challenges as well which need to be catered by continuous research and development, and innovations. The applications of natural fiber in composites have tremendous opportunities but there are threat from many high performance fibers which restricts natural fiber scope.

5 Criteria for Natural Fiber Selection

Natural fibers differ in their mechanical, physical and chemical properties and therefore selecting right fiber as reinforcement is very crucial. This will ultimately affect the functionality of the composites and the subsequent manufacturing processes. There are various criteria in selecting the reinforcement which are addressed in the sections below.

Density—Density is one of the important criteria is selecting natural fiber as it determines the weight of the composites. This is particularly important in an automotive application where low weight contributes towards reduced energy consumption and thereby supports sustainability. In this context, natural fibers are comparable with any synthetic fibers like glass, carbon, and aramid, where values of specific tensile strength and specific modulus of elasticity are considered important. Specific tensile strength and specific modulus of elasticity is the ratio between the mechanical properties (tensile strength and modulus of elasticity) and fiber density. The lower the value of density, the higher will be the specific values and thereby more favored to be used in composites.

Aspect ratio—The fiber length to diameter ratio is known as the aspect ratio. Continuous filaments are stronger and stiffer than short fibers and have a higher aspect ratio than short fibers (Lewin 2006). Short or discontinuous fibers (length > 100 × diameter) are used for bulk production as it is easier to fabricate complex parts and are isotropic in nature (Lewin 2006). Continuous fiber composites (from woven cloth and helical winding) have preferred orientation and single layers of different orientations are stacked together to attain desired strength and stiffness (Campbell 2010). Discontinuous fibers like chopped fibers and random mat have random orientation. High strength composites are produced with small diameter fibers as there are fewer surface defects, more flexibility but are costly to produce. However, it is important to note that if the fiber aspect ratio is too high, the fibers may get intertwined during processing and result in poor composite mechanical strength, due to poor dispersion (De and White 1996). The problem with high-performance synthetic fibers is that they are brittle and tends to break during processing whereas cellulosic/natural fibers are more flexible (Lee 1991). Hence, it is important to know fiber length and fiber length distribution to predict reinforcement strength.

Thermal conductivity—Thermal conductivity is an important criterion when it comes to industrial applications, especially in the automotive industry. In the case of natural fibers, the hollow tubular structure is responsible for thermal and acoustic insulation. Agoudjil et al. (2011), compared the values of thermal conductivity of date palm, coir, hemp and sisal and found out that hemp has the highest thermal conductivity followed by date palm, sisal, and coir. This makes natural fiber a good substitute for interiors in the automotive industry.

Chemical composition of the fiber—Cellulose, hemicellulose, and lignin are three major components of the cell wall of plant fiber. Different fibers have a different composition (Faruk et al. 2012; Madsen 2004) depending on the molecular composition and structure, which in turn determines mechanical, chemical and water absorption properties of the fiber. The more the cellulose content in the fiber, the less is the water absorption capacity of the fiber, hence making natural fiber a feasible alternative in the automotive sector.

Availability and cost of raw material—One of the reasons for selecting natural fiber over any synthetic fiber is abundant accessibility of natural fiber. However, the availability differs from type to type, region to region and different period of time. Also, the cost factor is important in this respect. The cost tends to fluctuate from time to time and therefore necessary steps should be taken to ensure that bulk quantity is purchased in advance for uninterrupted production. Even though the cost of natural fiber is comparatively less than glass or carbon, but there is intense competition within the varieties of natural fibers in terms of cost as illustrated in Fig. 6. It clearly explains that date palm is much cheaper than other varieties depicted in the figure.

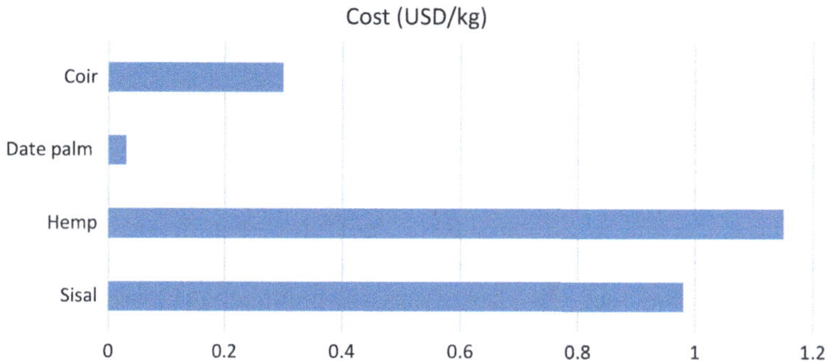

Fig. 6 Cost comparison of different natural fibers used in automotive industries (adapted from AL-Oqla and Sapuan 2014)

Mechanical properties of natural fiber—The value of elasticity, tensile strength, and elongation to break are mechanical properties important in the selection of the suitable reinforcing fibers in automotive applications. However, natural fibers have low mechanical properties but specific values like specific modulus of elasticity and specific tensile strength are comparable to the glass fiber. It is important to develop optimum reinforcements to obtain the desired mechanical properties. This is particularly important in hybrid bio-composites, in which there is a chance to manipulate biodegradable properties. This can be accomplished by right blend ratio of bast and leaf fibers (popularly known as engineered natural fibers), which gives right stiffness and toughness (Mohanty et al. 2002). The specific modulus of elasticity with respect to cost ratio is a critical factor in natural fiber selection. The more is the value of the ratio, the more desirable is the fiber type in most applications (AL-Oqla and Sapuan 2014).

6 Future of Date Palm Fibers

The recent interest in date palm fiber for composites has motivated favorable government policies pursuing light weight sustainable materials that has less reliable on petroleum resources. Although, there are limited applications of date palm fiber in composites, research and innovation will pave path for new opportunities and applications. Along with improving the mechanical properties of fibers using various chemical modifications, innovation in processing technologies may result in further expansion in the area of applications.

As we know, composites are made of reinforcement and matrix. Just replacing reinforcement in the composites does not solve our problem. In order to meet the emerging demands, bio-based resins have been introduced. Being at an early stage, there are tons of opportunities in the resin market. Although, cost is the pressing factor and thereby limits their use in the main stream composite applications. In order to increase their acceptance in industries, their price should be comparable to that of the petroleum-based ones. Research is being conducted at different research laboratories to develop new pathways to synthesize inexpensive biodegradable resins with better mechanical properties ranging from soy-based, corn, sugarcane, lignocellulose, whey and algae.

Natural insulation materials to advance energy efficiency of buildings; plant-derived products acting as carbon sinks to lock up carbon dioxide; recyclable or compostable materials to reduce the landfill crisis; and lightweight green automotive constituents increasing fuel efficiency of cars and plummeting carbon emissions are some of the materials of enormous interest in the future.

The success of date palm fiber depends mainly on the acceptability of the products in the main stream market. A lot of effort should go into educating the consumers about the long term benefits of using natural-fiber reinforced composites. This will play a key role in establishing and expanding date palm fiber market. The most important point to highlight in this regard is the utilization of a by-products, which otherwise goes waste, in various forms to support sustainability. A greater understanding of the date palm fiber by researchers and innovators will also contribute to a greater interest and will continue to result in more products in the future.

7 Conclusion

The advantages of using natural fiber composites are low specific weight, high strength and modulus, low cost renewable source of material, eco-friendly, easy recycle. There are huge unused quantities of agricultural residues around the globe.

It is important to analyze and evaluate all input and output of materials and energy of a product throughout their life cycle focusing mainly on the environmental factors associated with converting date palm fiber into products. This cradle-to grave approach each phase of production systems starting from raw materials, processing, manufacturing, distribution, use and reuse, and final disposal. With the new terminology evolving around circular economy, composite companies are now focusing on eliminating waste and use of renewal resources like high-performance natural-fiber. Date palm fiber composites is one of the promising alternative for composite industry.

Renewable resources offer a limitless supply of potentially sustainable raw materials for the production of biocomposites. Although in its infancy, there is a growing market for biocomposites products and with further technology development, a host of new applications can be anticipated which is not limited to date palm fiber but can also be extended to other natural fiber like flax, hemp, wood fibers, and so on. As discussed earlier, development of bio-polymers will further drive the date palm fiber market growth. Importance should be given to investment in research and development if a sustainable biocomposites industry is to be expanded. Inexpensive, environmentally friendly, and easy production of date palm fiber are attractive benefits for design development and applications. An effective collaboration among scientists, engineers, and designers is vital to achieve quality materials and produce acceptable design, which has benefits al the stakeholders including companies and customer. As a result, this will drive the economy, particularly of those countries where they are drown in abundance.

References

Agoudjil B, Benchabane A, Boudenne A, Ibos L, Fois M (2011) Renewable materials to reduce building heat loss: characterization of date palm wood. Energy Build 43(2–3):491–497. https://doi.org/10.1016/j.enbuild.2010.10.014

AL-Oqla FM, Sapuan SM (2014) Natural fiber reinforced polymer composites in industrial applications: feasibility of date palm fibers for sustainable automotive industry. J Clean Prod 66:347–354. https://doi.org/10.1016/j.jclepro.2013.10.05

Bledzki AK, Gassan J (1999) Composites reinforced with cellulose based fibres. Prog Polym Sci (Oxford) 24(2):221–274. https://doi.org/10.1016/S0079-6700(98)00018-5

Campbell FC (2010) Introduction to composites material. In: Structural composites material. ASM International

De SK, White JR (eds) (1996) Short fibre-polymer composites. Elsevier

Faruk O, Bledzki AK, Fink HP, Sain M (2012) Biocomposites reinforced with natural fibers: 2000–2010. Prog Polym Sci 37(11):1552–1596. https://doi.org/10.1016/j.progpolymsci.2012.04.003

Francucci G, Manthey NW, Cardona F, Aravinthan T (2014) Processing and characterization of 100% hemp-based biocomposites obtained by vacuum infusion. J Compos Mater 48(11):1323–1335. https://doi.org/10.1177/0021998313485266

John M, Thomas S (2008) Biofibres and biocomposites. Carbohyd Polym 71(3):343–364. https://doi.org/10.1016/j.carbpol.2007.05.040

Khot SN, Lascala JJ, Can E, Morye SS, Williams GI, Palmese GR, Wool RP et al (2001) Development and application of triglyceride-based polymers and composites. J Appl Polym Sci 82(3):703–723. https://doi.org/10.1002/app.1897

Lee SM (1991) International encyclopedia of composites, 2nd edn

Lewin M (2006) Handbook of fiber chemistry. Handbook of fiber chemistry. https://doi.org/10.1002/1521-3773(20010316)40:63.3.CO;2-C

Madsen B (2004) Properties of plant fibre yarn polymer composites. Thesis

Mohanty AK, Misra M, Drzal LT (2001) Surface modifications of natural fibers and performance of the resulting biocomposites: an overview. Compos Interfaces 8(5):313–343. https://doi.org/10.1163/156855401753255422

Mohanty AK, Misra M, Drzal LT (2002) Sustainable bio-composites from renewable resources: opportunities and challenges in the green materials world. J Polym Environ 10(1–2):19–26. https://doi.org/10.1023/A:1021013921916

Netravali AN, Chabba S (2003) Composites get greener. Mater Today 6(4):22–29. https://doi.org/10.1016/S1369-7021(03)00427-9

Prasad, R. C., & Ramakrishnan, P. (2000). Composites, science, and technology. New Age International. https://books.google.co.in/books?id=6QsoU6HtQAoC

Singh B, Verma A, Gupta M (1998) Studies on adsorptive interaction between natural fiber and coupling agents. J Appl Polym Sci 70:1847–1858

Van Vuure AW, Baets J, Wouters K, Hendrickx K (2015) Compressive properties of natural fibre composites. Mater Lett 149:138–140. https://doi.org/10.1016/j.matlet.2015.01.158

Zini E, Scandola M (2011) Green composites: an overview. Polym Compos 32:1905–1915. https://doi.org/10.1002/pc.21224

Date Palm Nano Composites Applications and Future Trends

Mohammad L. Hassan, Enas A. Hassan, and Wafaa S. Abo Elseoud

Abstract Since emerging the era of nanomaterials and their fascinating properties, an increasing interest in cellulosic nanomaterials (nanocelluloses) has been initiated all over the world due to the availability of wood and agricultural residues. Palm residues are important agricultural wastes rich in cellulose fibers. In fact, by virtue of their architecture, cellulose fibers is known to consist of nano-elements, so called elementary fibrils, embedded in a matrix of lignin (three dimensional polyphenolic polymer) and hemicelluloses (short-chain polysaccharides). These elementary fibrils have a width as low as 4 nm. Two kinds of nanocelluloses could be isolated so far from cellulosic fibers of different sources including palm residues: cellulose nanofibers and cellulose nanocrystals. These cellulosic nanomaterials have very interesting properties such as their relatively low density, high aspect ratio, high mechanical properties, biodegradability, non-toxicity, transparency in different polymer matrices, in addition to amenability for chemical modification of their hydroxyl groups at their surfaces to produce new nanomaterials. Motivated by these properties, researchers all over the world have used nanocelluloses in various applications such as papermaking, water purification, pharmaceutical products, biomedical materials, electrical and insulating materials, sensors, as well as in solar cells. This chapter is devoted to current state-of-the-art applications of nanocelluloses isolated from palm residues and expected future trends.

Keywords Date palm · Oil palm · Nanocomposites · Water treatment · Papermaking · Drug delivery · Pickering emulsions

---

M. L. Hassan (✉) · E. A. Hassan · W. S. Abo Elseoud
Cellulose and Paper Department, National Research Centre, Giza 12622, Egypt
e-mail: ml.hassan@nrc.sci.eg; mlhassan2012@gmail.com

E. A. Hassan
e-mail: ea.elgarhy@nrc.sci.eg; enasali69@yahoo.com

W. S. Abo Elseoud
e-mail: ws.abouelseoud@nrc.sci.eg; nrc_wafaa2014@yahoo.com

M. Midani et al. (eds.), *Date Palm Fiber Composites*, Composites Science and Technology,
https://doi.org/10.1007/978-981-15-9339-0_16

1 Introduction

There are different kinds of palm trees around the world. Those producing fruits have important economic value since their crops represent significant source of income for some countries. Date palm and oil palm are among these palm kinds; the first produces dates fruit with its high nutritional value while the second produces oil, which is widely used in cooking and making margarine. Huge amounts of palm residues left after harvesting the fruits; these residues are good source for polymeric materials that can be further utilized in value-added products. Namely, cellulose, hemicelluloses, and lignin are the common polymers in these palm residues. Each of these polymers has different properties and applications due to difference in chemical composition and architecture. Cellulose fibers isolated from palm residues represent important secondary product due to their use in strategic industries such as paper-making and cellulose derivatives industries. Furthermore, these fibers could be used during the last years in isolating nanomaterials, so called nanocelluloses, thanks to development of technologies such as ultrasonic, high pressure homogenization, ultra-fine grinding, cryo-crushing, and others, in addition to chemical and enzymatic methods. The produced nanomaterials, namely cellulose nanocrystals and cellulose nanofibers, have new interesting physical and mechanical properties as compared to the micro-scale cellulose fibers. Different parts of palm residues contain significant amounts of cellulose fibers, this includes residues from date palm and oil palm. For example, in case of oil palm residues (Fig. 1), the fronds contain up to 40–50% cellulose, the bunches contain 43–65% cellulose, while the trunk contain 29–37%

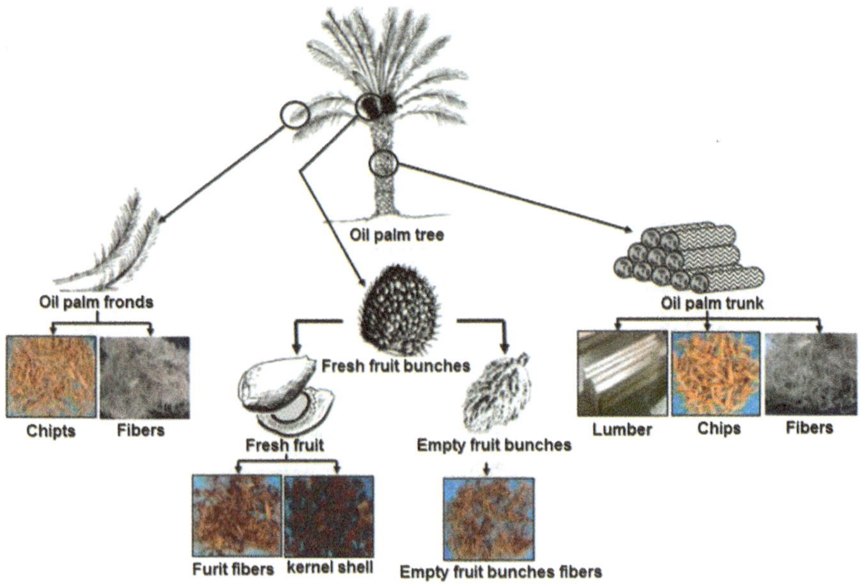

Fig. 1 Different parts of oil palm tree and residues (Dungani et al. 2018)

Fig. 2 Different parts of date palm tree and residues (Alshammari et al. 2019)

cellulose (Abdul Khalil et al. 2017). On the other hand, regarding date palm residues (Fig. 2), date palm leaves contain 35% cellulose, date palm fruit stalks contain 44% cellulose, date palm sheath contains 43.5% cellulose, and date palm trunk contains 40% cellulose (Alshammari et al. 2019). The following parts show the different potential applications of nanocelluloses isolated from palm residues.

2 Applications of Palm-Based Nanocelluloses

High cellulose content in residues from different kinds of palm trees (leaves, fronds, bunches, trunk, and stalks) motivated research in different places in the world to isolate cellulosic nanomaterials (cellulose nanofibers and cellulose nanocrystals) and to apply them in different areas. Cellulose nanocrystals are isolated from cellulose fibers by controlled acid hydrolysis following by washing to remove the excess acid and applying ultrasonic treatment (Mariano et al. 2014). On the other hand, cellulose nanofibers are isolated by mainly mechanical disintegration of the fibers into nanofibers using different technologies such as high-pressure homogenizers, ultra-fine grinding, cryo-crushing, and combination of chemical or enzymatic pretreatments with these technologies (Shaghaleh et al. 2018; Wang et al. 2019).

The early use of cellulosic nanomaterials was mainly focused on their use as reinforcing biodegradable fillers thanks to their high aspect ratio and strength. Later on, due to the naturally occurring functional groups and the amenability for functional modification, wider applications started to appear as shown in the following parts.

2.1 Applications of Palm-Based Nanocelluloses as Reinforcing Fillers in Polymers

Cellulose nanofibers (CNF) and cellulose nanocrystals (CNC) isolated from different parts of palm trees of different species were used with many natural and synthetic polymers as reinforcing elements. This includes their use with hydrophilic and hydrophobic polymers. In the latter case, modification or adding compatibilizers was necessary to get nanocomposites with good mechanical properties. The main idea for the remarkable nanocelluloses role as reinforcing elements is the formation of network within the polymer matrix at critical concentration; this network add strength to polymer martrix at room temperature or even when it is soften by temperature lower than degradation temperature of cellulose. Interfacial interaction between nanocelluloses and polymer matrix is quite important too to ensure good dispersion of nanocelluloses (Mondal 2018) (Fig. 3).

Natural rubber latex is widely used in different applications such as medical gloves and products, threads, adhesives, tires, cushion, automotive industry, packaging, and insulation (Boonkerd 2017). But natural rubber latex suffers from low mechanical properties (low tensile strength and modulus). Since natural rubber comes in form of water-based emulsion, it could be easily mixed with CNF or CNC. Use of cellulosic nanomaterials adds biodegradability to rubber products and also impart them new mechanical properties. Cellulosic nanomaterials from different parts and kinds of palm trees were studied to modify and improve the properties of natural rubber. For example, cellulose nanocrystals and nanofibers extracted from oil palm bunches and

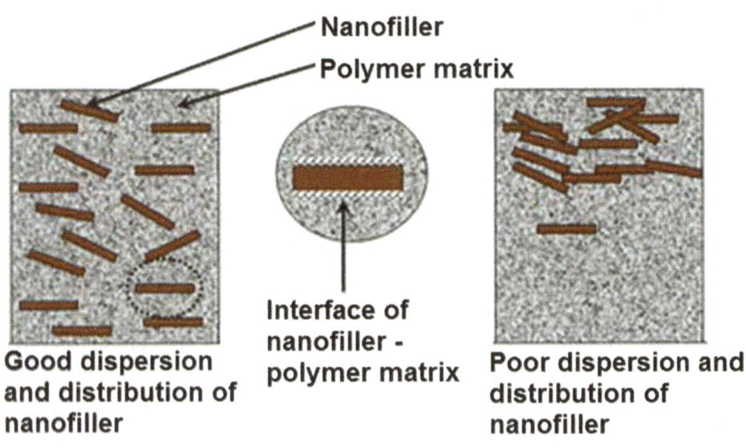

Fig. 3 Dispersion and distribution of nanocellulose fillers in polymer matrices (Mondal 2018)

the rachis of date palm tree were used as reinforcing elements in natural rubber latex films prepared by casting (Fiorote et al. 2019; Ladhar et al. 2017; Bendahou et al. 2009, 2010). Significant improvement in mechanical properties of natural rubber at room temperature and above its glass–rubber transition temperature could be achieved by addition of nanocrystals and nanofibers. The nanofibers resulted in better improvement in mechanical properties of rubber matrix than the nanocrystals due to the higher aspect ratio and possibility of entanglements of the former. The presence of residual lignin, extractive substances and fatty acids at the surface of CNF was also suggested to promote higher adhesion level with the polymeric matrix. In addition to the improvement of mechanical properties of natural rubber due to addition of cellulose nanocrystals isolated from date palm rachis, dielectric properties were found to be improved too (Agrebi et al. 2017). Rubber nanocomposites containing cellulose nanocrystals and nanofibers also showed interesting electrical conductivity properties which qualify the prepared nanocomposites for use as battery separators (Ladhar et al. 2015a, b).

Polylactic acids are biodegradable polymers with interesting properties such as being produced from renewable resources. It is used in making biodegradable plastic films and products for food and biomedical applications as well as in 3D printing due to its melting at convenient temperatures (Divakara and Shetty 2019). However, polylactic acids have low glass transition temperature that makes it unsuitable for use with hot products (Nguyen et al. 2018). Cellulosic nanomaterials are good candidates for reinforcing polylactic acid thanks to the presence of the polar carboxylic groups in the later. Therefore, nanocellulose isolated from palm residues were used to improve some properties of polylactic acid polymers. For example, cellulose nanocrystals isolated from oil palm empty fruit bunch were used at relatively low loadings (1–5%) to improve mechanical properties of polylactic acid films (Simangunsong et al. 2018; Haafiz et al. 2016). However, thermal stability of the nanocomposites decreased with increasing the nanocrystals loading (Haafiz et al. 2016), and addition of nanocrystals did not affect water vapor permeability of the nanocomposites films (Simangunsong et al. 2018). Nanocrystals from the same source were used after surface oxidation to introduce carboxylic groups, to improve mechanical properties of polylactic acid and poly-(3-hydroxybutyrate-co-3-hydroxyvalerate), and to decrease oxygen permeability across polymer films (Dasan, et al. 2017a). Opposite results were recorded in case of using cellulose nanocrystals isolated from TEMPO-oxidized oil palm empty fruit bunch with polylactic acid (Indarti et al. 2016), where no improvement in mechanical properties (tensile strength and modulus) could be achieved in spite of the increasing crystallinity of polylactic acid upon addition the nanocrystals. Poor compatibility between the nanocrystals and polylactic acid was concluded under the conditions used in that work. Cellulose nanofibers isolated from oil palm empty fruit bunches were used with graphene oxide and thyme essential oil to prepare polylactic acid nanocomposite films for active packaging applications. The role of the nanocrystals was found to mainly decrease strain of the films (Salehudin and Muhamad 2018).

Starch is an important commercial natural polymer with many applications. There are different kinds and properties of starch based on source of isolation (Chisenga et al. 2019; Hsieh et al. 2019; Šárka and Dvořáček 2017; Liu et al. 2017; Shevkani et al. 2017). In addition, modified starches are widely available with wider applications than native starch (Vanier et al. 2017; Singh et al. 2010). Use of starch in composites is widely studied in different applications such as in food (Altuna et al. 2018), drug delivery (Gopinath et al. 2018), biodegradable bio-based material for packaging applications (Niranjana and Prashantha 2018), pharmaceuticals (Singh et al. 2010), and in oil field (Zhang 2001). To improve mechanical properties of starch, cellulose nanofibers and nanocrystals isolated from palm residues were widely studied. For example, cellulose nanofibers isolated from sugar palm residues (Arengapinnata sugar palm frond, trunks, and bunches) by high pressure homogenization were used with starch to improve several properties of starch matrix (Atikah et al. 2019; Ilyas et al. 2018, 2019, 2019a). Low nanofibers loadings (up to 1%) were used in the prepared nanocomposites. Not only mechanical properties were improved but also improved water resistance and faster biodegradability of the starch nanocomposites in soil were achieved. Cellulose nanofibers isolated from oil palm empty fruit bunches by mechanical treatment using ultrafine grinder could improve mechanical properties of polyvinyl alcohol/native starch with different amylose content (Lisdayana et al 2018; Fahma 2017a). Addition of 3% of the nanofibers brought about the maximum increase in mechanical properties. CNF isolated from the same source but using unbleached pulp, i.e. containing lignin, was used for improving mechanical properties of starch films and also add resistance to water sorption properties (Ago et al. 2016); the starch/cellulose nanofibers nanocomposites could be used as alternative to the fossil-oil based and non-biodegradable polystyrene foams. Regarding the use of cellulose nanocrystals, those extracted from oil palm empty fruit bunch were used at low weight ratio (2 wt. %) to improve tensile strength and modulus of starch (Salehudin et al. 2014). Higher ratios of cellulose nanofibers in the starch matrix resulted in decreasing the mechanical properties, probably due to interruption of hydrogen bonding between starch chains. In another work, cellulose nanocrystals isolated from the same residues were used in loadings from 0.5 to 5% as reinforcing material for tapioca starch crosslinked with citric acid (Owi et al. 2017). Mechanical properties were improved up to 3% loading of the nanocrystals and declined thereafter. Cellulose nanocrystals isolated from oil palm mesocarp fibers were used as reinforcing filler for cassava starch at loadings up to 10% (Campos et al. 2017); optimum loading was at 6% of nanocrystals.

Polyvinyl alcohol is water soluble synthetic polymer and has the ability to form films with good mechanical properties, resistant to oil, and good air barrier properties. It is non-toxic, biocompatible and biodegradable too. These unique properties of PVC polyvinyl alcohol reflected in its wide use in different areas including electrochemical devices (Mokhtar et al. 2016), cement materials (Thong et al. 2016), biomedical products (Baker et al. 2012) and in drug systems (Pluta and Karolewicz 2001). It was also used in textiles (Akbari et al. 2016) and flexible electronics (Huynh and Kutner 2015). To enhance its properties, cellulosic nanomaterials from palm residues and others were used as reinforcing materials. For example, cellulose nanofibers isolated from oil palm empty fruit bunch unbleached pulp were used as reinforcing agent in polyvinyl alcohol/chitosan composites (Solikhin et al. 2018). The isolated nanofibers had lignin and highly amorphous. Nanofibers loading up to 0.5 wt.% resulted in increasing mechanical properties of the films while at higher loadings the mechanical properties declined due to strong aggregation of the nanofibers and formation of non-smoothed wrinkled films.

TEMPO-oxidized cellulose nanofibers isolated from oil palm empty fruit bunches were used at loading from 0.5 to 6 wt. % to reinforce polyvinyl alcohol (Asad et al. 2018). The optimum improvement in mechanical properties was achieved at 4% of the nanofibers. Addition of the nanofibers to polyvinyl alcohol improved its thermal stability; above 2% loading, reduction of its melting point was recorded. Cellulose nanofibers from oil palm empty fruit bunches were used to reinforce polyvinyl alcohol and the effect of sonication time of polyvinyl alcohol/cellulose nanofibers mixture before casting was studied. The maximum increase in mechanical properties was achieved at sonication time up to 9 min. Cellulose nanofibers isolated from same residues were used at 3% loading with nanosilica to reinforce polyvinyl alcohol (Ching et al. 2015). Addition of 0.5% nanosilica with cellulose nanofibers resulted in improving mechanical properties of polyvinyl alcohol films. Cellulose nanofibers isolated from same residues were also used to reinforce polyvinyl alcohol/starch blend films. Addition of 5% (w/v) cellulose nanofibers showed the best mechanical properties, water resistance, and biodegradability (Lani et al. 2014).

Acrylate polymers are a wide category of materials since they have different structures depending on the monomeric units used in their synthesis (Mittal 2016). Their properties are widely varied too depending on the nature of monomers used. Therefore, their use also varies to include many areas. Cellulose nanofibers and nanocrystals from palm residues were used with different acrylate polymers as reinforcing elements. For example, cellulose nanofibers and nanocrystals isolated from rachis of date palm tree were used with polyacrylic acid latex consisting of butyl acrylate and styrene (Boufi et al. 2014). The higher aspect ratio of cellulose nanofibers resulted in higher mechanical properties than in case of using cellulose nanocrystals. But it was noticed that the transparency of nanocomposites decreased is spite of the smaller diameter of nanofibers used indicating their agglomeration in the films. Cellulose nanocrystals from rachis of the date palm tree were also used with poly(styrene-co-2-ethyl hexylacrylate) copolymer and methacryloxypropyl triethoxy silane compatibilizer for making stable suspension and then films by casting (Ben Mabrouk et al. 2011). Addition of both of cellulose nanocrystals at less than 3 wt.% and the coupling agent at 1 wt. % resulted in enhancement the mechanical properties at the rubbery state of acrylate polymer. The prepared nanocomposites showed also interesting electrical properties.

Polyurethanes are another important class of polymers. They are prepared by reaction of polyol compounds with diisocyanate compounds. Since a wide variety of polyol and diisocyanate compounds are available and can by synthesized. Many different kinds of polyurethanes with different properties and applications exist (Akindoyo et al. 2016). Products vary from flexible and rigid foams (Kausar 2018), coatings and membranes (Joshi et al. 2018), biomedical products (Wang et al. 2012), and for water treatment (Lemos et al. 2007). Use of cellulosic nanomaterials from palm residues as biodegradable natural reinforcing materials for different kinds of polyurethanes was studied. For examples, cellulose nanocrystals isolated from date palm rachis were used in ratios from 0.4 to 2.4% as reinforcing elements to improve mechanical properties of rigid polyurethane prepared from 80:20 polyether and palm-kernel oil based polyester polyols (Septevani et al. 2018). Addition of cellulose nanocrystals at 0.4% improved heat insulation properties of polyurethane matrix (Septevani et al. 2017). Cellulose nanofibers isolated by high intensity homogenizer and cellulose nanocrystals from date palm rachis were used to improve mechanical of polycaprolactone diol-based polyurethanes (Benhamou et al. 2015). Cellulose nanofibers imparted polyurethane better mechanical and thermal properties than nanocrystals and also showed better dispersion in polyurethane matrix.

Epoxy thermoset resins are important polymers with wide applications and uses such as in paints, adhesives, coatings, sealants, flooring, and making plastic composites (Jin et al. 2015). However, cured epoxy resins suffer some shortcomings for variety of advanced applications such as its notably poor thermal and viscoelastic properties. Cellulosic nanomaterials can enhance the properties of epoxy resins since the later are rich in polar functional groups which have excellent compatibility with hydroxyl groups of cellulose. For example, cellulose nanocrystals isolated from oil palm empty fruit bunch fibers at loadings from 1 to 5% were used to enhance dynamic mechanical properties of epoxy composites and resulted in light weight and thermally stable (lower thermal expansion coefficient) composite structural materials at 3% nanocrystals loading (Saba et al. 2017, 2019). In a different study, cellulose nanofibers isolated from the same palm residues were used at lower loadings (up to 0.75wt.%) to reinforce mechanical properties and thermal stability an epoxy matrix (Ireana et al. 2016); the optimum ratio of cellulose nanofibers was 0.5%. Nanoparticles isolated from oil palm empty fruit bunch after chemical treatment with stannous chloride and bromine water followed by using cryogenizer and high energy ball milling were used with epoxy resin to reinforce a non-woven mat made from kenaf fibers. Montmorillonite and organically-modified montmorillonite were used for comparison too. Addition of cellulose nanoparticles at 3 wt.% loading improved thermal stability of the nanocomposites and was equivalent to the addition of montmorillonite but less than in case of using the organically-modified montmorillonite. The improved thermal stability of the nanocomposites in case of using cellulose nanoparticles was due to the thermal stability of the cellulose nanoparticles (Saba et al. 2015). Addition of cellulose nanoparticles, montmorillonite and organically-modified montmorillonite to the epoxy matrix resulted also in increasing mechanical properties of the prepared nanocomposites as concluded from dynamic mechanical analysis results (Saba et al. 2016a, b).

Regarding the use of polyethylene and polypropylene, the most popular plastic used worldwide, cellulose nanofibers from oil palm mesocarp fibers were isolated in a specially designed extruder and mixed at the same extruder with polyethylene, i.e., one-pot process, to prepare nanocomposite films with similar mechanical properties to those prepared by the conventionally followed two-step protocol for making cellulose nanofibers/polyethylene nanocomposites, i.e., isolation of cellulose nanofibers first using different technologies, then mixing of nanofibers with polyethylene in an extruder (Yasim-Anuar et al. 2019).

There is a recent interest in isolating nanofibers from unbleached pulps not only for economic and environmental benefits but also for producing nanofibers with different surface properties due to presence of lignin. Cellulose nanofibers with lignin have different surface properties than those isolated from bleached pulps, and thus improved compatibility with hydrophobic polymers could be expected. Freeze-dried cellulose nanofibers with 9% lignin were isolated from empty palm fruit bunch fibers and used for reinforcing polypropylene in the presence of soy protein isolated, hydroxypropyl cellulose, and maleated polypropylene as coupling agents (Ferrer et al. 2016). Maximum improvement in mechanical properties was 15% upon addition of 1% of the nanofibers in the presence of soy protein isolate as a coupling agent.

In addition to the aforementioned polymers matrices, others were tested with nanocelluloses isolated palm trees residues. For example, cellulose nanocrystals isolate from oil palm shells were used as reinforcing elements in seaweed matrix at up to 30% nanocrystals loading (Abdul Khalil et al. 2017). Maximum reinforcing was achieved at 20% loading but the hydrophobicity and elongation at break of the films were decreased. Cellulose nanofibers isolated from oil palm empty fruit bunches were used after its acetylation as reinforcing elements in polychloroprene rubber matrix at nanofibers loading up to 5 wt.% (Fahma et al. 2014, 2014a). Polychloroprene rubber is used for gaskets, cable jackets, tubing, seals, O-rings, tire-sidewalls, gasoline hoses and weather-resistant products, electrical insulators, coatings and adhesives (Martín-Martínez 2005). Due to the presence of the hydrophobic acetyl groups of the modified nanofibers, improvement in tensile strength and modulus of polymer matrix took place as compared to the non-modified nanofibers. In addition, the nanofibers resulted in higher storage modulus of polychlorprene polymer matrix above its glass transition temperature. The optimum ratio of the acetylated nanofibers was at 3 wt.%. Cellulose nanofibers isolated from oil palm empty fruit bunches by chemi-mechanical method (Gea et al. 2018) were used to improve mechanical properties of polycaprolactone, one of the biodegradable synthetic polymers with different applications in packaging (Lasprilla-Botero et al. 2018; Yun et al. 2017) and biomedical and pharmaceutical applications (Espinoza et al. 2020; Wang et al. 2018). Nanofibers loadings up to 60% were used in polycaprolactone polymer and the optimum ratio that gave the highest mechanical properties was 20%.

2.2 *Applications of Palm-Based Nanocelluloses in Water Treatment*

Cellulose nanofibers and nanocrystals enjoy properties made them fit well in applications related to water treatment. The high surface area of cellulosic nanomaterials and abundant functional groups as well as the wide possibility of introducing different functional groups make them very promising materials for water treatment applications. In addition, the possibility of forming network with tunable porosity make

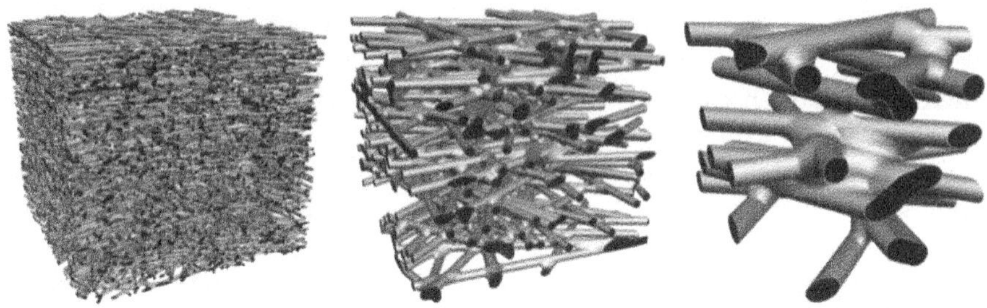

Fig. 4 Illustration of the correlations between pore size and fiber diameter at a constant porosity of 80% in a fixed volume (Ma et al. 2011)

cellulose nanofibers or nanocrystals good candidates for ultra- and nanofiltration membranes (Fig. 4, Ma et al. 2011).

In area of water treatment, cellulose nanocrystals isolated from oil palm empty fruit bunch were used as a biosorbent for textile effluent contaminant remediation from methylene blue dye. Adsorption capacity of the dye by the nanocrystals reached 50.9 mg/g at nanocrystals dosage as low as 0.066 mg/ml (Shanmugarajah et al. 2019). Cellulose nanocrystals isolated from the same source were used for cadmium ion removal from aqueous solution at different pH value; the maximum removal was at pH7 (Lim et al. 2016).

Cellulose nanocrystals from date palm leaves were used with polyamide to form electro-spun fibers for making high-efficiency membranes for vegetable and diesel oil extraction from oil/water mixtures (Sobolčiak et al. 2017). Addition of nanocrystals resulted in significant increase in mechanical properties of polyamide films; increase in Young's modulus by 224% and the tensile strength by 110%, as well as increasing the hydrophilicity of polyamide were achieved by addition of 1% cellulose nanocrystals.

Cellulose nanofibers isolated from date palm fruit stalks by ultrafine grinding were used to make thin film membranes by simple vacuum filtration with porosity suitable for ultrafiltration applications for removing oil from oil-in-water emulsions and bacteria (Hassan et al. 2017, 2017a) (Figs. 5, 6 and 7).

Another approach for preparation of nanocellulosic composites for water treatment application was via in-situ synthesis of zinc oxide nanoparticles in the presence of cellulose nanocrystals isolated from oil palm empty fruit bunches (Lefatshe et al. 2017). The presence of ZnO nanoparticles with cellulose nanocrystals imparted the prepared nanocomposite films photocatalytic activity and capability of degrading methylene blue dye in water solution. The nanocomposite films showed also antibacterial activity against Gram-positive *Staphylococcus aureus* and Gram-negative *Escherichia coli* (Fig. 8).

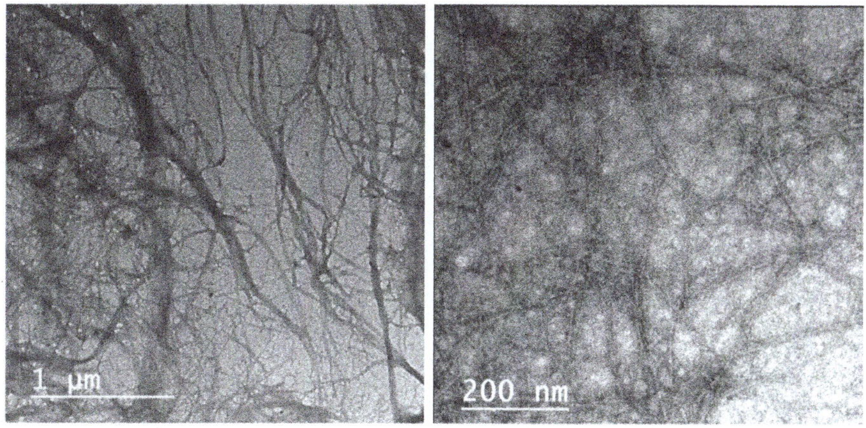

Fig. 5 TEM of cellulose nanofibers (**left**) and TEMPO-oxidized cellulose nanofibers (**right**) isolated from date palm fruit stalks (Hassan et al. 2017a)

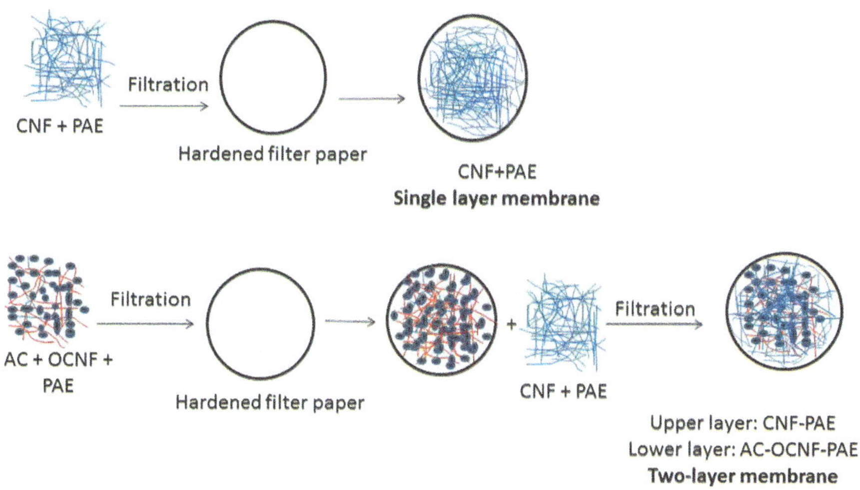

Fig. 6 Preparation of single layer date palm cellulose nanofibers (CNF) and two-layer CNF/AC (activated carbon) membranes on hardened filter paper (Hassan et al. 2017a)

2.3 *Applications of Palm-Based Nanocelluloses in Papermaking*

Paper industry uses a wide variety of additives for improving specific properties of paper products and/or imparts them new properties. The properties of nanocelluloses qualify them for different roles in papermaking industry. The most favorable property of nanocelluloses is that they are of the same kind of main paper material (cellulose

Fig. 7 SEM image of oven-dried CNF/AC membrane after filtration of *E. coli* bacteria suspension (Hassan et al. 2017a)

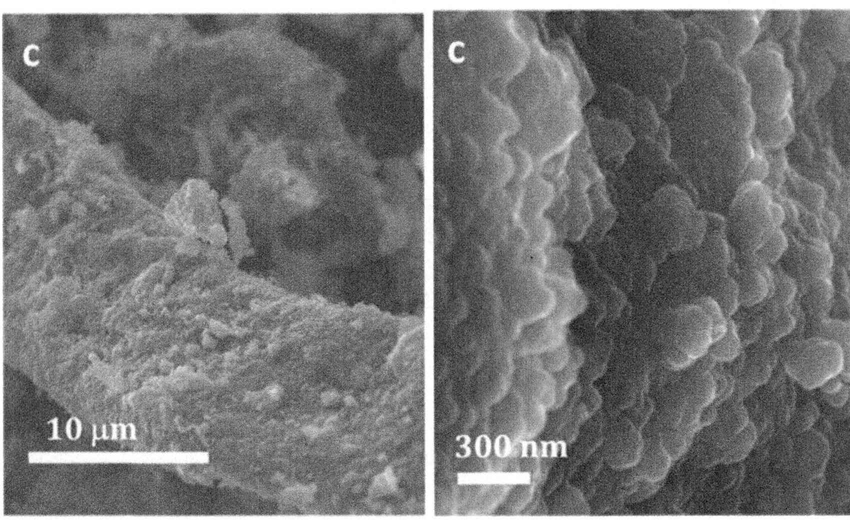

Fig. 8 SEM micrographs of ZnO/CNC nanocomposite at low and high magnification (Lefatshe et al. 2017)

fibers). Therefore, adding cellulosic nanomaterials to paper sheet will not negatively affect the main properties of paper such as flexibility, strength, printability, and biodegradability.

An important process during papermaking is refining of pulp fibers before paper sheets making. The refining step is usually done by mechanical action using what

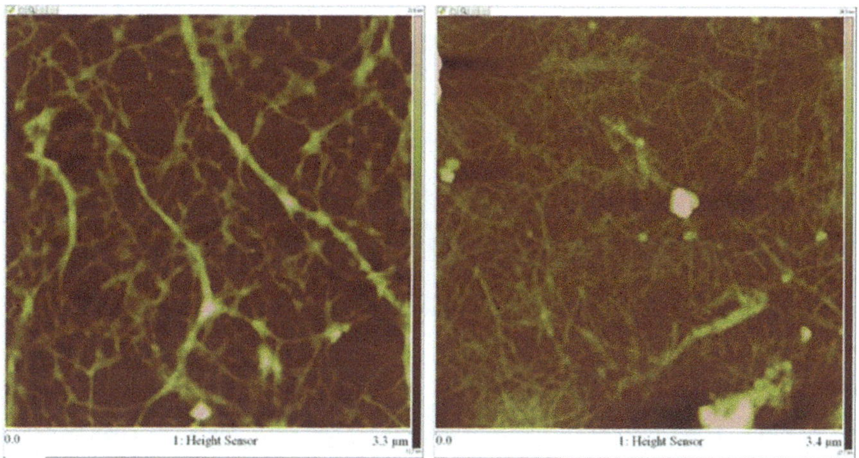

Fig. 9 Atomic force microscopy (AFM) images of palm cellulose nanofibers (left) and TEMPO-oxidized cellulose nanofibers (right) (Hassan et al. 2015)

is known as disc refiners, which results in generating tiny fibrils at the surface of the fibers. This remarkably increases strength properties of paper sheets but on the expense of increasing energy consumption. Cellulose nanofibers and TEMPO-oxidized cellulose nanofibers isolated from date palm fruit stalks were used to replace the refining step during papermaking from softwood and bagasse pulps (Hassan et al. 2015). The results showed that both kinds of nanofibers could improve mechanical properties of paper sheet similar to that obtained as a result pulp refining in case of bagasse pulp (short fibers). In addition, wet strength property was significantly improved to much more extent than that resulted from refining. The oxidized nanofibers were more effective than the nanofibers, probably due to the presence of carboxylate groups, which result in stronger hydrogen bonding between them and the pulp fibers. On the other hand, in case of softwood pulp (long fibers), refining was more effective than adding cellulose nanofibers in improving the different paper sheet properties (Figs. 9 and 10).

2.4 *Applications of Palm-Based Nanocelluloses in Drug Delivery*

Use of natural-based nanomaterials in pharmaceutical area of research is highly demanded since they are non-toxic. There are different roles that nanocelluloses can play in drug delivery applications. These roles are based on the presence of hydroxyl functional groups at the surface of nanocelluloses and their rod like structure. The hydroxyl functional groups can be utilized in forming physical or chemical bonds with different drugs; these bonds could be sensitive to the media inside human bodies

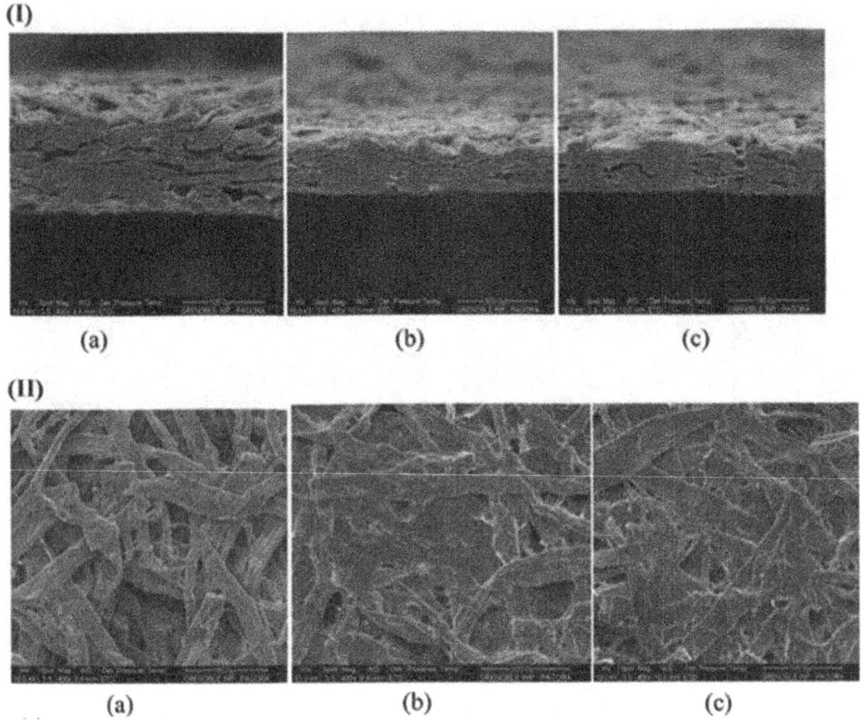

Fig. 10 SEM of paper sheet edges (I) and surfaces (II) prepared using: **a** softwood fibers, **b** softwood + 20% cellulose nanofibers, and **c** softwood + 20% TEMPO-oxidized cellulose nanofibers (Hassan et al 2015)

and the loaded drugs could be released. In addition, other functional groups could be introduced at the surface of nanocelluloses. Furthermore, the rod-like structure of nanocelluloses could act as physical barrier to release of drug and thus control their rate of release inside the body.

Regarding the use of nanocelluloses derived from palm residues in the area of drug release, cellulose nanocrystals isolated from palm kernel cake was used in alginate matrix crosslinked with calcium ions to regulate release of chlorhexidine digluconate, which is used to eradicate broad spectrum of bacteria that cause infection of dentinal tubules (Evelyna et al. 2019) (Fig. 11).

Cellulose nanocrystals and TEMPO-oxidized cellulose nanocrystals isolated from date palm fruit stalks were used in chitosan nanoparticles to regulate the release of anti-hyperglycemic drug Repaglinide (Abo Elseoud et al. 2018). Chitosan nanoparticles were formed via ionic gelation, and the chitosan/cellulose nanocrystals or chitosan/oxidized cellulose nanocrystals systems had high encapsulation efficiency of the drug reached 98%. Presence of the oxidized nanocrystals caused more sustained release of the drug than the non-modified nanocrystals, probably due to the

Fig. 11 TEM images of cellulose nanofibers isolated from palm kernel cake (left) and SEM images of nanocellulose-alginate nanocomposites microcapsules (right) (Evelyna et al. 2019)

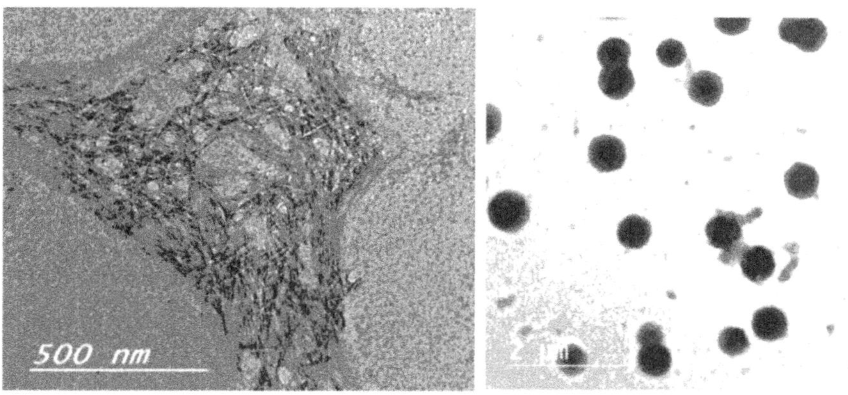

Fig. 12 TEM images of palm oxidized cellulose nanocrytals and their nanocomposites with chitosan nanoparticles (Abo Elseoud et al. 2018)

stronger hydrogen bonding between the drug and the oxidized nanocrystals (Figs. 12 and 13).

2.5 *Applications of Palm-Based Nanocelluloses in Pharmaceuticals and Biomedical Products*

In addition to use of nanocelluloses in pharmaceutical area due to their non-toxicity, they could be also used in biomedical applications due to the same reason and also due to their biocompatibility. In that area of research, threads based on cellulose nanofibers isolated from oil palm empty bunches were used for wound healing application (Yunus et al. 2019). The cellulose nanofibers threads were compared to others made from silk, polyvinyl alcohol, polyvinyl alcohol/bacterial cellulose, and polyvinyl alcohol/cellulose nanofibers. The study, which was carried out on

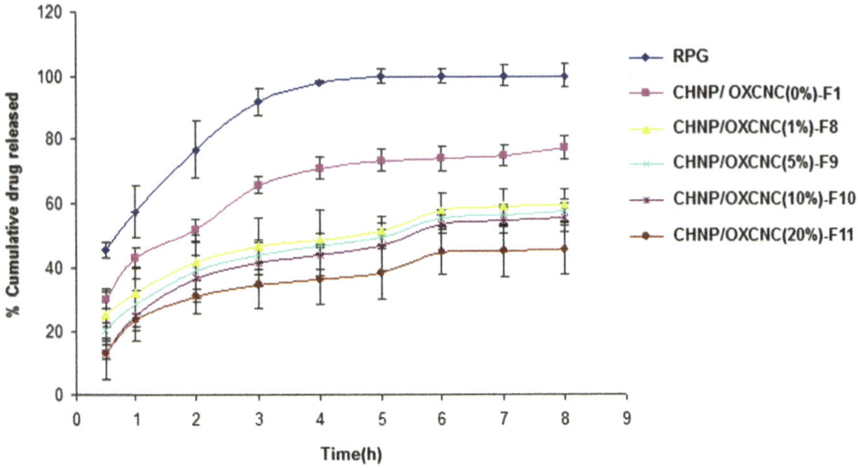

Fig. 13 In vitro release profiles of Repaglenide (RPG) from RPG-loaded chitosan nanoparticles/oxidized cellulose nanocrystals (CHNP/OXCNC) nanocomposites (Abo Elseoud et al. 2018)

male wistar rats showed proliferation and migration of white blood cells. In addition, granulation of tissue components in wound healing process increased in sutures area compared to the control. Although the response of white blood cells and tissue reaction was different for the different kinds of threads used, the wound healing process completed without significant differences in the healing duration.

2.6 Applications of Palm-Based Nanocelluloses as Dispersants in Emulsions

Emulsions are widely used in different industries including food, agricultural, chemical industries, cosmetics, inks, paints, as well as in pharmaceuticals. Cellulose nanocrystals, as derived from natural polymer, are good candidates as solid dispersants in liquid emulsions in different industries. They could be used to stabilize other solid nanoparticles in suspension. For example, cellulose nanocrystals isolated from oil palm empty fruit bunch were found to be good dispersants for oil-in-water Pickering emulsion (Foo et al. 2017). The nanocrystals could be also used to stabilize magnetite nanoparticles at their surface when the magnetite nanoparticles were *in-situ* synthetized in the presence of nanocrystals (Low et al. 2017). Cellulose nanocrystals/magnetite nanoparticles could be successfully used to form olein-in-water Pickering emulsion with good stability at different pH values and for times for up to 2 weeks (Fig. 14).

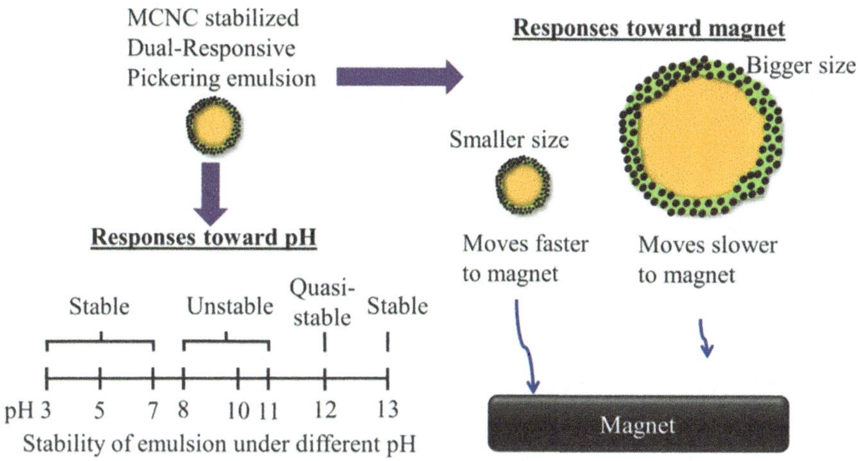

Fig. 14 Use of magnetite/cellulose nanocrystals (MCNC) for stabilization of oil-in-water emulsions at different pH values (Low et al. 2017)

3 Conclusions and Future Outlook

Palm residues represent good source for isolating cellulosic nanomaterials with attractive properties. Nanocelluloses isolated from different palm residues were successfully applied as reinforcing elements with different natural and synthetic polymers. Nanocelluloses from palm residues also added some desirable properties to polymers matrices such as increased resistance to water vapor permeability, thermal stability, and faster biodegradability. Nanocelluloses from palm residues were also successfully used as sorbents for removal of contaminants from water, and for making thin-film ultrafiltration membranes. Nanocelluloses from palm residues could be used in papermaking industry to improve mechanical properties and water resistance of paper products. They could also be used to save energy during papermaking since their addition could improve mechanical properties of paper made from short fibers without the need to do refining step. Nanocellulose from palm residues could be also used in controlled drug release delivery systems with or without modification of their surfaces; their addition to polymer matrices could regulate release of the active drug materials. Nanocelluloses form palm residues could be used to stabilize Pickering emulsions, which are used in different industrial applications.

Since the properties of nanocelluloses and their nanocomposites depends on the source of lignocellulosic materials used in their isolation, there are many applications still need to be explored for nanocelluloses isolated from palm residues, especially date palm ones which were explored to much less extent than those from oil palm. Nanocelluloses from palm residues still need to be tested in different applications related to tissue engineering, flexible electronics (conductors, semiconductors, insulator, and super capacitors), solar cell, hydrogels, food products, and others.

References

Abdul Khalil HPS, Tye YY, Ismail Z et al (2017) Oil palm shell nanofiller in seaweed-based composite film: Mechanical, physical, and morphological properties. BioResources 12(3):5996–6010

Abo-Elseoud WS, Hassan ML, Sabaa MW et al (2018) Chitosan nanoparticles/cellulose nanocrystals nanocomposites as a carrier system for the controlled release of repaglinide. Int J Biol Macromol 111:604–613

Ago M, Ferrer A, Rojas OJ (2016) Starch based biofoams reinforced with lignocellulose nanofibrils from residual palm empty fruit bunches: water sorption and mechanical strength. ACS Sustain Chem Eng 4(10):5546–5552

Agrebi F, Ghorbel N, Ladhar A et al (2017) Enhanced dielectric properties induced by loading cellulosic nanowhiskers in natural rubber: Modeling and analysis of electrode polarization. Mater Chem Phys 200:155–163

Akbari M, Tamayol A, Bagherifard S et al (2016) Textile Technologies and Tissue Engineering: A Path Toward Organ Weaving. Adv Healthc Mater 5(7):751–766

Akindoyo JO, Beg MDH, Ghazali S et al (2016) Polyurethane types, synthesis and applications-a review. RSC Adv 6(115):114453–114482

Alshammari BA, Saba N, Alotaibi MD, Alotibi MF et al (2019) Evaluation of Mechanical, Physical, and Morphological Properties of Epoxy Composites Reinforced with Different Date Palm Fillers. Materials. https://doi.org/10.3390/ma12132145

Altuna L, Herrera ML, Foresti ML et al (2018) Synthesis and characterization of octenyl succinic anhydride modified starches for food applications. A review of recent literature. Food Hydrocoll 80:97–110

Asad M, Saba N, Asiri AM et al (2018) Preparation and characterization of nanocomposite films from oil palm pulp nanocellulose/poly (Vinyl alcohol) by casting method. Carbohydr Polym 191:103–111

Atikah MSN, Ilyas RA, Sapuan SM et al (2019) Degradation and physical properties of sugar palm starch/sugar palm nanofibrillated cellulose bionanocomposite. Polimery/Polymers 64(10):680–689

Baker MI, Walsh SP, Schwartz Z et al (2012) A review of polyvinyl alcohol and its uses in cartilage and orthopedic applications. J Biomed Mater Res B Appl Biomater 100B(5):1451–1457

Ben Mabrouk A, Kaddami H, Magnin A et al (2011) Preparation of nanocomposite dispersions based on cellulose whiskers and acrylic copolymer by miniemulsion polymerization: Effect of the silane content. Polym Eng Sci 51(1):62–70

Bendahou A, Kaddami H, Raihane M et al (2009) Nanocomposite materials based on date palm tree cellulose whiskers. Rev Roum Chim 54(7):571–575

Bendahou A, Kaddami H, Dufresne A (2010) Investigation on the effect of cellulosic nanoparticles' morphology on the properties of natural rubber based nanocomposites. Eur Polym J 46(4):609–620

Benhamou K, Kaddami H, Magnin A et al (2015) Bio-based polyurethane reinforced with cellulose nanofibers: A comprehensive investigation on the effect of interface. Carbohydr Polym 122:202–211

Boonkerd, K. (2017) Development and modification of natural rubber for advanced application. In: Kobayashi T (ed) Applied environmental materials science for sustainability. Hershey, PA

Boufi S, Kaddami H, Dufresne A (2014) Mechanical performance and transparency of nanocellulose reinforced polymer nanocomposites. Macromol Mater Eng 299(5):560–568

Campos AD, SenaNeto ARD, Rodrigues VB et al (2017) Bionanocomposites produced from cassava starch and oil palm mesocarp cellulose nanowhiskers. Carbohydr Polym 175:330–336

Ching YC, Rahman A, Ching KY et al (2015) Preparation and characterization of polyvinyl alcohol-based composite reinforced with nanocellulose and nanosilica. BioResources 10(2):3364–3377

Chisenga SM, Workneh TS, Bultosa G et al (2019) Progress in research and applications of cassava flour and starch: a review. J Food Sci Technol 56(6):2799–2813

Dasan YK, Bhat AH, Ahmad F (2017) Polymer blend of PLA/PHBV based bionanocomposites reinforced with nanocrystalline cellulose for potential application as packaging material. Carbohydr Polym 157:1323–1332
Dasan YK, Bhat AH, Faiz A (2017a) Development and material properties of poly(lactic acid)/poly(3-hydroxybutyrate-co-3-hydroxyvalerate)-based nanocrystalline cellulose nanocomposites. J Appl Polym Sci 134(5):44328
Divakara Shetty S, Shetty N (2019) Investigation of mechanical properties and applications of polylactic acids—a review. Mater Res Express 6(11):112002
Dungani R, Aditiawati P, Aprilia S, Yuniarti K, Karliati T, Suwandhi I, Sumardi I (2018) Biomaterial from Oil palm aaste: properties. Characterization and applications. https://doi.org/10.5772/intechopen.76412
Espinoza SM, Patil HI, San Martin Martinez E (2020) Poly-ε-caprolactone (PCL), a promising polymer for pharmaceutical and biomedical applications: Focus on nanomedicine in cancer. Int J Polym Mater Po 69(2):85–126
Evelyna A, Astifanni TK, Ruth I et al (2019) Preparation of nanocellulose-alginate nanocomposites for chlorhexidine digluconate drug carrier. IOP Conf Ser Mater Sci Eng 547(1):012046
Fahma F, Hori N, Iwata T et al (2014) Preparation and characterization of polychloroprene nanocomposites with cellulose nanofibers from oil palm empty fruit bunches as a nanofiller. J Appl Polym Sci 131(8):40159
Fahma F, Takemura A, Saito Y (2014a) Acetylation and stepwise solvent-exchange to modify hydrophilic cellulose whiskers to polychloroprene-compatible nanofiller. Cellulose 21(4):2519–2527
Fahma F, Hori N, Iwata T et al (2017) PVA nanocomposites reinforced with cellulose nanofibers from oil palm empty fruit bunches (OPEFBs). Emir J Food Agr 29(5):323–329
Fahma, F., Sugiarto, Sunarti, TC et al (2017a) Thermoplastic Cassava Starch-PVA Composite Films with Cellulose Nanofibers from Oil Palm Empty Fruit Bunches as Reinforcement Agent. J Appl Polym Sci 2745721
Ferrer A, Hoeger IC, Lu X et al (2016) Reinforcement of polypropylene with lignocellulose nanofibrils and compatibilization with biobased polymers. J Appl Polym Sci 133(34):43854
Fiorote JA, Freire AP, Rodrigues DS et al (2019) Preparation of composites from natural rubber and oil palm empty fruit bunch cellulose: Effect of cellulose morphology on properties. BioResources 14(2):3168–3181
Foo ML, Tan KW, Wu TY et al (2017) characteristic study of nanocrystalline cellulose and its potential in forming pickering emulsion. Chem Eng Trans 60:97–102
Gea S, Surga M, Rahayu S et al (2018) The analysis of thermal and mechanical properties of biocomposite polycaprolactone/cellulose nanofiber from oil palm empty fruit bunches. AIP Conf Proc 2049:020063
Gopinath V, Saravanan S, Al-Maleki AR et al (2018) A review of natural polysaccharides for drug delivery applications: Special focus on cellulose, starch and glycogen. Biomed Pharmacother 107:96–108
Haafiz MKM, Hassan A, Khalil HPS et al (2016) Exploring the effect of cellulose nanowhiskers isolated from oil palm biomass on polylactic acid properties. Int J Biol Macromol 85:370–378
Hassan ML, Bras J, Mauret E et al (2015) Palm rachis microfibrillated cellulose and oxidized-microfibrillated cellulose for improving paper sheets properties of unbeaten softwood and bagasse pulps. Ind Crops Prod 64:9–15
Hassan E, Hassan M, Abou-zeid R et al (2017) Use of bacterial cellulose and crosslinked cellulose nanofibers membranes for removal of oil from oil-in-water emulsions. Polymers 9(9):388
Hassan M, Abou-Zeid R, Hassan E et al (2017a) Membranes based on cellulose nanofibers and activated carbon for removal of Escherichia coli bacteria from water. Polymers 9(8):335
Hsieh C-F, Liu W, Whaley JK et al (2019) Structure, properties, and potential applications of waxy tapioca starches—a review. Trends Food Sci Technol 83:225–234
Huynh T-P, Kutner W (2015) Molecularly imprinted polymers as recognition materials for electronic tongues. Biosens Bioelectron 74:856–864

Ilyas RA, Sapuan SM, Atiqah A et al (2019) Sugar palm (Arengapinnata (Wurmb.) Merr) starch films containing sugar palm nanofibrillated cellulose as reinforcement: Water barrier properties. Polym Compos 41(2):459–467

Ilyas RA, Sapuan SM, Ibrahim R et al (2019a) Effect of sugar palm nanofibrillated cellulose concentrations on morphological, mechanical and physical properties of biodegradable films based on agro-waste sugar palm (Arengapinnata (Wurmb.) Merr) starch. J Mater Res Technol. doi: https://doi.org/10.1016/j.jmrt.2019.08.028

Ilyas RA, Sapuan SM, Ishak MR et al (2018) Development and characterization of sugar palm nanocrystalline cellulose reinforced sugar palm starch bionanocomposites. Carbohydr Polym 202:186–202

Indarti E, Roslan R, Marwan, et al (2016) Polylactic acid bionanocomposites filled with nanocrystalline cellulose from TEMPO-oxidized oil palm lignocellulosic biomass. BioResources 11(4):8615–8626

IreanaYusra AF, Abdul Khalil HPS, Hossain MS et al (2016) Characterization of plant nanofiber-reinforced epoxy composites. BioResources 10(4):8268–8280

Jin F-L, Li X, Park S-J (2015) Synthesis and application of epoxy resins: A review. J Ind Eng Chem 29:1–11

Joshi M, Adak B, Butola BS (2018) Polyurethane nanocomposite based gas barrier films, membranes and coatings: a review on synthesis, characterization and potential applications. Prog Mater Sci 97:230–282

Kausar A (2018) Polyurethane composite foams in high-performance applications: a review. Polymer Plast Tech Eng 57(4):346–369

Ladhar A, Arous M, Kaddami H et al (2015a) Ionic hopping conductivity in potential batteries separator based on natural rubber-nanocellulose green nanocomposites. J Mol Liq 211(5048):792–802

Ladhar A, Arous M, Kaddami H et al (2015b) L.C.AC and DC electrical conductivity in natural rubber/nanofibrillated cellulose nanocomposites. J Mol Liq 209:272–279

Ladhar A, Arous M, Kaddami H et al (2017) Correlation between the dielectric and the mechanical behavior of cellulose nanocomposites extracted from the rachis of the date palm tree. (2017) IOP Conf Ser Mater Sci Eng 258(1):012001

Lani NS, Ngadi N, Johari A et al (2014) Isolation, characterization, and application of nanocellulose from oil palm empty fruit bunch fiber as nanocomposites. J Nanomater 702538

Lasprilla-Botero J, Torres-Giner S, Pardo-Figuerez M et al (2018) Superhydrophobic bilayer coating based on annealed electrospun ultrathin poly(ε-caprolactone) fibers and electrosprayed nanostructured silica microparticles for easy emptying packaging applications. Coatings 8(5):173

Lefatshe K, Muiva CM, Kebaabetswe LP (2017) Extraction of nanocellulose and in-situ casting of ZnO/cellulose nanocomposite with enhanced photocatalytic and antibacterial activity. Carbohydr Polym 164:301–308

Lemos VA, Santos MS, Santos ES et al (2007) Application of polyurethane foam as a sorbent for trace metal pre-concentration—a review. Spectrochim Acta B 62(1):4–12

Lim YH, Chew IML, Choong TSY et al (2016) NanoCrystalline cellulose isolated from oil palm empty fruit bunch and its potential in cadmium metal removal. MATEC Web Conf 59:04002

Lisdayana N, Fahma F, Sunarti TC et al (2018) Thermoplastic starch–PVA nanocomposite films reinforced with nanocellulose from Oil Palm Empty Fruit Bunches (OPEFBs):eEffect of Starch Type. J Nat Fibers

Liu G, Gu Z, Hong Y et al (2017) Structure, functionality and applications of debranched starch: a review. Trends Food Sci Technol 63:70–79

Low LE, Tey BT, Ong BH et al (2017) Palm olein-in-water Pickering emulsion stabilized by Fe3O4-cellulose nanocrystal nanocomposites and their responses to pH. Carbohydr Polym 155:391–399

Mariano M, El Kissi N, Dufresne A (2014) Cellulose nanocrystals and related nanocomposites: review of some properties and challenges. J Polym Sci B Polym Phys 52(12):791–806

Martín-Martínez JM (2005) Polychloroprene rubber adhesives: applications and properties. Second Edition, Handbook of Adhesion, pp 329–332

Ma H, Burger C, Hsiao BS et al (2011) Ultra-fine cellulose nanofibers: new nano-scale materials for water purification. J Mater Chem 21(21):7507–7510

Mittal V (ed) (2016) Functional polymer blends: synthesis, properties, and performance. Florida, Boca Raton

Mokhtar M, Majlan EH, Talib MZM et al (2016) A short review on alkaline solid polymer electrolyte based on polyvinyl alcohol (PVA) as polymer electrolyte for electrochemical devices applications. Int J Appl Eng Res 11(19):10009–10015

Mondal S (2018) Review on nanocellulose polymer nanocomposites. Polym-Plast Technol Eng 57(13):1377–1391

Nguyen HTH, Qi P, Rostagno M et al (2018) The quest for high glass transition temperature bioplastics. J Mater Chem a 6(20):9298–9331

Niranjana Prabhu T, Prashantha K (2018) A review on present status and future challenges of starch based polymer films and their composites in food packaging applications. Polym Compos 39(7):2499–2522

Owi WT, Lin OH, Sam ST et al (2017) Tapioca starch based green nanocomposites with environmental friendly cross-linker. Chem. Eng. Trans 56:463–468

Pluta J, Karolewicz B (2001) The application of polyvinyl alcohol in the technology of modern drug form. Polim Med 31(1–2):11–17

Saba N, Paridah MT, Abdan K et al (2015) Preparation and characterization of fire retardant nano-filler from oil palm empty fruit bunch fibers. BioResources 10(3):4530–4543

Saba N, Paridah MT, Abdan K et al (2016) Thermal properties of oil palm nano filler/kenaf reinforced epoxy hybrid nanocomposites. AIP Conf Proc 1787:050020

Saba N, Paridah MT, Abdan K et al (2016a) Dynamic mechanical properties of oil palm nano filler/kenaf/epoxy hybrid nanocomposites. Constr Build Mater 124:133–138

Saba N, Paridah MT, Abdan K et al (2016b) Effect of oil palm nano filler on mechanical and morphological properties of kenaf reinforced epoxy composites. Constr Build Mater 123:15–26

Saba N, Jawaid M, Paridah MT et al (2017) Physical, structural and thermomechanical properties of nano oil palm empty fruit bunch filler based epoxy nanocomposites. Ind Crops Prod 108:840–843

Saba N, Jawaid M, Alothman OY et al (2019) Evaluation of dynamic properties of nano oil palm empty fruit bunch filler/epoxy composites. J Mater Res Technol 8(1):1470–1475

Salehudin MH, Muhamad II (2018) Effect of graphene oxide and cellulose nanofiber towards mechanical properties of polylactic acid based active packaging using response surface methodology. Malaysian J Anal Sci 22(6):984–998

Salehudin MH, Salleh E, Muhamad II et al (2014) Starch-based biofilm reinforced with empty fruit bunch cellulose nanofibre. Mater Res Innov 18(S6):322–325

Šárka E, Dvořáček V (2017) New processing and applications of waxy starch (a review). J. Food Eng 206:77–87

Septevani AA, Evans DAC, Annamalai PK et al (2017) The use of cellulose nanocrystals to enhance the thermal insulation properties and sustainability of rigid polyurethane foam. Ind Crops Prod 107:114–121

Septevani AA, Evans DAC, Martin DJ et al (2018) (2018) Hybrid polyether-palm oil polyester polyol based rigid polyurethane foam reinforced with cellulose nanocrystal. Ind Crops Prod 112:378–388

Shaghaleh H, Xu X, Wang S (2018) Current progress in production of biopolymeric materials based on cellulose, cellulose nanofibers, and cellulose derivatives. RSC Adv 8(2):825–842

Shanmugarajah B, Chew IM, Mubarak NM et al (2019) Valorization of palm oil agro-waste into cellulose biosorbents for highly effective textile effluent remediation. J Clean Prod 210:697–709

Shevkani K, Singh N, Bajaj R et al (2017) Wheat starch production, structure, functionality and applications—a review. Int J Food Sci Tech 52(1):38–58

Simangunsong DI, Hutapea THA, Lee HW et al (2018) The effect of nanocrystalline cellulose (NCC) filler on polylactic acid (PLA) nanocomposite properties. J Eng Technol Sci 50(4):578–587

Singh AV, Nath LK, Singh A (2010) Pharmaceutical, food and non-food applications of modified starches: acritical review. Elec J Env Agricult Food Chem 9(7):1214–1221

Sobolčiak P, Tanvir A, Popelka A et al (2017) The preparation, properties and applications of electrospun co-polyamide 6,12 membranes modified by cellulose nanocrystals. Mater Des 132:314–323

Solikhin A, Hadi YS, Massijaya MY et al (2018) Properties of Poly(Vinyl Alcohol)/Chitosan nanocomposite films reinforced with oil palm empty fruit bunch amorphous lignocellulose nanofibers. J Polym Environ 26(8):3316–3333

Thong CC, Teo DCL, Ng CK (2016) Application of polyvinyl alcohol (PVA) in cement-based composite materials: a review of its engineering properties and microstructure behavior. Constr Build Mater 107:172–180

Vanier NL, El Halal SLM, Dias ARG et al (2017) Molecular structure, functionality and applications of oxidized starches: a review. Food Chem 221:1546–1559

Wang L, Ghubayra AR, Bauer A J-P et al (2012) Polyurethane for biomedical applications: a review of recent developments. The design and manufacture of medical devices, pp 115–151

Wang L, Ghubayra AR, Bauer AJ-P et al (2018) Recent advances in electrospun poly(E-caprolactone)-based materials and their biomedical applications. Latest developments, challenges and future directions, nano/micro-structured materials for energy and biomedical applications, pp 77–130

Wang J, Liu X, Jin T et al (2019) Preparation of nanocellulose and its potential in reinforced composites: a review. J Biomater Sci Polym Ed 30(11):919–946

Yasim-Anuar TAT, Ariffin H, Norrrahim MNF et al (2019) Sustainable one-pot process for the production of cellulose nanofiber and polyethylene/cellulose nanofiber composites. J Clean Prod 207:590–599

Yunus M, Fahma F, Abidin Z et al (2019) Cellulose based surgical threads from oil palm empty fruit bunches in wound healing on Male Wistar rats. Indian Vet J 96(6):21–23

Yun X, Wang Y, Li M et al (2017) Application of permselective poly(ε-caprolactone) film for equilibrium-modified atmosphere packaging of strawberry in cold storage. J Food Process Pres 41(6)e13247

Zhang L-M (2001) A review of starches and their derivatives for oilfield applications in China. Starch/Staerke 53(9):401–407

The manufacturer's authorised representative in the EU is Springer Nature Customer Service Centre GmbH, Europaplatz 3, 69115 Heidelberg, Germany. If you have any concerns regarding our products, please contact ProductSafety@springernature.com

Printed and bound by CPI Group (UK) Ltd, Croydon, CR0 4YY
30/09/2026
02225287-0015